AF831352

Metal Surface Engineering

Metal Surface Engineering: Developed Tribology offers updated and advanced information on tribology for ferrous and non-ferrous materials, and discusses the practical application and performance of surface engineering. It also covers wear prevention and low-friction techniques in mechanical, electrical, and electronic devices and components of equipment used in the automotive and aerospace industry to improve efficiency and extend equipment life. Furthermore, this book offers analytical approaches and symbolic models that demonstrate the control of materials under different conditions, helping to develop new industrial products and applications. In addition, this book also covers such advanced technologies as advanced coatings, thermal techniques, and diversification.

This is a comprehensive reference guide for students studying ferrous and non-ferrous materials, helping them to better understand the methods and techniques used in processing and reducing wear and friction. This book serves as a helpful reference for academics, scientists, and researchers by providing theoretical and experimental foundations for the study of networks and wear, and by supporting the development of new solutions in surface engineering.

Metal Surface Engineering

Developed Tribology

Riyadh A. Al-Samarai and Yarub Al-Douri

CRC Press
Taylor & Francis Group
Boca Raton London New York

CRC Press is an imprint of the
Taylor & Francis Group, an **informa** business

First edition published 2025
by CRC Press
2385 NW Executive Center Drive, Suite 320, Boca Raton FL 33431

and by CRC Press
4 Park Square, Milton Park, Abingdon, Oxon, OX14 4RN

CRC Press is an imprint of Taylor & Francis Group, LLC

© 2025 Riyadh A. Al-Samarai and Yarub Al-Douri

Reasonable efforts have been made to publish reliable data and information, but the author and publisher cannot assume responsibility for the validity of all materials or the consequences of their use. The authors and publishers have attempted to trace the copyright holders of all material reproduced in this publication and apologize to copyright holders if permission to publish in this form has not been obtained. If any copyright material has not been acknowledged please write and let us know so we may rectify in any future reprint.

Except as permitted under U.S. Copyright Law, no part of this book may be reprinted, reproduced, transmitted, or utilized in any form by any electronic, mechanical, or other means, now known or hereafter invented, including photocopying, microfilming, and recording, or in any information storage or retrieval system, without written permission from the publishers.

For permission to photocopy or use material electronically from this work, access www.copyright.com or contact the Copyright Clearance Center, Inc. (CCC), 222 Rosewood Drive, Danvers, MA 01923, 978-750-8400. For works that are not available on CCC please contact mpkbookspermissions@tandf.co.uk

Trademark notice: Product or corporate names may be trademarks or registered trademarks and are used only for identification and explanation without intent to infringe.

ISBN: 9781032945736 (hbk)
ISBN: 9781032954103 (pbk)
ISBN: 9781003584735 (ebk)

DOI: 10.1201/9781003584735

Typeset in Times
by KnowledgeWorks Global Ltd.

Access the Support Material: http://www.routledge.com/9781032945736

Contents

Preface

One of the most important methods for reducing wear and friction and achieving superior mechanical, chemical and tribological properties is to improve the performance with surface treatment. Product efficiency and productivity have increased and will continue to increase as a result of improvements in surface treatment procedures. Over time, these processes have undergone significant evolution, moving from traditional to advanced technologies. The use of biomaterials, smart materials and hybrid materials in advanced surface engineering procedures is one of the areas of focus. This book provides detailed explanations of many surface finishing techniques and their uses. Several technologies have been developed and used for surface treatment of products made of various materials and modifications are examined in detail. The main purpose of this book is to provide readers with a comprehensive overview of the surface. Engineering technologies including hybrid and smart materials, nanomaterials, industry, smart manufacturing and technologies are used in each industry's unique applications. In addition, it discusses several topics related to surface engineering, as well as process-specific information. Computational and simulation analysis, artificial intelligence and machine learning, process control, design and quality modification for engineering interoperability are some of the aspects discussed in detail.

The specialists have uniquely compiled archives of the latest research from a variety of sources, creating a platform that offers readers access to both surface engineering technologies and advanced research technologies used in these procedures. This easy read provides readers with a comprehensive understanding of surface treatment techniques, materials, applications and optimization styles. The readers of this textbook are the academics, scientists, researchers, students and engineers working in manufacturing field. This will provide with a guidance to help them selecting the correct processes for many types of materials used in surface engineering. The purpose of this book is to help the scientific community identify steps needed for further research in related fields. This offers a unique perspective on the potential for incremental improvements in surface engineering technology, explaining the latest developments in production technologies. To keep pace with the demands of industrial production, these processes must be updated with new materials and technologies as industry advances during the Fourth Industrial Revolution. This book includes a section devoted to the optimization and application of surface engineering methods. Tribology is an interdisciplinary field of study that includes additive chemistry, lubricants, surface engineering (including surface modification, surface coating and surface topography analysis), mechanical engineering and materials science. Tribology, or the "science of friction", was once considered a new and extremely complex field because it combines several concepts such as wear, lubrication and friction. Around 3500 BC, the study of tribology was first applied at a primitive level to move stones using water as a lubricant to reduce friction and protect the surface of material, ultimately resulting in less wear.

Today, tribology is a fairly broad field, including metals, polymers, ceramics and various composite materials with a wide range of applications. To keep the scientific community and industry abreast of the latest developments, this book offers a comprehensive updated information on all material pairs used in tribological contact. Prominent researchers from business, academia, government and private research institutions around the world will find value in the knowledge offered in this application-oriented book. In addition, the latest tribology research worldwide is presented here. The current state of the industry, trends, potential future directions, opportunities, etc. are covered in detail, making it user-friendly for novice researchers as well. Moreover, the main goal is to provide a thorough and systematic overview of recent advances in nanotribology and lubrication.

This book consists of ten chapters covering the following topics: Using a preferred selection index approach, Chapter 1 analyzes current developments in metal surface engineering, including methodologies, characterization strategies and applications. Chapter 2 uses geometric surfaces introduced as a new concept. In addition, surface technology is a vital and technologically advanced area in the manufacturing and consumer products industries. This includes both high-tech and low-tech methods that often modify the surfaces of components or materials to create a new surface as presented in Chapter 3. By texturing the surface of a material, it is possible to change its wear and friction properties. This significantly reduces wear and friction due to conformal flat contact, as can be seen in Chapter 4.

In tribology, severe corrosion is the main cause of failure of AISI 316 stainless steel. One well-known method for improving the mechanical properties of contact surfaces is the application of surface textures. In this study, the steel surface is processed using laser ablation to create a microtexture. Various solid lubricants such as CaF_2, h-BN, WS_2 and graphite are mixed with microfibers are shown in Chapter 5. Various AM surfaces have been used to illustrate surface textural features and their influence on the tribological behavior of surfaces. Chapter 6 reviews the behavior of surfaces and other tribological elements of natural and artificial systems under a relative motion in the field of tribology, which studies wear, friction and lubrication. Research and review articles are included in the tribology, surface engineering that discuss the criticality of wear resistance for how the surface layers of different materials create their properties using different methods and techniques as described in Chapter 7. In surface engineering and manufacturing, the surface finishing technique plays a significant role in providing details on how scientists are using non-communicable diseases to find new materials with new properties for practical use. Nanofabrication methods can be divided into three categories as presented in Chapter 8. Health and safety issues in emerging surface engineering techniques and subsurface phenomena are the main drivers of modern interdisciplinary research to have a major impact on technology and science as discussed in Chapter 9, followed by developments in surface coating technologies. In addition to review surface treatment and finishing techniques, Chapter 10 presents recent developments in surface engineering and development of nanostructured materials based on intelligent surface properties, which is growing rapidly. One of the main tasks is a reliable and accurate description of the electrical structure.

Riyadh A. Al-Samarai
University of Samarra, Iraq

Yarub Al-Douri
Al-Bayan University, Iraq
Piri Reis University, Turkey
University of Malaya, Malaysia

Author Biographies

Prof. Dr. Riyadh A. Al-Samarai was born in Salah Al-Din in 1970. Al-Samarai is a graduate of the University of Technology in Baghdad, holding BSc & MSC in mechanical engineering. He obtained PhD in mechanical engineering-Applied Mechanics from the University Malaysia Perlis, Malaysia. He held several positions: Director of Missions and Social Relations, Assistant Dean for Scientific Affairs and Head of the Department of Electromechanical Engineering at Samarra University. Al-Samarai is the winner of the Iraqi Scientific Excellence Award 2024. He has launched MSC program in electromechanical engineering at Samarra University. He has received many medals and awards, including the Creativity Award for Scientists in Malaysia, many medals in international scientific forums and conferences, Iraqi Forum for Intellectuals and Academics (IFIA), and the Best Research Paper Award at the Asian Universities Conference in Malaysia, with other awards. Al-Samarai is the editor-in-chief of the *Samarra University Journal of Engineering and Science.* Al-Samarai has 32 years of research experience, 23 years of university teaching experience, 16 years of administrative experience and more than 60 publications including citations > 350, h-index = 7 and research grants of US$1 million.

Prof. Dr. Yarub Al-Douri is a Fellow of the European Academy of Sciences. Al-Douri is winner of the prestigious Khalifa International Award for Date Palm and Agricultural Innovation 2024. He is one of the Middle-East, North of Africa and Southeast Asia's most renowned scientists known for his contributions in renewable energy and nanotechnology. Al-Douri is a scientist, researcher, administrator and an excellent lecturer who has students all over the world. Al-Douri has Doctorat D'etat in materials science (2000), MSc in physics (1995) and BSc in physics (1991). He has been appointed full-professor, visiting professor, adjunct professor, consultant expert, associate professor, assistant professor, research fellow (a), scientific collaborator and post-doc in UAE, KSA, Iraq, Malaysia, Turkey, Pakistan, Algeria, Yemen, Singapore, Germany and France, respectively. Al-Douri initiated Nanotechnology Engineering MSc Program and Nano Computing Laboratory, in addition to founding Applied Materials Laboratory. Al-Douri was Head of Department of Physics, Head of Department of Nanomaterials and Department of International Networking and Collaboration, additionally to Secretary of Department of Physics. He has received 78 national and international prizes and awards including IAAM Scientist Award by International Association of Advanced Materials, Sweden 2022. He is also the winner of World's Top 2% Scientist Career-Long Citation Impact by Stanford University, USA 2020, World's Top 2% Scientists by Stanford University, USA 2024, 2023, 2022, 2021 & 2020; OeAD Award, Austria 2020; Japan Society for the Promotion of Science (JSPS) Award 2019, Asian Universities Alliance (AUA) Award 2019, Iraqi Forum for Intellectuals and Academics Award 2019, Best Researcher Award at Cihan University Sulaimaniya 2021 & 2019, Best Paper at Global Conference on Energy and Sustainable Development, Coventry, UK 2015, Distinguished Researcher Award at University Malaysia Perlis 2011–2015, Gold Award at ITEX Kuala Lumpur 2013, TWAS-UNESCO Associateship 2009–2012 & 2012–2015 and others. He has more than 31 years' experience of research, teaching, administrative and editorial board management, organizing events, research grants and consultations, in addition to more than 894 publications with renowned international publishers, and US$5.1M research grants. Al-Douri has notable citations > 14.4K and h-index > 66. He has supervised more than 25 students. Al-Douri is associate editor of *Nano-Micro Letters* (Q1, IF= 31.6); editor-in-chief of *Experimental and Theoretical Nanotechnology* (Scopus-indexed); editor-in-chief of *World Journal of Nano Science and Engineering*, editor and peer-reviewer of 34 international journals, member of 13 international scientific bodies, organizer 37 international conferences, chairperson of 11 international conferences. He appointed 38 internal and external examiner for post-graduates students, presented 41 training courses and public lectures, and has been invited 64 times as keynote, invited and guest speakers to many reputed universities in Europe, Asia and Africa that ended up in generating more than 9 MOUs and MOAs. His research field focuses on renewable energy, nanotechnology, nanoelectronics, nanomaterials, modelling and simulation, semiconductors and optical studies. Al-Douri has contributed to academic and research life for upgrading universities to the international level. Finally, Al-Douri is a public figure among international media in the UK, Singapore, Malaysia, Qatar and UAE.

1 Characteristics of Metal Surface Engineering

1.1 INTRODUCTION

Surface engineering of metals studies the microstructure, properties of surface layers and mechanisms of development of classical and modern surface treatment methods. The first part of this chapter provides an overview of fundamentals of surface engineering, history of its development and proposes a two-class classification scheme for surface layers [1–4]. The main capabilities and useful properties of surface layers and coatings are discussed, as well as how these properties interact with other properties and how important they are for proper operation and selection. An innovative classification of methods for creating a surface layer is presented. Recent developments in this area are elaborated. These methods are determined by a direct or radial contact of particles or heating medium with surface being treated. This field has further expanded with recent developments in surface technologies and characterization techniques leading to innovative results and inventive applications. Surface treatment methods are divided into two categories: physical and chemical procedures. There are three types of vapor deposition: chemical vapor deposition (CVD), physical vapor deposition (PVD) and laser surface treatment, which involve applying heat or energy to a surface to alter it. These methods provide a flexibility in tuning surface characteristics to meet specific needs such as hardness, adhesion and surface roughness. However, chemical methods, including chemical etching, sol–gel coating, chemical coating and electroplating, offer the potential for controlled application and chemical surface modification to improve functionalization, biocompatibility and corrosion resistance.

The effectiveness of surface-finishing procedures is understood and evaluated using various characterization approaches. Surface shape, crystallographic structure and phase composition are examined using microstructure techniques such as X-ray diffraction (XRD), transmission electron microscopy (TEM) and scanning electron microscopy (SEM). Atomic force microscopy (AFM), scanning tunneling microscopy (STM) and surface profilometry are few examples of surface topography analysis tools that can accurately measure and visualize surface roughness and texture. Atomic movement and phase transitions that affect physical properties are areas of interface engineering, including surface engineering and advanced manufacturing. Because process conditions determine the structure, scientists and engineers have powerful tools to create artifacts with a wide range of industrial and scientific uses. In some cases, surface technology and

DOI: 10.1201/9781003584735-1

additive manufacturing are linked. In these cases, materials are joined layer by layer to create pieces of three-dimensional (3D) model data. Surface technology is the process of improving the surface characteristics of a component without affecting the characteristics of base substrate material [5, 6].

Magnesium is a substance that has great potential in automotive components and biomedical device manufacturing. Magnesium's light weight and excellent strength make it an attractive material to use in many industries. However, magnesium remains highly reactive and flammable in physiological environments. Controlling the corrosion susceptibility and availability of magnesium in the presence of other metallic elements or aqueous conditions remains challenging. Apart from more specific problems in each industry, including production methods, these are the two main disadvantages of magnesium materials that prevent their widespread use. When applying surface technology to tribological applications, the two main goals are to modify the frictional properties of a component and increase its resistance to wear and damage. Sometimes both goals are achieved simultaneously. There are three main categories of methods that can be used to create surface of component. The first category includes operations such as transformation which somehow changes the surface without changing its composition, surface melting and hardening.

The actions of the second group somehow change the surface by changing the composition of the surface layer. These processes can lead to changes such as transformation (e.g., carburization of steel, which changes maximum hardness and hardenability) or destruction of current crystal structure through solid solution formation or lattice destruction. However, because of the reaction between the materials added to the process and the support elements, these processes can also directly result in the developing of new phases that differ from the support phases. Sometimes these new phases form precipitates of surface-engineered layers within the matrix (e.g., vanadium or aluminum nitrides are formed when a suitable steel is nitride, but in other situations the new phases separate into their layer).

The biomedical industry uses magnesium specifically in bone implants due to its strength properties similar to those of organic skeletons. In addition, magnesium is suitable for biodegradable implants or medical devices due to its strong corrosive or destructive properties. Modern engines and automobiles also use cast magnesium alloys as components because magnesium's high strength and light weight allow for increased acceleration, faster driving and lower energy consumption. A semi-passive hydroxide coating, which practically does not prevent corrosion processes, is created by magnesium.

1.2 SURFACE-ENGINEERED LAYER DELAMINATION

Understanding how a surface-engineered component differs from a homogeneous component in the context of tribological contact is critical before delving into surface engineering procedures as it will clarify the key differences between

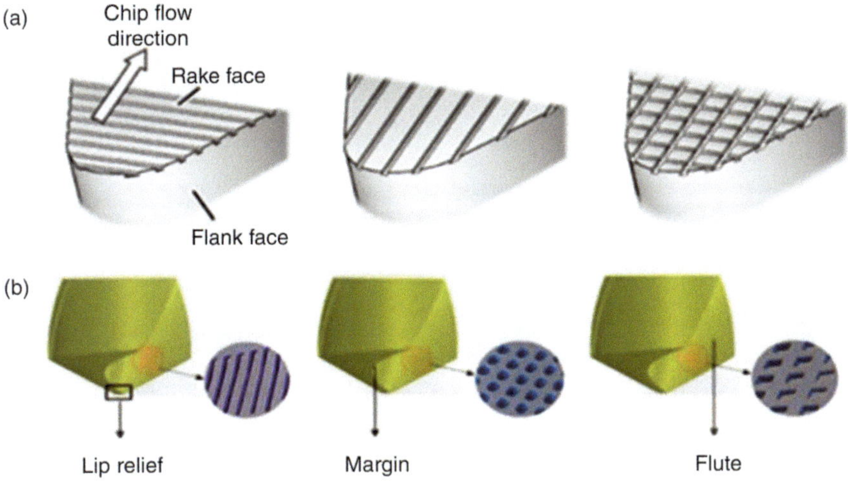

FIGURE 1.1 Microtextures on rake face of drill (bottom) (a); microgrooves created on rake face of rotating cutting tool (b). The arrow indicates the direction of chip flow [13] [open access].

different types of processes. Microgrooves are created on the rake face of rotating cutting tool; an arrow indicates the direction of chip flow; microtextures on the rake face of drill bit (Figure 1.1). After one, three and five days, the morphology and roughness of corrosion products are examined by AFM [13].

Numerous characteristics can be identified that determine the tribological response of a homogeneous material (in terms of wear and friction). The biggest barrier to the widespread use of magnesium is its tendency to oxidize heavily, although metallurgical procedures such as alloying can increase the degree of film negativity. These procedures include anodizing (which deposits an aluminum oxide coating on an aluminum alloy) and boriding (which can form layers of borides on iron, titanium or cobalt-based alloys). The third category, sometimes called coating procedures, involves methods of applying material to the surface. Hence, the original material is not changed – an artificially created layer is applied to the surface. A wide range of possible methods and materials are available in this category. Processes in this group include painting, CVD, surfacing and electroplating [7, 8]. The thickness of corrosion products gradually increases with increasing of corrosion period, from 132.1 nm after one day to 384.4 nm after three days and 984.7 nm after five days (Figure 1.2a). Moreover, a noticeable change in the morphology of products is observed (Figure 1.2c). After one day of corrosion, Q235 steel matrix is severely damaged and most of irregular corrosion products are formed on the surface (Figure 1.2b). After three days, a thick layer and uniform octahedral crystalline particles are formed (Figure 1.2b). The octahedral particles collected for three days comprise cubic

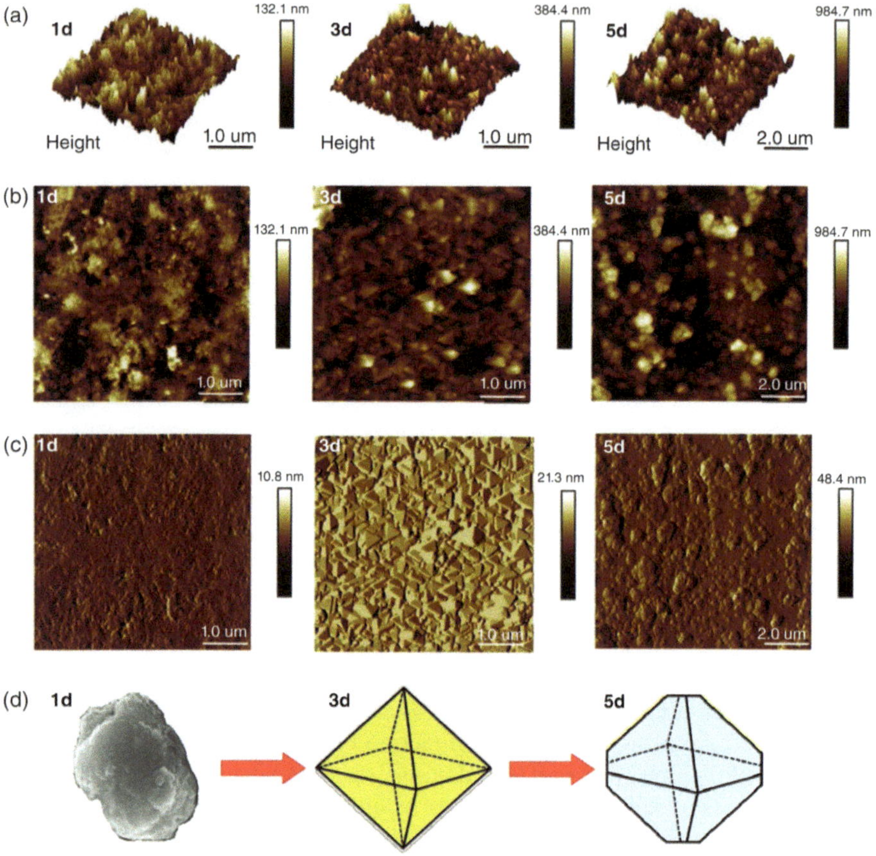

FIGURE 1.2 AFM images (a–c) and schematics (d) show different corrosion times of corrosion products on the surface of Q235 steel [10] [open access].

crystals as corrosion products. After five days, some corrosion resistance is observed (Figure 1.2b). Figure 1.2d shows the structural development of corrosion products [10].

You can create the surface of component to improve its responsiveness. The artificial surface layer has a set of qualities that contribute to a better tribological reaction. However, interface characteristics, which can be categorized, influence the system in addition to characteristics of substrate and surface-engineered layers, making it scattered or clear. The response of surface component to tribological challenge depends on the behavior of both the material beneath the surface and the surface material itself. Examples of such problems include rolling contact in a bearing, slipping between gear teeth or abrasion from a rough hard counter-body. The layer designed on the surface may fail if the circumstances are bad enough be eliminated, by delamination (separation at interface) rather than gradual wear. The bond strength of substrate to coating, degree of tribological conditions, thickness of coating layer (all of these taken together control the stresses at interface), and any additional factors that contribute to coating delamination will determine whether coating delamination will occur. Conversely, the likelihood of separation between the surface layer and underlying material is low when the interface is diffuse.

Laser melting deposition (LMD) process can simultaneously process two or more dissimilar materials to produce parts, composite parts or parts whose surfaces are formed from composite materials that can be manufactured in a single step. This is because the LMD machine creates parts by precisely following the CAD model. The processing parameters that influence the characteristics of laser-clad metal parts are described in detail. This chapter demonstrates LMD's ability to produce complex unique parts, improve surface quality, repair expensive components and reduce purchase-to-flight ratios in the aerospace component manufacturing process. It also emphasizes the use of nontraditional like CNC machines. On another hand, the separation of surface layer and underlying material is less likely when the interface is diffuse. It is difficult to characterize or quantify the interface strength of systems with sharp interfaces. When an interface is in tension rather than shear, it typically exhibits significantly different features. Uniaxial strength (σi) or surface energy (Gi) is often used to characterize interface strength. A significant amount of work goes into enhancing interfacial strength in most surface-finishing procedures, such as cleaning and roughening the outer surface before coating as used in vapor deposition and electroplating processes and sandblasting, which is often used before thermal spraying or application bond layer, which is an intermediate layer that forms a strong bond with each material, even if there is no strong bond between the substrate and coating. Although stress gradients are common in coatings, the assumption is as follows::

- The coating represents the strain energy density (energy per unit volume) of the coating under biaxial compression.
- The coating is thinner than size of component.
- The stresses in the coating are constant throughout its thickness. The state of surface technology has changed dramatically due to recent developments. Improved mechanical and tribological properties can be obtained through precise control of microstructure and composition at nanoscale, which has led to emergence of nanostructured coatings [10–12] as an important area of research. Due to their exceptional hardness, corrosion resistance and low friction, these coatings are in great demand in various industries.

To effectively achieve good mechanical properties and/or restore the surface quality of metals that have withstood prolonged use under severe operating conditions, the use of material coating methods is recommended [9–14]. The need for surface treatment is most important as surface wear due to mechanical damage, stress or fatigue has significant financial implications. In some other situations, reengineering is more cost-effective than replacing the entire machine [15, 16].

1.3 CHANGING THE SURFACES OF COMPONENTS WITHOUT CHANGING THEIR COMPOSITION

1.3.1 Transformation Hardening

Due to its excellent mechanical properties and wear resistance, Cr/Cr-DLC/DLC #4 DLC coating with Cr base layer and multilayer Cr/C intermediate layer appears to be a strong contender. In contrast to single diamond like carbon (DLC) and CrN coatings with large cracks and delaminated regions, scientists [12] have developed Cr-containing DLC coating with transfer layer and multilayer CrN/DLC architecture. This allows for a sharp temperature gradient to be created, austenitizing the surface while leaving most of the material ferrite (and preventing it from converting to martensitic). Due to the localization of heat near the surface, care must be taken to prevent melting. During rapid cooling, such as quenching, austenite transforms into martensitic. Depending on the carbon concentration, austenite typically occurs in iron-carbon alloys at temperatures above the critical eutectoid temperature (723°C) and below 1500°C. Normal cooling rates cause the carbon in the austenite to diffuse as it cools, forming carbon-rich cementite and leaving behind carbon-poor ferrite. Pearlite is a layer of ferrite and cementite that can form depending on the composition of alloy. However, if rapidly cooled, the carbon becomes supersaturated with carbon and transforms into a highly stressed, body-centered tetragonal form known as martensitic. This coating is characterized by hard Cr layers combined with lubricating amorphous DLC layers and provides better fracture toughness during indentation. In addition, the interfaces of multilayer transfer structure help to limit crack propagation. The strong CrN load support layers, interfaces and lubricating DLC layer are responsible for the low wear rate of CrN/DLC/Cr-DLC coating. The dent morphology, cross-sectional profiles and 3D images of worn tracks sliding with AISI 440C ball are shown in Figure 1.3.

Mass conversion hardening of steels is widely used and usually undergoes an austenitizing process in furnace before hardening to martensitic. In industry, martensitic hardening, especially surface hardening is used to improve the wear resistance and fatigue strength of parts subjected to loads under unloaded surfaces (e.g., bending) and loaded surfaces (Hertzian compression) [17]. Most load-bearing components in the automotive and aerospace industries are strengthened by significantly improving load-bearing capacity, allowing the least amount of material to be used (lightweight). The main difference with surface hardening is the rapid application of heat to the surface, which limits the ability of heat to diffuse into the bulk and away from the surface [12].

This happens because carbon does not have time to diffuse. In martensitic, each carbon atom still exists as an interstitial impurity. The relative amounts of each are determined by the cooling rate of cementite, martensitic and ferrite. Consequently, it determines the mechanical properties of resulting steel, including toughness, tensile strength and hardness.

High power input for a short time will promote the formation of shallow martensitic layer. The transient hardening power density depends on the heat source. Typical methods include induction hardening using high-frequency electrical

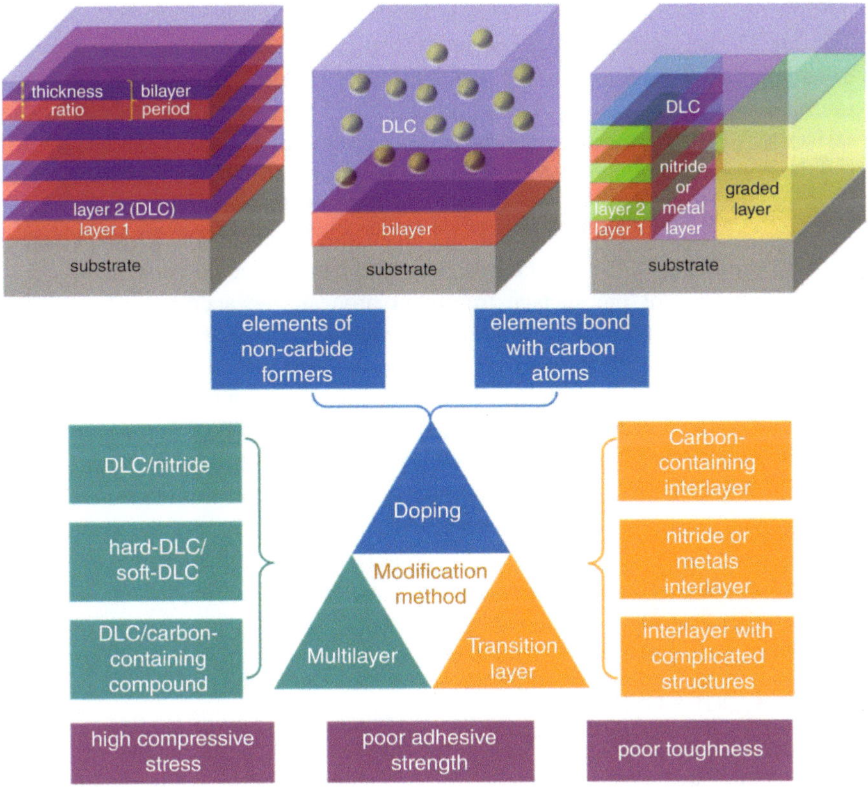

FIGURE 1.3 SEM systematic diagram of modification methods of DLC coating and limitations that need to be resolved [12] [open access].

induction heating, oxy-acetylene or oxy-propane flame hardening, and electron beam hardening by electron beam or laser hardening, respectively. After the surface is austenitized, the workpiece is cooled using external hardening in both flame and induction hardening processes. This is usually done using water jets or water bath. Typically, both approaches achieve hardening depths in the range of 0.25–6 mm. Hardening is the relative ability of ferroalloy to form martensitic. The depth below the hardened surface at which metal exhibits certain hardness, such as 50 high-carbon chromium (HRC), or certain percentage of martensitic in the microstructure, is a standard for evaluating hardness. While martensitic can reach up to 72 HRC, pearl steel can only reach maximum hardness of 43 HRC. If carbon levels exceed 0.2–0.3%, fresh martensitic becomes extremely brittle. Most applications cannot use it due to its extreme fragility. The tempering procedure, which involves gently heating the hardened steel, eliminates this brittleness (albeit with some loss of hardness) [18, 21].

The impact of advances in surface engineering has been seen in the energy sector. Increasing the catalytic activity of materials has been achieved through modifications to the surface to enable more efficient energy conversion activity [22, 23]. The aerospace and defense sectors utilize surface engineering approaches to enhance the functioning and dependability of critical components that operate in challenging environmental conditions [12]. In the automobile industry, wear-resistant coatings and corrosion protection technologies play a key role in ensuring the dependability and safety of engine components and vehicle parts [13, 14]. High-performance devices are being developed in the meantime, thanks to surface changes in the semiconductor and electronics sectors [15]. Despite tremendous progress, the field of surface engineering faces numerous obstacles.

The intricacy of the surface treatment process, the accuracy of deposition parameters and large-scale scalability of manufacturing processes all require more research and development. Furthermore, the economic feasibility of these techniques as well as technology's smooth integration with other manufacturing processes are prerequisites for surface finishing's practical application. In the realm of surface engineering, emerging trends and technologies show a lot of potential for future. Sophisticated computer modeling and simulation techniques can aid in the design and optimization of engineered surfaces of the desired quality. Furthermore, the studies on novel materials like grapheme and other two-dimensional materials show promise for the development of intricate coatings and functionalized surfaces with unique properties.

1.3.2 Curve Control

Bézier curves are essential for computer graphics and geometric modeling. They offer the customer easy-to-use tools for modifying the shape of parametric polynomial curve in addition to many useful features. By assigning the shape parameter to each control point and taking into account rational Bézier curves, which have the added advantage of being able to accurately describe each conic section, even greater control and flexibility can be achieved. In this chapter, we explore the editing capabilities that arise from the barycentric representation of rational Bézier curve. In particular, we demonstrate how to switch between the Bézier shape and barycentric shape, discuss the effects of changing the constituent parts of barycentric shape (nodes, interpolation points and weights) and examine the relationship between the degree of elevation of Bézier shape and insertion of point in barycentric shape. In addition, we explore the useful properties of barycentric shape in curve evaluation. The combination of moderate hardness (450 V) and high-volume fraction of retained austenite (89%) resulted in excellent cavity wear resistance. Due to high martensitic convertibility of laser-melted samples using different processing parameters, it is found that the cavitation corrosion resistance of samples improved with an increasing volume percentage of retained austenite present in UNS S42000 steel. However, cavitation wear resistance increases to maximum value with an increased hardness and then decreases as hardness further increases. This indicates that martensitic convertibility is more important than hardness during cavitation wear. Laser cladding specimens have smaller pits than as-received UNS S42000 hardened steel due to potential anisotropy.

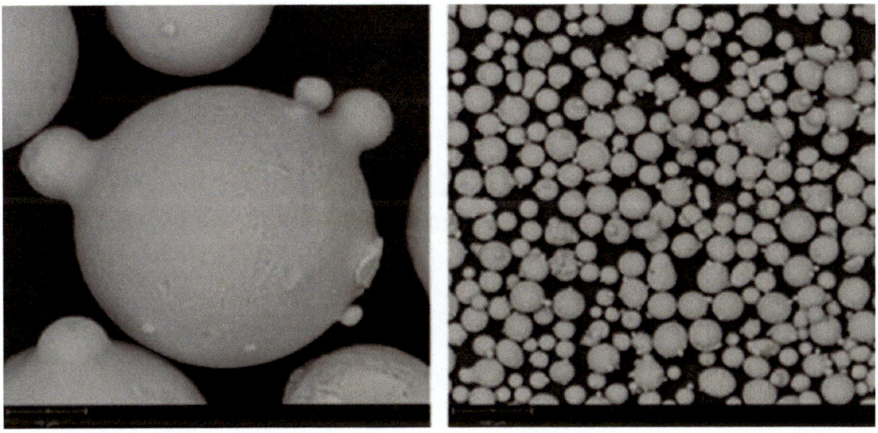

FIGURE 1.4 Powdered stellate at various magnifications. Thirty microns is the average particle size [16] [open access].

Melting or polishing of carbide particles and presence of retained austenite resulted in improved wear resistance, as evidenced by the fact that the etching potential increases linearly with the amount of retained austenite.

By changing process and power, the rate of cooling after a power failure can be adjusted by changing the temperature profile over time. While simple melting leaves composition unchanged, rapid solidification improves the microstructure, which in turn significantly increases the hardness of machined surface compared to the bulk. A Diamond Jet Hybrid (DJH) 2600 is used to create the HVOF coatings. The spray distance is 25 mm, oxygen gas pressure is 22 bar, carrier gas flow rate is 4 L/minute, fuel flow rate is 280 L/minute and oxygen flow rate is 830 L/minute. Figure 1.4 shows the satellite powders used in this study [16].

Surface melting reacts effectively with cast irons, which form a fine-grained structure similar to that produced by traditional die casting. The wear surfaces of crankshafts, camshafts and cam followers are often subjected to TIG melting process that combines strong core material with an extremely hard surface.

1.3.3 Surface Texturing

Surface texturing is the process of creating topographic pattern on a surface, usually at scale much smaller than tribological contact in the form of grooves or dimples. These characteristics can change lubricant flow, lubricant film thickness, or how debris is captured or pushed out of sliding contact, as well as other ways that can affect the contact. Dimpled textures that remind people of golf balls are gaining more and more attention and seem to have a great potential. However, some argue that texturing is harmful or unhelpful in certain test scenarios, suggesting that a full understanding of how texturing affects tribological interactions remains unclear [11]. Clearly, texturing effects must be considered in addition to

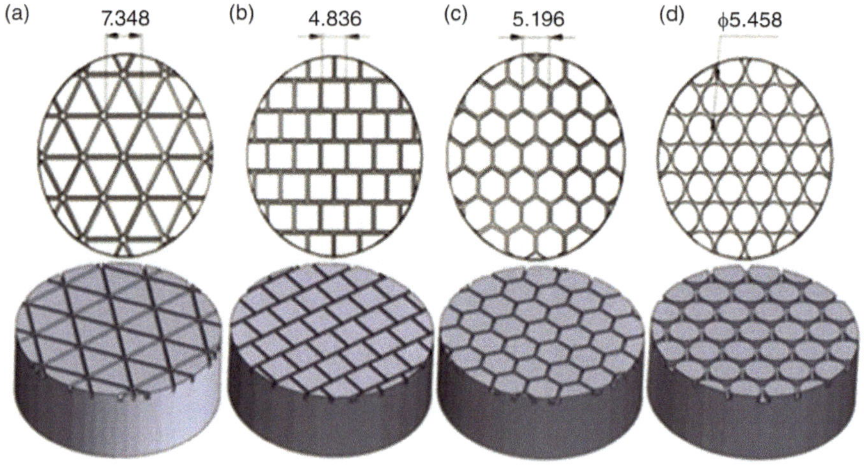

FIGURE 1.5 Four different cross-groove texture shapes shown in schematic illustration of (a) triangle texture, (b) square texture, (c) hexagonal texture and (d) rounded texture [11] [open access].

more typical aspects of tribological contact. Figure 1.5 illustrates a schematic diagram of sample.

All main shape areas are the same, and groove width is 0.5 mm to investigate the effect of transverse groove texture shape on lubrication properties. The samples are copied. On a steel ball, the low-friction DLC is covering glided. Since texture impedes the formation of transfer film on the steel counter-body, the friction increases when the balls slide over a similarly textured surface in absence of lubrication.

A study is conducted on DLC-coated tools for milling aluminum alloys, as shown in Figure 1.6. M-1: 51/60 (sp3 content ratio), M-2: 44:60, M-3: 36:60, and YG6: substrate are alternating layers of high sp3 and low sp3 in Figure 1.6. Panel (a) shows friction coefficient versus sliding times curves. Panel (b) shows average wear rate and friction coefficient of the samples. Panel (c) shows the hardness of the samples, wear rate, and H/E ratio. Panel (d) shows cross-sections of wear track and surface profiles. In contrast, creating a lubricant transfer layer at contact point helped protect the tool surface and reduced the friction force [12].

There are several methods for accomplishing this kind of surface texturing, including laser, lithography, electrical discharge machining (EDM) and mechanical indentation engraving. The primary factors are the topographic features' dimensions and forms, their area fraction and any pattern orientation about the contact's shape and relative motion. Figure 1.7 illustrates how DLC-coated tools are used for turning. The diamond-coated insert has shown a significant advantage over uncoated and DLC-coated inserts in cutting aluminum alloy. Its tool life is significantly longer than uncoated insert, 55 times longer when cutting Al-Si alloy and 27 times longer when cutting aluminum bronze. However, the tool life of DLC-coated insert is 1.5 times longer when cutting mid-Si-Al alloy and 7 times

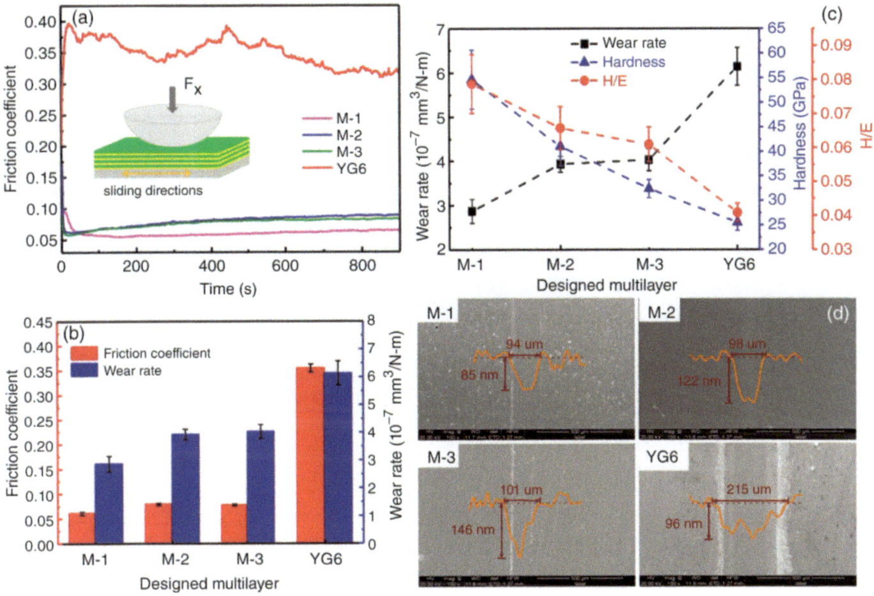

FIGURE 1.6 DLC multilayer construction of alternating high-sp3-fraction and low-sp3-proportion layers, M-1: 51/60 (sp3 content ratio), M-2: 44:60, M-3: 36:60, YG6: substrate. (a) Friction coefficient curves against sliding times. (b) Average friction coefficient and wear rate of samples. (c) Wear rate, hardness and H/E ratio of the samples. (d) Surface and cross-sectional profiles of wear tracks [12] [open access].

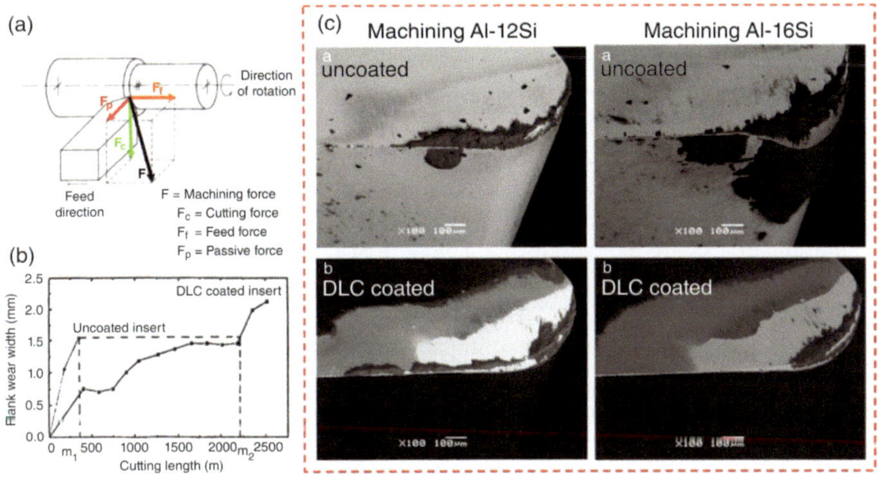

FIGURE 1.7 (a) Schematic representation of the turning operation, (b) wear curves of implants, and (c) adhesion of aluminum to tools. (a-c) Formation of built-up edge (BUE) due to aluminum adhesion on uncoated tools, increasing cutting forces. (c-b) Reduced adhesion and minimized BUE formation on DLC-coated tools, enhancing machining performance [12] [Open access].

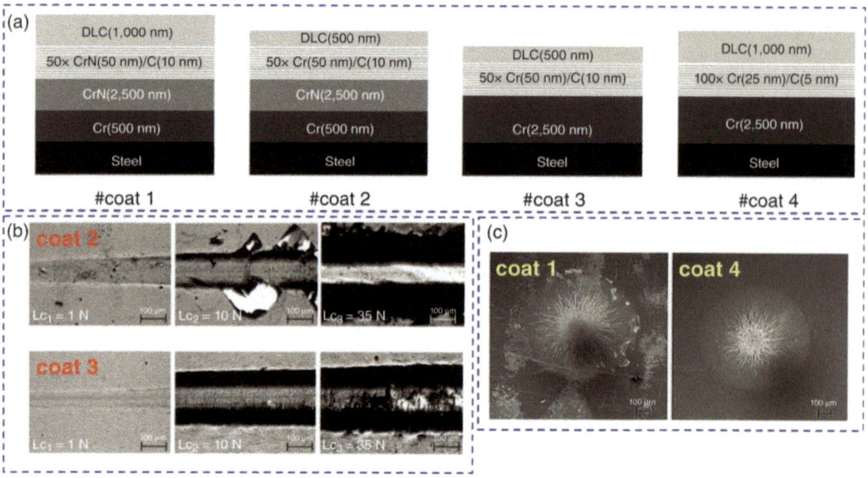

FIGURE 1.8 The adhesion evaluation of DLC coating with a complex multilayer transition layer [51]. (a) Coating design for the various deposition conditions, (b) scratch track optical observation of sample #coat 2 and sample #coat 3, (c) SEM micrographs of the Rockwell test impacts of sample #coat 1 and sample #coat 4 [12] [open access].

longer when cutting aluminum bronze compared to uncoated insert when cutting high-Si-Al alloy [12].

When using DLC-coated and uncoated tools for micro-milling of Inconel 718 superalloy, the coated tools significantly reduce the cutting forces and improved the surface smoothness of workpiece. In addition, compared with uncoated tool, the DLC-coated tool generally has excellent anti-burr performance, as can be seen in Figure 1.8 that shows the experimental results.

As shown in Figure 1.9, when cutting AISI 52.100 (HRC 56–58) hardened steel, the DLC-coated tool with 35 at.% Si shows the strongest wear resistance compared to two DLC-coated tools with different Si concentrations (2 and 58 at.% Si) and uncoated ceramic tools. The wear mechanism is adhesive fatigue, and 2 at.% Si-DLC coated tool shows high tendency to adhere to workpiece. In addition, cracks are likely to form on 58 at.% Si-DLC tool, with transverse cracks being the most common type, which propagated into the ceramic substrate [12]. The tool containing 35 at.% Si-DLC demonstrates an excellent anti-adhesion to the workpiece material and strong impact toughness to prevent crack formation.

As shown in Figure 1.10c, the dry turning performance of hydrogenated multilayer DLC and WC/C coatings when applied on Al_2O_3/TiCN composite ceramic cutting tools for machining AISI 52100 is a high-carbon chromium steel (62 HRC). When cutting steel, coating peeling occurred in both WC/C and DLC coatings, although it is less severe on WC/C multilayer coated tool. This is due to the fact that WC/C coating has stronger adhesion at substrate interface (51N) than the DLC coating (44N). The WC/C coating exhibited less peeling than DLC coating, which is accompanied by better degree of thermal stability [11]. Meanwhile,

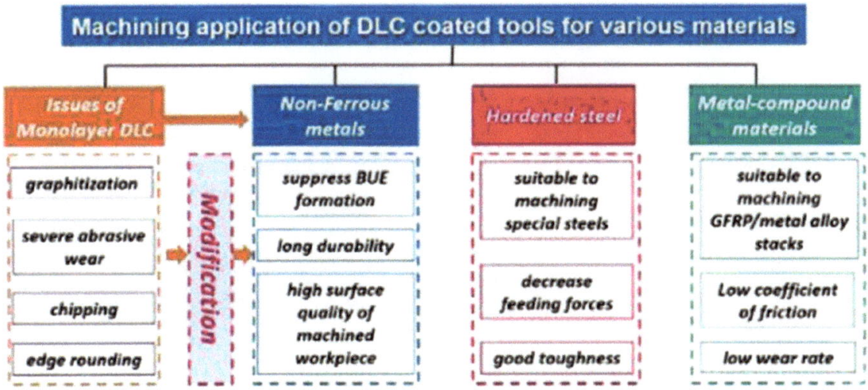

FIGURE 1.9 Systematic diagram of machining application of DLC-coated tools for various materials [12] [open access].

compared with uncoated tools, the two types of coated tools have exhibited superior wear resistance and cutting ability. According to reference [11], the morphology of CVD diamond films consists of deep valleys and octahedral asperities, while the surface topography of DLC films is flatter [12].

Compared with CVD diamond films, DLC films showed lower coefficient of friction (CoF) and higher adhesion to WC-Co substrate. While the DLC-coated tool continuously maintains nearly constant wear throughout the turning process,

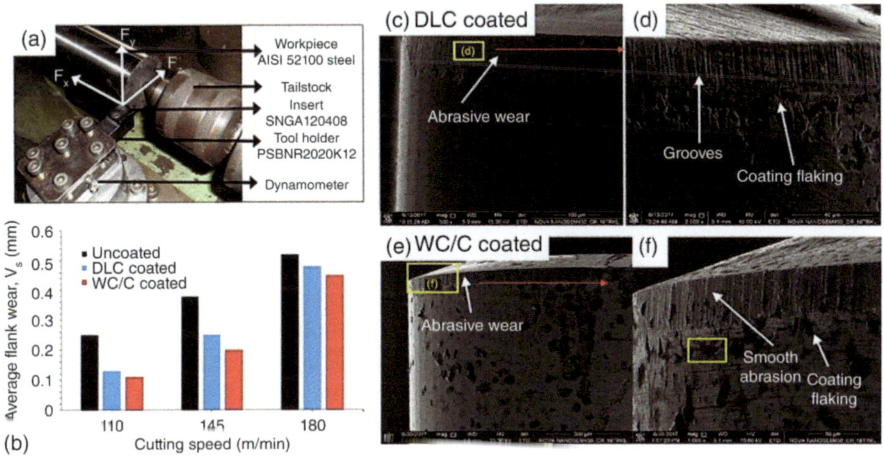

FIGURE 1.10 (a) Experimental configuration for turning. Dry turning performances of hydrogenated DLC and WC/C multiplayer coatings. (b) Cutting tool average flank wear. SEM micrographs of flank surfaces of cutting tools coated with DLC (c) and WC/C (e). The enlarged representations of matching yellow rectangle areas in (c, e) are (d, f), respectively [12] [Unrestricted access].

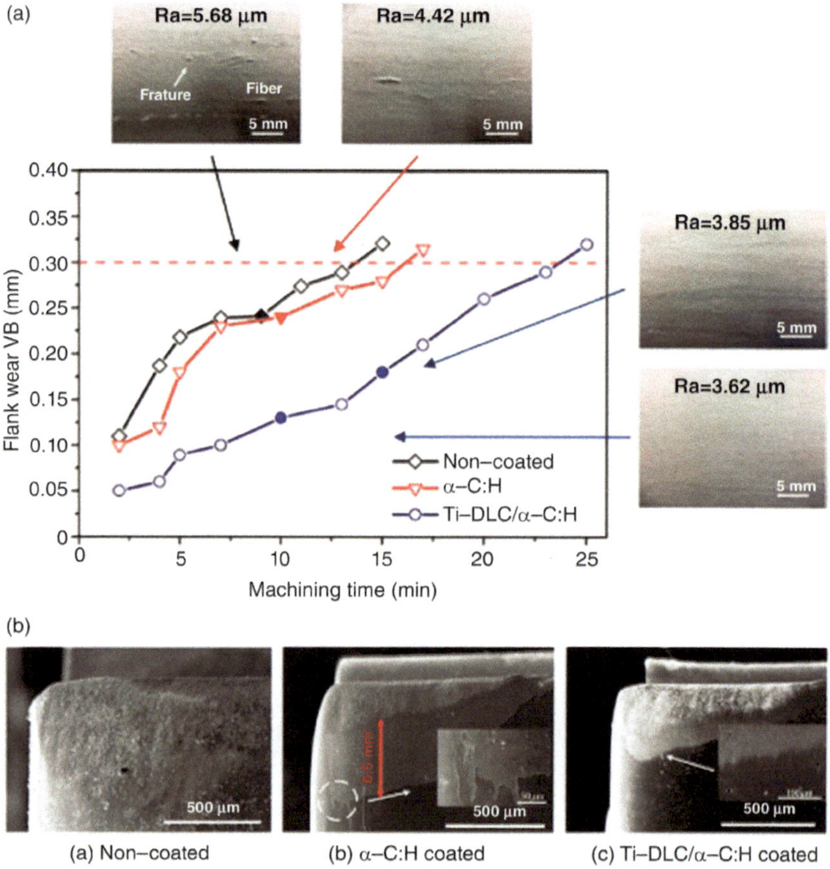

FIGURE 1.11 Multilayer Ti-DLC/a-C: H film in the turning of fiberglass (a) machined surface quality and associated changes in flank wear. (b) Images of flank wear. (c) effect of Ti-DLC/a-C: H coating on lateral wear of tools in GFRP machining, except for a-C[12] [open access].

the flank abrasive wear of CVD diamond tools maintains moderate degree before peeling off from the substrate. The Ti-DLC/a-C:H multilayer film with abrasive wear dominant wear mode in GFRP turning without obvious diffusion wear had the lowest wear rate, longest tool life and best-machined surface quality as well as oxidation wear [18]. The flank wear of cutting tools with Ti-DLC/a-C:H coating, a-C:H coating and without coating is shown in Figure 1.11.

When drilling stacks, the DLC-coated tool showed lower peak thrust force and delamination factor value than the TiAlN-coated drills, which depend on the thrust force and tool wear. In addition, as shown in Figure 1.12, the difference between DCFRPin and DCFRPout with TiAlN-coated tool is more noticeable than DLC-coated tool. In the machining process, the tool coating, which is located between the tool substrate and workpiece, is critical for achieving excellent cutting performance, a strong thermal barrier and well-finished surface [12].

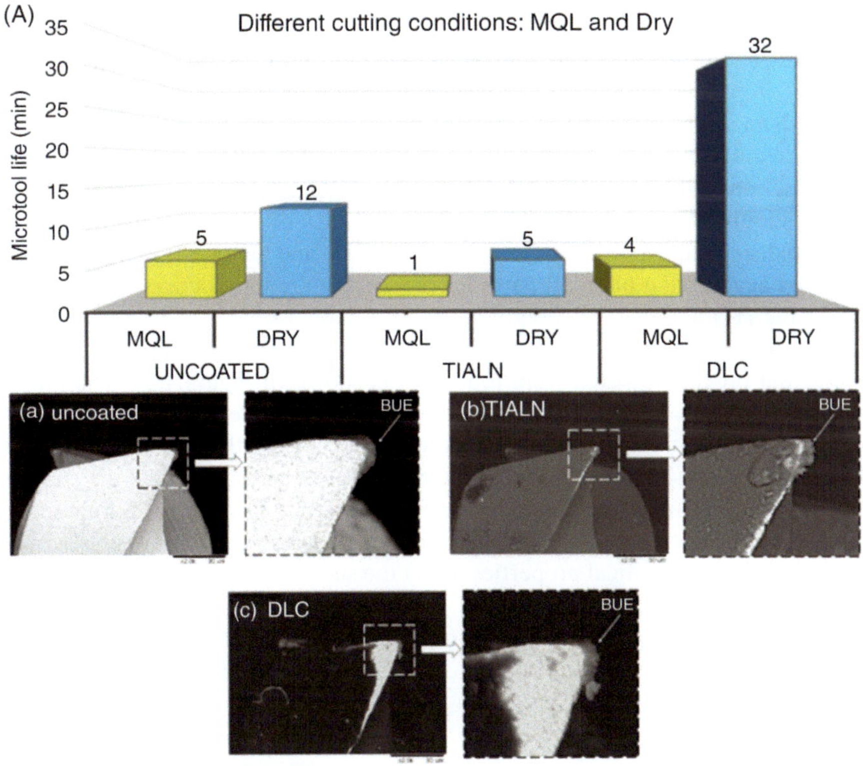

FIGURE 1.12 Dry micro-milling of the Ti-6Al-4V alloy [76]. (A) Microtools' working life comparison. Wear morphologies of cutting edge and BUE of (a) uncoated tool, (b) TiAlN coated tool, (c) DLC coated tool under dry conditions [12] [open access].

TiAlN coated tools are protected from abrasive wear [17]. The TiAlN nano-composite-coated tool has shown 2.25 times higher flank wear than DLC-coated tool. In absence of built-up edge (BUE) during CFRP/AA7075 stack drilling, the main wear mechanisms of DLC-coated tool are chipping, edge rounding and flank wear. Due to low friction coefficient, high hardness and good chemical inertness, DLC coating is used as a type of cutting tool coating material for cutting various materials. Unfortunately, many poor mechanical properties such as high internal compressive stress and poor adhesion to substrate seriously limit the use of DLC coatings in manufacturing and machining.

1.4 TECHNIQUES FOR METAL SURFACE ENGINEERING

To achieve the desired qualities and properties, metals must undergo surface treatment processes to change their surface properties. This section discusses the basic physical and chemical methods of metal surface technology. Surface treatment processes include electroplating, surface hardening, oxidation, chemical steam, thermal spraying, nitriding (CVD) and PVD.

1.4.1 Physical Techniques

Microstructure mechanisms, characteristics and surface layer characterization are key topics of discussion as mechanical and materials engineers review current advances in surface engineering techniques and applications. Functional surfaces in mechanical systems, surface engineering through friction stir processing and friction cladding, material deposition using the cold laser deposition process, laser additive manufacturing for metal surface modification, surface modification and structure control to control liquid phase, and nanoscale adhesion behavior and aggregation of small particles are among the topics covered [18]. Flame spraying [17] and plasma spraying [16] are a few examples of thermal spraying techniques. An outline of wire laser additive manufacturing method and pictures of actual procedure taken from top-down is shown in Figure 1.13. The melting pool and molten metal of wire are simultaneously created by laser. The metallurgical bond with the substrate can be made by passing the molten wire into the pool. The bead-shaped solid body is created by the wire feeder on the substrate and movement of laser processing head. Numerous factors, including surface quality, deposition shape and quality, the final microstructure of deposited layer and resulting mechanical properties affect the success of weighted linear additive model (WLAM) process.

This method uses physical evaporation and condensation of starting materials to create a thin layer of coating. Materials, including ceramics, alloys and

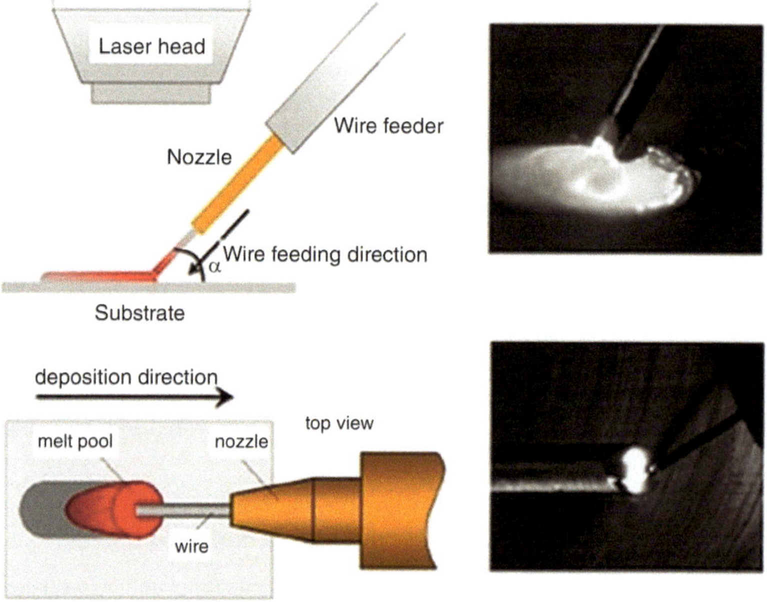

FIGURE 1.13 Illustration of additive manufacturing process using wire laser along with top-down photos of the actual workflow [13] [open access].

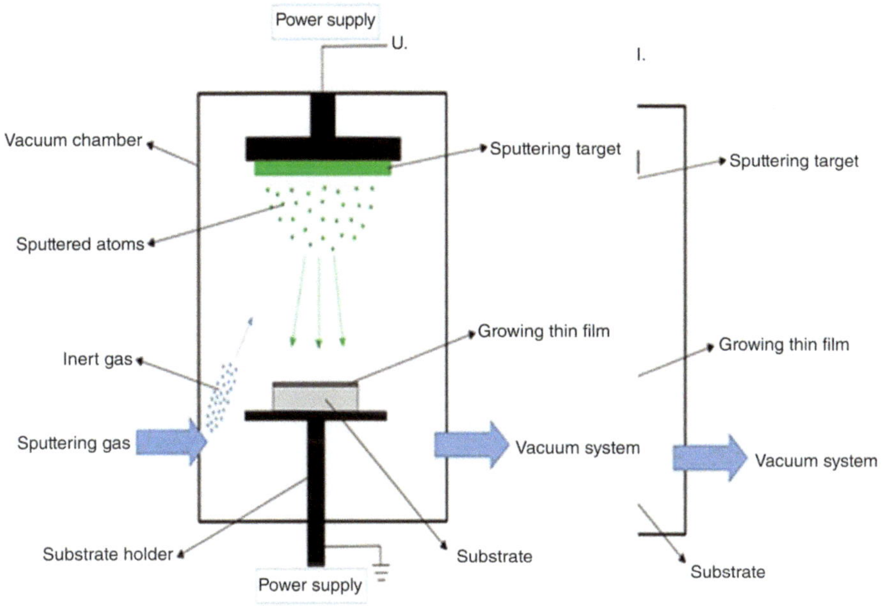

FIGURE 1.14 Physical vapor deposition coating [13] [open access].

metals, are deposited using techniques such as spraying and evaporation. PVD coatings are suitable for wear-resistant and attractive applications due to their excellent adhesion, hardness and chemical stability [23]. The PVD coating process is shown in Figure 1.14. CVD is a chemical process that allows vaporized precursor molecules to react to form thin coatings on metal surfaces. On a heated substrate, the precursor molecules are destroyed to form a hard coating. CVD coatings have precise control of composition and thickness, conformal coating and high adhesion [21–23].

The formation of complex structure is facilitated by computer numerical control (CNC) sequencing which provides a single or integrated combination of motion involving the part, electron beam or wire feeder. The main advantage of electron beams additive manufacturing (EBAM) compared to other directed energy deposition (DED) technologies is that it prevents surface oxidation, which improves the purity of formed layers. Better and faster beam control with electromagnetic lenses is another advantage of this approach; it reduces the difficulty of increasing the deposition rate of electrically conductive materials, especially in highly reflective alloys such as copper and aluminum, and increases the sensitiv ity index. The delicate nature of Index can be defined as the volume of product divided by the size of actual deposit (Figure 1.15).

Using a concentrated electron beam in a high vacuum chamber, EBAM can produce and maintain a molten pool. It is important to note that metal wire is the main feed material used in this procedure. The functional safety assessment

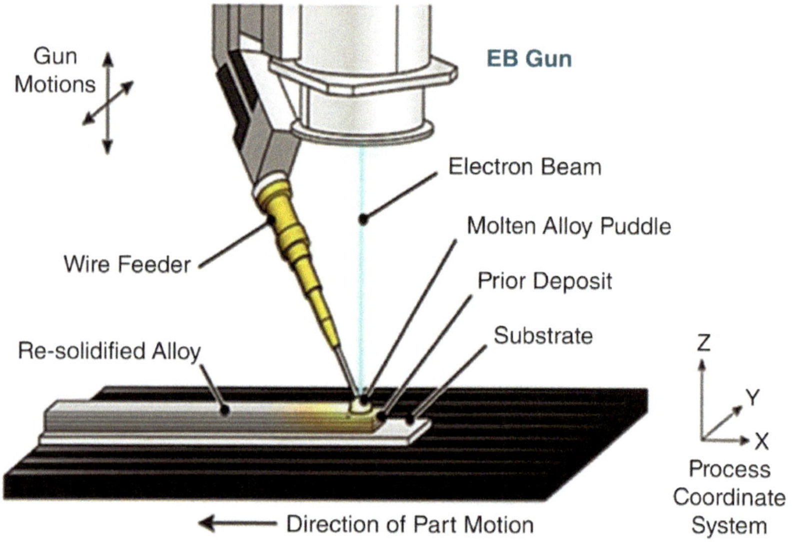

FIGURE 1.15 Rudimentary procedure of additive manufacturing with electron beam [13] [open access].

method (FSAM) process is shown in Figure 1.16. Enough heat is generated in the workpiece due to friction and plastic deformation to allow the layers to join. Material transport, heat generation and consolidation of material from the front of tool to back of tool ensure the creation of the joint. The geometry of tool used in FSAM determines not only the process parameters but also, to some extent, the macroscopic and microscopic characteristics of the manufactured parts [23, 24].

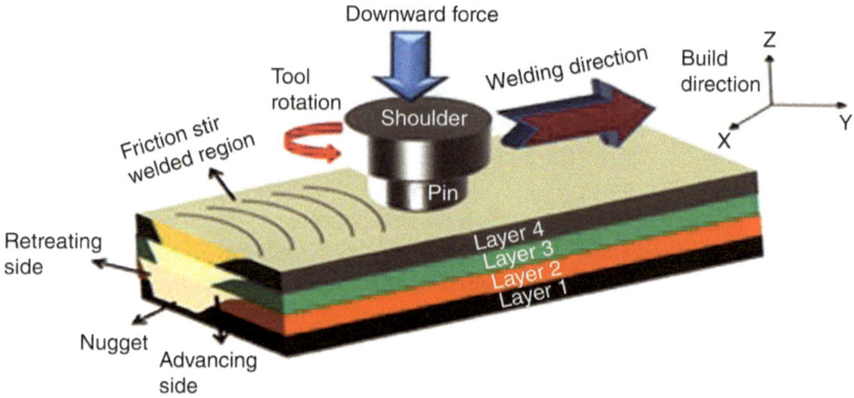

FIGURE 1.16 Schematic of laser cladding [13] [open access].

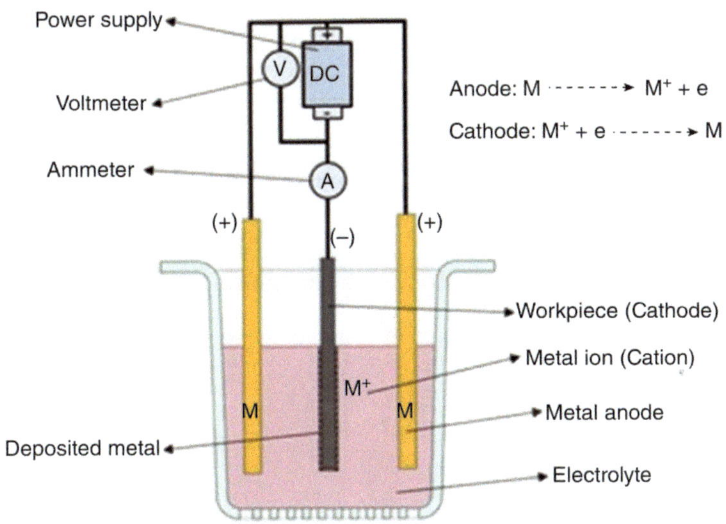

FIGURE 1.17 Electroplating process [13] [open access].

1.4.2 Chemical Techniques

Galvanic coating is the widely used electroplating method that uses an electrochemical process to deposit a metal coating on a conductive substrate. Electrodeposition of appropriate metal is initiated by applying an electric current to substrate while it is immersed in an electrolyte solution containing metal ions, as shown in Figure 1.17.

The sol–gel treatment method was originally documented by Eppelmann in the mid-18th century. It is previously used in the production of building and decorative materials, but has since undergone significant development and has been used in the production of glass, ceramics, catalysts, coatings, composites and fibers. In general, the sol–gel process entails the transition of system from liquid (usually colloidal) phase, known as "sol", to a solid (gel) phase. Surface qualities, such as conductivity, appearance and corrosion resistance, can be improved by electroplating. It is widely used in many industries, including automotive, electronics and jewelry design [21, 22]. Chemical coating is unlike electroplating, which requires an external power source to apply the metal coating to the substrate. Chemical plating (shown in Figure 1.18) is a non-electrochemical method [11, 12]. Due to unique properties of sol–gel materials (e.g., optical transparency, high surface porosity and chemical inertness), this method has gradually expanded to the encapsulation of specific components for biorecognition. A biosensor is a device that uses a biological recognition element to provide selective quantitative or semi-quantitative analytical information. This is accomplished using a transducer and processor to convert the biochemical event into a detectable signal. An 8-10 A transducer converts changes in physical and chemical properties during the detection of analyte into a corresponding analyte signal, the selective

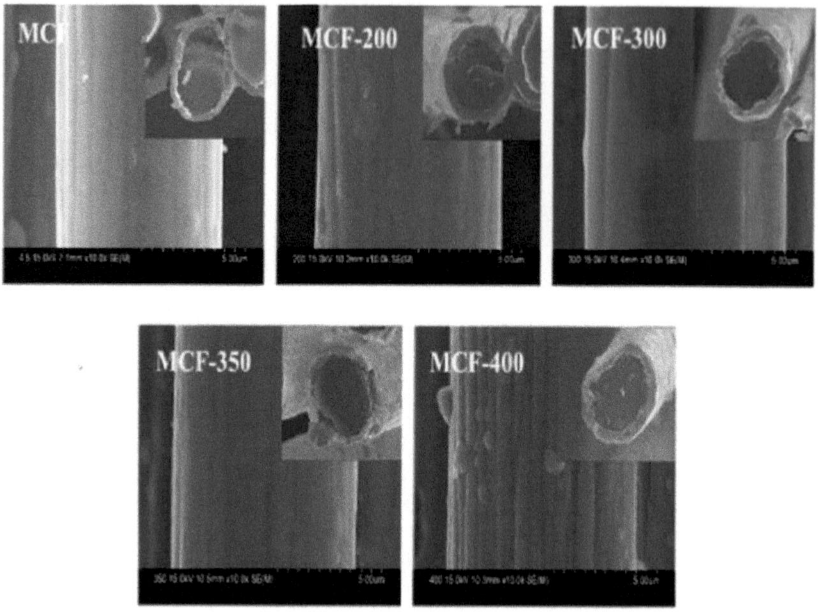

FIGURE 1.18 SEM images of MCF as a function of heat treatment temperature [12] [open access].

biological recognition element. Bioreceptor constitutes a biosensor, usually a self-contained analytical device capable of detecting the analyte. Biosensors can be classified based on the concept of biorecognition (e.g., enzymatic, immunoaffinity recognition or whole-cell biosensor) or transport element (e.g., electrochemical, optical, piezoelectric or thermal sensor or magnetic biosensor).

Advantages of sol–gel coatings include chemical resistance, optical clarity and strong adhesion. They are used in protective coatings, optics, sensors and so on [13–15]. Figure 1.19 shows a diagram of sol–gel coating procedure. Chemical etching, sometimes called chemical milling, is the process of using chemical etchants to specifically remove material from a metal surface. Sol–gel materials provide a suitable and adaptable basis for the production of biosensors.

Since the initial adoption of term "biosensor" [13], many interesting biosensor designs have been proposed in several research papers over the past few decades. However, since immobilization is a critical step in biosensor development, simple, affordable and reliable bioreceptor immobilization methods are still needed for biosensor development.

The sol–gel method is used in the production of variety of biosensors for a wide range of applications. A comprehensive literature review reveals that a large number of recent publications unequivocally show the growing interest in the sol–gel method for developing biosensors [16]. It is critical to remember that the choice of surface treatment process depends on several variables such as the quality required, substrate type, scalability and cost.

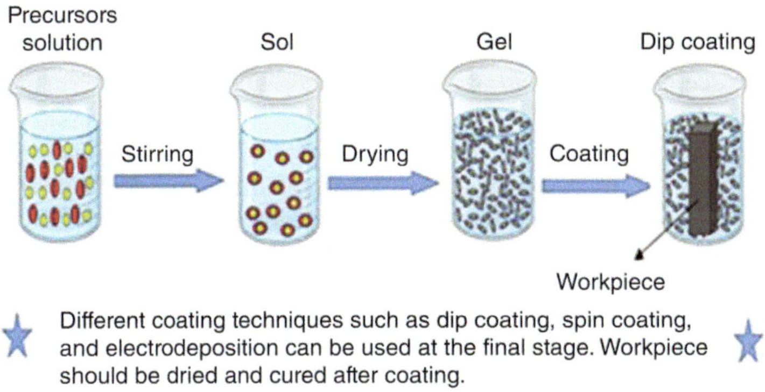

FIGURE 1.19 Process flow diagram for sol–gel dip coating [13] [open access].

Each strategy has its advantages and disadvantages, so the best approach should be chosen after careful analysis of specific application's needs and performance standards. Over few past years, additional study has been carried out to improve these methods. For example, developments in the field of thermal spray coatings have made it possible to create nanostructured coatings with improved tribological and mechanical characteristics [17, 18].

1.5 CHARACTERIZATION METHODS FOR SURFACE ENGINEERING

A beam of ions is used in many surface inspection techniques, such as secondary ion mass spectrometry (SIMS), to strike the surface and remove atoms of sample material. Mass spectrometry is used to ionize, identify, and measure these atoms. Ion scattering spectroscopy (ISS) examines the surface coating. A specific amount of beam particles will be reflected elastically if an ion beam is directed at the sample surface. The mass of these atoms can be determined by measuring the energy of scattered particles. The function of emission angle and strength of scattered ions gives information about the surface crystal structure. The difference results from the intensity of scattered beams from absorbing atoms casting a shadow on substrate atoms. The positions are occupied by surface atoms that can be determined by applying scattering theory [2, 3].

1.5.1 Engineering Modeling

Modern technologies are developing rapidly which naturally increases the demands on the mathematical modeling process since real-life situations require complex coupled models. In addition, the model selection is necessary since several models can usually be used to model a certain physical phenomenon. It seems that the quality of final coupled model depends on the model selection. In this sense, understanding the function of individual models in a complex coupled model and studying the practical significance of various model relationships are some of most important responsibilities of a modeler. Figure 1.20 shows 2D microscopic, phase contrast, darkfield and 3D profilometric images that are used to analyze the surfaces of ultrasonic test (UT), static compression test (SCT) and dynamic compression test (DCT) samples. The UT samples have more green areas. The DCT samples show an increase in red areas. The DCT samples show a noticeable change in height. In both electron and light microscopy, methods that remove un-scattered rays from the image are called darkfield microscopy. This image gives an accurate representation of sample being measured in a dark field, as does a darkfield microscope. Therefore, the field surrounding the sample is often dark. Figure 1.20 illustrates that the DCT sample appears to have blacker and whiter areas. Evaluation of mathematical models solely based on their physical assumptions, i.e., without taking into account a specific engineering example or performing calculations using abstract mathematical tools are needed to create the model. So, some methods have been proposed for applying abstract mathematics to practical mathematical modeling. These include graph theory [3], abstract Hilbert spaces [4, 5], abstract algebraic approach [6, 7], predicate logic [8, 9], type theory [10, 11] and category theory [12, 13].

Uncertainties are commonly used to describe various problems that reduce the quality of final coupled model in engineering applications. As stated in reference [1], the following categories of uncertainty associated with modeling can be

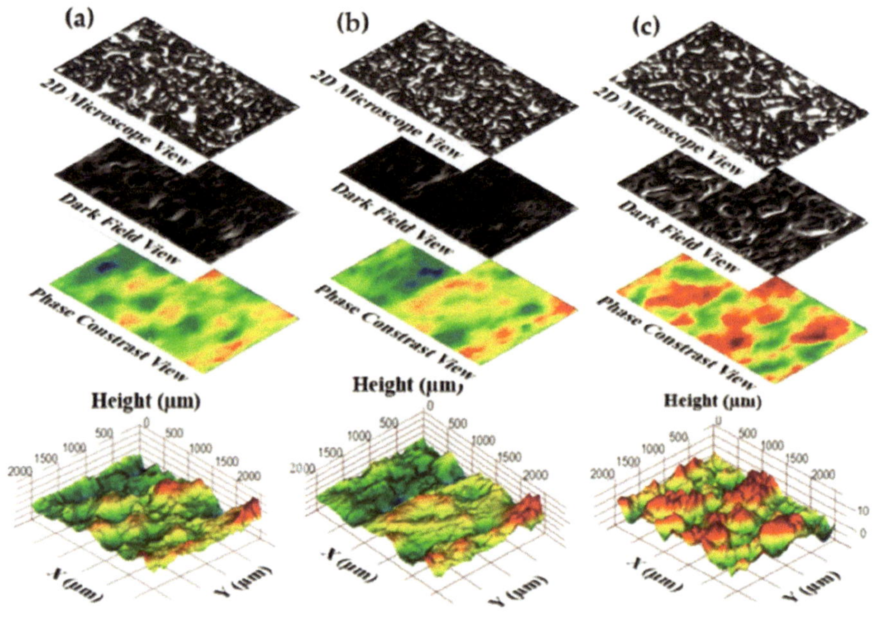

FIGURE 1.20 Surface photographs of samples obtained using (a) UT (500 μs-6 A), (b) SCT (300 μs-6 A), and (c) DCT (500 μs-6 A)[7] [open access].

distinguished: (i) model input data; (ii) numerical approximation and (iii) model shape. As examples in references [2, 3] and references therein show, computational and statistical methods can be used to determine and handle the first two categories; however, additional attention is needed for the third type. Uncertainty in model shape suggests that a conceptual modeling error has occurred or that fundamental physical assumptions of the models have been violated. Methods need to be developed to correct conceptual modeling errors because impact of these deficiencies on the overall modeling process is much more significant. Typical surfaces used in AFM study are shown in Figure 1.21. The difference between polishing with 0.25-μm paste and with 1-μm paste is shown. Despite the obvious waviness and scratches of surfaces, it is still possible to determine a baseline that corresponded to the polished pyrite surface free of foreign particles and oxide. From this baseline, the peak heights are calculated. In this study, the depth of valleys concerning the baseline is not considered. Various features of photographs are corrected and analyzed using the open-source image analysis tool Gwyddion v2.48. According to topographic study, the pyrite in Figure 1.21a has an average height of 0.4 nm in contrast to pyrite in Figure 1.21b, whose

FIGURE 1.21 Polished {100} pyrite surface coated with 1 μm and 0.25 μm diamond paste, respectively. show 3D surfaces in parts (a) and (b), the scanning size is 10 × 10 μm², Parts (c) and (d), respectively.

average height is 0.35 nm. Both surfaces are smooth, but the pyrite polished with the thickest paste has more local depressions, which are probably caused by scratches during the polishing process. By all measures, pyrite polished with the finest paste is smoother [8].

This study aims to explore and improve the category theory-based modeling approach first presented in reference [13]. Several factors encourage the abstract description of mathematical models using category theory: (i) the abstract nature of category theory allows a wide range of objects and structures to be described in a general way; (ii) the practical interpretation of abstract constructs using category theory-based modeling methodology is simple, making the methodology useful in engineering practice; (iii) the scalability of category theory is an integral part, which suggests that descriptions of more complex things and structures can be made using the same ideas as descriptions of their constituent elements; (iv) finally, there are diverse applications of category theory, from modeling of dynamic systems [14] to ontological representation of knowledge [15].

The advantages of category theory are now recognized and appreciated by both non-mathematicians and those interested in applications. Grain boundaries, material defects and crystal structures are shown using TEM. This is especially useful for characterizing interfaces, nanoparticles and thin films. A deeper understanding of the properties and functionalities of materials is made possible by the important information that TEM offers regarding their characteristics at the atomic level. This is very useful for thin film analysis as it allows precise testing of uniformity, thickness and crystal structure. Moreover, because TEM provides high-resolution images to study size, shape and internal structure, it is a critical for studying nanoparticles and surfaces. TEM's companion focused ion beam (FIB) technology enables precise sample preparation, allowing in situ nanoscale modification and cross-sectional imaging. FIB allows the control of grinding extraction of thin lamellas and the creation of nanostructures, increasing the potential of TEM studies. The diffraction pattern produced by exposing a polycrystalline material to XRD can reveal details of the arrangement of atoms and the crystal lattice. The crystallinity of phases found in coatings or surface-modified layers can be assessed using XRD.

By comparing diffraction results, phases in a sample can be identified by matching their structure to databases of known crystallographic information. In addition, XRD can reveal information about the quality of crystal developed, as well as any stress or structural defects in the material. XRD is a valuable tool for characterizing crystalline phases formed during surface modification or coating applications. This information is a critical to maximizing the performance of engineered surfaces and understanding the correlation between structure and properties.

1.5.2 Surface Topography Analysis

Cell fate can also be influenced by surface topography. This occurs mainly through control of elongation, tension and shape. As demonstrated, Managed Service Providers (MSPs) on smaller nanotubes (~30 nm) underwent adhesion without differentiation, whereas MSCs on larger nanotubes (~100 nm) are targeted to osteocyte development through cytoskeletal contractions. According to another study in 2015, MSCs grown on substrates with ridges and grooves located close together (high surface area) rather than far apart (low surface area) prefer an osteogenic fate. MSC fate is thought to be influenced by additional topographic surface properties such as pattern frequency, cell alignment defined by isotropic patterns and roughness (Ra = 1–2 μm promotes osteoclast development). Figure 1.22 demonstrates the result of exposing a polished unoxidized pyrite surface to a 10^{-3} M solution. For the first time, adsorbate spots are regularly formed on the surface; this is not observed on the original surface. These spots have an average peak height of 8 nm, an average radius of about 130 nm and an average peak of 5 nm, which is 10 times higher than average without PAX. These spots have a truncated cone shape with elevation angles (radius to height) of about 3 to 4 degrees. With the addition of advanced modes like SCPM and MFM, the capabilities of AFM have been extended to include the investigation of electrical and magnetic properties. STM is a method that lets you look at a surface and get atomic-level images [8].

Texture, surface characteristics and roughness are among the variables that can be quantified using AFM. In addition, more sophisticated modes such as magnetic force microscopy (MFM) and scanning Kelvin probe microscopy

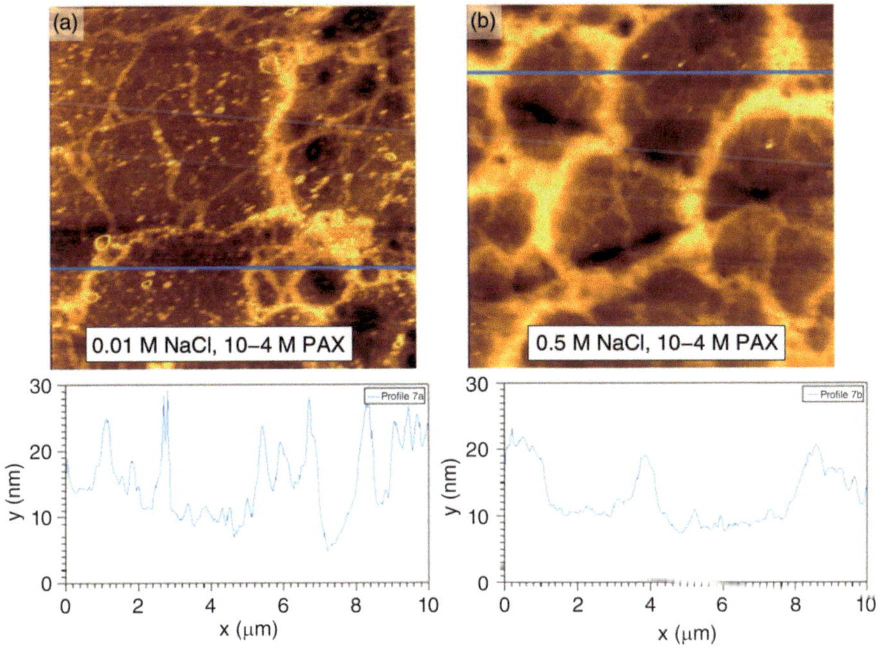

FIGURE 1.22 Surfaces {100} of oxidized pyrite exposed to PAX and sodium chloride at pH 8.5, scan size is 10 × 10 μm², and corresponding height profiles (blue lines in the images). (a) 0.01 M NaCl, 10^{-4} M PAX; (b) 0.5 M NaCl, 10^{-4} M PAX [8] [open access].

(SKPM) can reveal details about magnetic and surface potentials, respectively. Due to its versatility, AFM is a useful technique for surface characterization in a variety of academic and industrial contexts. Exposing oxidized pyrite samples to 10^{-3} M PAX solution produced the results shown in Figure 1.23. Figure 1.23a depicts a bright oxide percolation network with a few small isolated shiny spots, as well as a finer, denser network associated with a percolation backbone that can be easily identified. The percolation backbone intersected by line 1 in the upper part of image consists of two pillars, one with height of 16 nm and base width of 1000 nm, and another with height of 14 nm and base width of 1800 nm according to the height profiles of Figure 1.23a. In contrast, the backbone intersected by line 2 is shown in the lower part of the image.

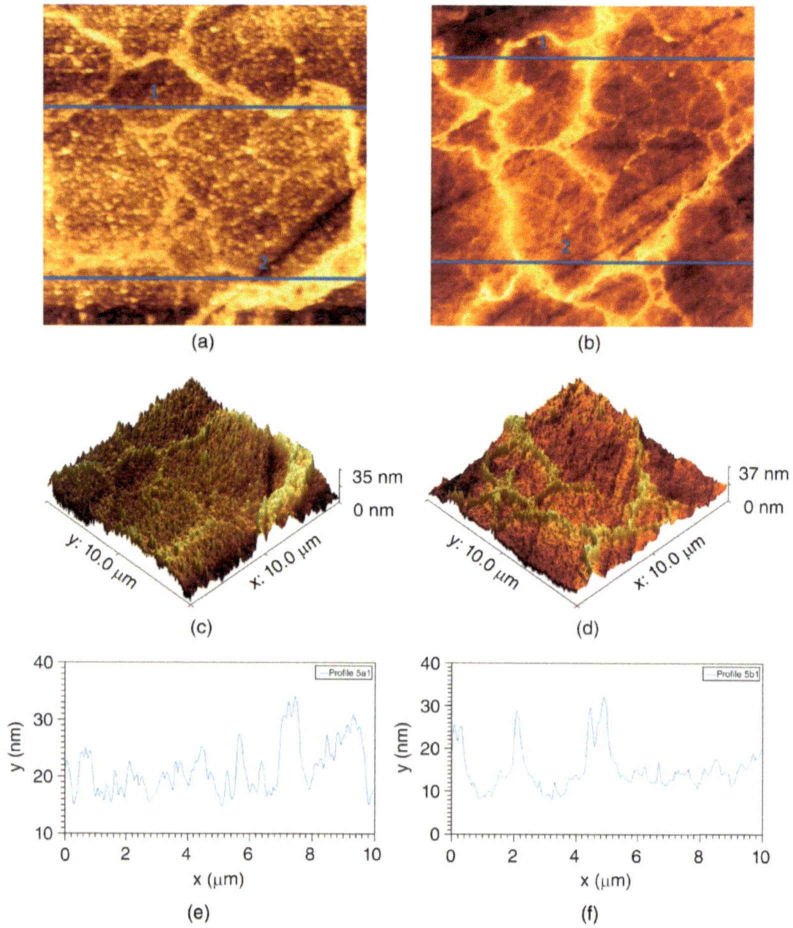

FIGURE 1.23 At pH 8.5, surface of oxidized pyrite is exposed to 10^{-3} M PAX. Blue lines 1 and 2 in (a) and (b) correspond to these profiles. The scan size is 10 × 10 µm² (c, d) topographic maps of corresponding photographs. (e, f) Elevation profiles with highest peaks corresponding to pillars, respectively [8] [open access].

The sharp metal tip of probe scans the surface of sample as a tiny tunnel current runs between the surface and material. STM tracks the distance between the tip and sample to produce an atomic-resolution 3D image of surface. MSC osteogenesis is influenced by porosity, which includes the size and frequency of pores throughout the substrate. Materials that mimic the high porosity of trabecular bone promote the development of osteogenic function. A growing body of research suggests that osteogenesis can be induced by substrates that change geometric shape to promote cellular elongation.

Surface profiling is a technique used to evaluate changes in specimen height and determine surface roughness. It scans the surface with a stylus or uses optical methods to record the height profile and quantitative information on characteristics including average roughness (Ra), maximum height (Rmax) and spatial wavelength characteristics. In the field of surface processing, this method is a widely used for process optimization, surface characterization and quality control. Prolometry helps ensure a given uniformity and quality of resulting surfaces by measuring surface roughness. Through its assistance in assessing the effectiveness of coatings and surface treatments, researchers and engineers can improve material performance and optimize operations in areas where precise control of surface quality is critical. When it comes to creating different textures and surface configurations, such as complex geometries and continuous and discontinuous surfaces, laser ablation is an ideal method for the microfabrication of tools consisting of hard and superhard materials (Figure 1.24).

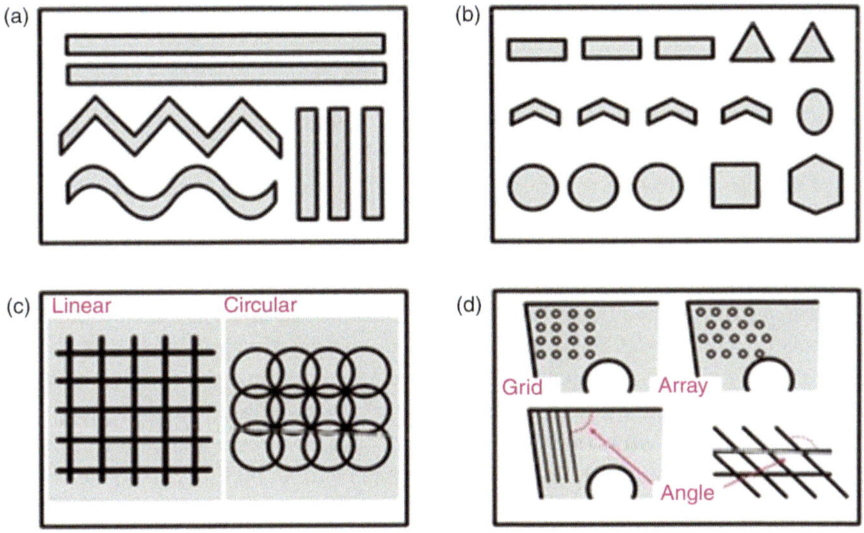

FIGURE 1.24 Popular microtexture configurations and designs, including (a) continuous textures, (b) discrete textures, (c) hatching characteristics and (d) texture orientation [13] [open access].

1.6 TRIBOLOGICAL AND MECHANICAL PROPERTIES

Hardness test is a test that determines how resistant a substance is to piercing or indentation. Vickers, Brinell and Rockwell hardness tests are three popular methods for measuring hardness. The mechanical properties of material such as its toughness, hardness and elastic modulus are assessed during these tests and critical to determining the durability and wear resistance of materials intended for surface engineering. In practical applications, various commonly used hardness tests such as Brinell, Vickers and Rockwell tests are used.

Each test determines the hardness of material using specific indentation methods and measurement specifications. Hardness testing evaluates material's resistance to deformation and external influences through controlled indentation or penetration. Tribology is the study of wear, lubrication and friction on engineering surfaces to accurately determine surface interactions and recommend improvements for specific applications. As a result, the study of tribology is interdisciplinary and includes mechanics, chemistry, physics and materials science. This chapter describes how ion implantation can affect adhesive contact between clean surfaces. This directly affects the important technological areas of metal forming and extrusion, where there are strong local adhesion forces and plastic deformation forces. Die wear behavior has been greatly improved by ion implantation, which may be the most significant advance in die processing since cemented tungsten carbide is first used. This section also provides a summary of using ion implantation in manufacturing technology. This discusses two types of wear tests that are well-suited to measure wear of adhesives. These are the crossed cylinder test, and the pin and disc test. The goal is to calculate the total wear removed as a function of speed and load over a specified period.

Real contact and sliding conditions are simulated using methods such as reciprocating sliding tests, pin-on-disc tests and ball-on-disc tests. These tests quantify variables such as surface degradation, wear rate and coefficient of friction. They provide valuable data on how well surface treatment techniques reduce wear, improve lubrication and improve the tribological performance of materials [21].

These tests provide a realistic assessment of material wear resistance and lubrication effectiveness by simulating the mechanical interactions and pressures encountered during sliding or friction. The results of these tests provide measurements that are useful in selecting the best coatings, lubricants or surface treatments to improve the tribological performance of materials under real-world operating conditions [22]. Researchers can obtain detailed information about friction characteristics such as adhesion, surface interactions and coefficient of friction at the nanoscale by examining the lateral force signal.

It is a critical to understand the basic differences in scale and contact mechanics when comparing macro- and nanoscale friction measurements using frequency modulation force microscopy (FFM). The forces and surface interactions that determine friction behavior at nanoscale can be very different from those at the macroscale. The influence of nanoscale friction from a number of elements such as adhesion, surface roughness, lubrication effects and interactions at the atomic level. On another hand, the normal force, surface roughness and type of sliding material are the main determining factors of macroscale friction [16–18]. While, the friction measurements at both the nano- and macro-scale provide an important insight into the fundamental principles of friction and essential for development.

Researchers can evaluate how surface technology coatings or modified layers respond to external forces using scratch and indentation tests. These tests measure the hardness of a material, determining how resistant it is to deformation due to controlled dents or scratches. They also measure the bond strength of modified layers or coatings, determining how well they can withstand applied loads without breaking or delaminating. These tests provide a decisive characterization of mechanical properties of surface materials produced. They support the process of selecting appropriate materials, improving coating methods and determining the durability of specific surfaces under different operating conditions. Scratch and dent test measurements help determine how well a material resists wear, withstands mechanical stress, and can continue to function.

Using these characterization approaches, researchers can gain a comprehensive understanding of structural, topographical and mechanical properties of surface-modified metals. This knowledge is useful. Evaluate, refine and ensure the quality of surface treatment procedures. Better materials and surface treatment methods can be developed with the help of performance data collected through these methods, which establishes correlations between processing parameters, microstructure and functional characteristics. Appropriate characterization methods will depend on specific objectives and characteristics of surface survey. Because each approach has its pros and cons, it is common practice to fully characterize manufactured surfaces using a combination of complementary approaches.

1.6.1 Softness and Continuity and Surface Technologies

Soft set theory is introduced in 1999 as a new method for solving problems with partial information. Many fields such as "function smoothness", "Riemann integration", "measurement theory", "probability theory", "game theory" and others have used this idea [1]. Soft set theory depends critically on the nature of parameter sets, which provide a broad framework for modeling uncertain data. In a relatively short time, it has contributed significantly to the development of soft set theory. The theoretical foundations of soft set theory are studied in detail in reference [2]. They have developed operators and operations between soft sets specifically for this use. Subsequently, other mathematicians recreated and put forward new varieties of operators and operations between the provided soft sets, thanks to the efforts by authors in reference [3] for a review of recent contributions using soft operators and operations. The structure of soft topology, which is an extension of general topology, is established in 2011 [4]. Thereafter, soft topology is added to a set of general topological concepts. In 2013, it described soft continuity of functions. Subsequently, other varieties of soft openness and soft continuity of functions are developed in [6–16] and other literature. Many disciplines of mathematics, most notably the theory of real functions, investigate various forms of generalized continuity. In many industries, surface treatments are necessary to improve material performance, protect it from degradation and create advanced functionality. Surface Technologies' Catalysts and Coatings divisions provide surface chemical solutions. Its offerings include automotive original equipment manufacturers (OEMs), refinish coatings and surface treatments, catalysts, battery materials and precious and base metals services. These industries it serves include the automotive and chemical industries.

The goal of current research is to create anti-corrosion coatings that are more durable and perform better while being environmentally friendly [16, 17]. When the coating is damaged, corrosion inhibitors or microcapsules containing healing substances are released, which successfully restore the protective layer and stop additional corrosion. Figure 1.25 shows (a) the stages of manufacturing the cutting tool insert; (b) the laser machining process used to prepare the cutting edge [30] and (c) the laser processing method used to create the grooved chip breaker design [31].

Through customized products, technologies and solutions, we improve our customers' applications and processes and provide them with regional support. By leveraging our diverse portfolio of technologies and strengthening our position as a leading supplier of battery materials and surface coating solutions. In the aerospace sector, where efforts are constantly being made to improve the performance and durability of aircraft components, surface treatment techniques are essential. Significant advances in coating technology resulting from recent

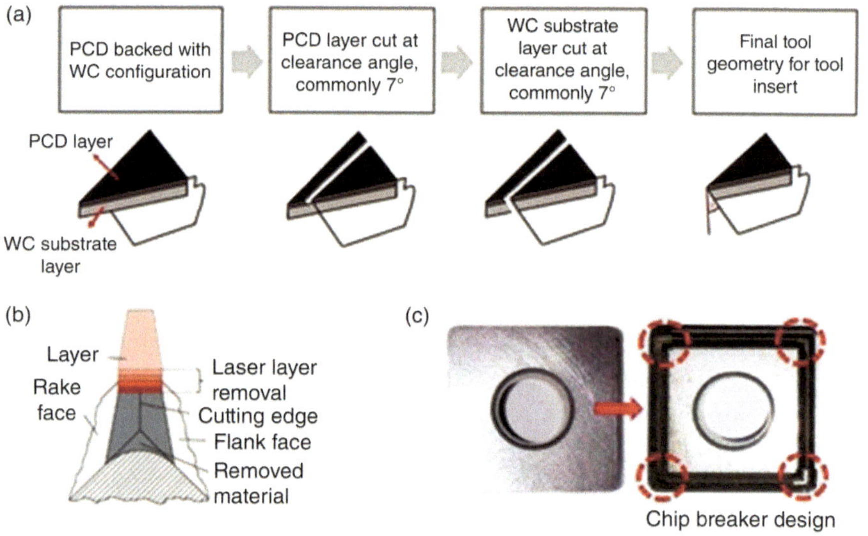

FIGURE 1.25 (a) Manufacturing steps of cutting insert adapted; (b) laser machining for cutting edge preparation; and (c) laser machining for design of groove chip breaker [13] [open access].

research and development in the field have improved resistance to high temperatures, wear and corrosion.

The development of enhanced thermal barrier coatings (TBC) is one of noteworthy areas of research. To reduce heat transfer and protect base materials from thermal degradation, TBC is applied to turbine blades and engine parts. Through the exploration of new materials and application methods, recent research has focused on improving the durability and thermal stability of TBC [10–13]. For example, studies have examined the use of ceramic matrix composites as TBCs because they have better thermal performance than traditional materials [14]. These TBC improvements result in longer component life, increased engine efficiency and reduced maintenance costs. The goal of recent research has been to create new wear-resistant coatings with low friction coefficients, increased hardness and resistance to abrasive wear [22, 23]. For example, the use of DLC coatings has shown promising results in reducing wear and extending the service life of critical aircraft components [13]. Due to their exceptional hardness and reduced friction, DLC coatings are ideal for applications where it is important that something be resistant to wear. In addition, anti-corrosion coatings are necessary to protect aircraft parts from aggressive substances and the environment [15].

Current research and surface treatment have advanced significantly as a result of developments that meet the unique requirements of military applications. Research on surface treatment methods for military equipment has focused on hardening methods [18]. The goal is to improve the strength and hardness of materials used in armored vehicles and weapons so they can withstand harsh environmental conditions without breaking or deforming. Innovative heat treatment methods have been explored to optimize the mechanical properties of critical components by achieving precise and localized hardening, such as induction hardening [10] and laser surface hardening [21]. By increasing the wear resistance and durability of military equipment, these methods make it possible to guarantee its reliable operation even under stressful conditions [11].

Numerous studies have also been conducted on protective coatings to improve the corrosion resistance of military weapons. Military equipment is exposed to corrosive substances in harsh environments such as deserts or sea, which can degrade its performance and compromise safety. The development of multifunctional coatings that provide corrosion protection [12] along with other benefits such as camouflage or radar signature reduction has been the focus of recent research. For example, corrosion resistance and stealth can be achieved using nanostructured coatings, which increases the operational efficiency of military equipment.

Specialized coatings have attracted the attention of experts who specialize in surface treatments for military vehicles to meet specific operational needs. These coatings provide low reflectivity, reduced infrared signals and improved adhesion for concealment [13, 14]. Technological developments in materials and application methods have made it possible to develop coatings that increase stealth and reduce the visibility of equipment in different ranges of electromagnetic spectrum. This can significantly improve the effectiveness and survivability of military assets in combat conditions.

1.6.2 Automotive Industry

Coatings are resistant to wear of engine parts. Surface treatments are necessary to improve engine performance and extend component life in the automotive industry. To reduce wear, minimize friction and improve wear-resistant coatings, recent research and development is aimed at maximizing fuel economy [15, 16]. DLC coatings have become an excellent means of improving the wear resistance of engine components. High hardness, minimal friction and excellent adhesion to the substrate are just some of the special qualities of DLC coatings [19]. To improve their tribological performance, recent studies have explored various variants of DLC coatings, including tetrahedral amorphous carbon [21] and amorphous hydrogenated DLC [17, 18]. Because of its special tribological properties, tetrahedral amorphous carbon is a highly desirable type of carbon for a variety of applications, especially in enhancing wear resistance and reducing friction in mechanical systems [21]. These coatings are proven to have exceptional abrasion resistance, which reduces friction losses and improves internal combustion engine performance.

The design, development, production, marketing, sales, maintenance, repair and customization of automobiles are all part of the wide range of businesses and organizations that constitute the automotive industry. In terms of income, this is one of the largest sectors in the world (from 16% in France to 40% in Slovakia) [22, 23]. Many techniques have been explored, including surface pretreatment, the use of intermediate materials and optimized deposition procedures [12–14]. For example, the use of interlayers such as titanium nitride (TiN) or chromium nitride (CrN) has been demonstrated to improve adhesion, resulting in better wear resistance and longer component life [15–17].

In addition, interest in single-crystalline coatings has increased in recent years [13]. Because single-crystalline coatings lack grain boundaries and have strong crystalline order, they have better mechanical properties than polycrystalline coatings. They are attractive for wear-resistant applications because they provide better adhesion, reduced friction, and greater wear resistance. Researchers can optimize the performance of single-crystalline coatings under challenging operating conditions and tune their properties to suit specific applications by precisely controlling their crystallographic orientation protection against corrosion of car components. In the automotive sector, corrosion is a major issue that compromises the durability and appeal of vehicles. Automotive parts now have effective corrosion protection that ensures longevity in a variety of situations, thanks in large part to surface treatment procedures [20].

Automotive components are increasingly interested in zinc-nickel alloy coatings as an anti-corrosion solution [18]. For example, to improve the adhesion and corrosion researchers have studied the use of plating bath containing trivalent chromium ions for chromium coatings [21, 22]. This particular bath, which uses trivalent chromium ions, is an environmentally friendly replacement for the conventional hexavalent chromium bath [22, 23]. Trivalent chromium coatings reduce the environmental impact of standard chrome plating methods, while providing greater corrosion resistance, adhesion and visual appeal. To further improve the corrosion protection of automotive parts, researchers have also explored nano-composite coatings incorporating organic inhibitors or nanoparticles [16, 17].

There have been significant advances in metal surface finishing in recent years, driven in large part by the demand for improved functionality, protection and material performance in several industries. This study included a comprehensive analysis of approaches, characterization strategies, uses, challenges and potential development paths in the field of metal surface engineering. Metal surface engineering techniques involve a variety of physical and chemical procedures. Various physical methods are available to modify surface quality, including thermal spray coating. Effective methods for applying coatings and changing surface chemistry are chemical processes, including chemical etching, sol–gel coating, electroplating and electroplating.

These techniques allow modification of surface properties such as abrasion resistance, electrical properties, biocompatibility and corrosion resistance to suit the needs of specific applications. Characterization methods are important in assessing the quality and performance of surface engineering materials. Microstructure characterization techniques that shed light on surface morphology, crystal structure and phase composition include XRD, TEM and SEM. AFM and STM are surface topography analysis techniques that can accurately measure and image surface features at the nanoscale.

The fatigue properties of aircraft engine components have been studied and successfully improved using mechanical surface treatment (MST), including shot peeling (SP), ultrasonic surface rolling (USR), laser impact treatment (LSP), low ductility polishing (LPB) and deep cold rolling. (DCR) [6–10]. The main reasons for this gain are the gradient microstructural changes caused by MST, high compressive residual stresses and strain hardening in the subsurface layer of localized plastic deformation. A strengthened subsurface layer can eliminate crack sources from the surface to the subsurface and slow down crack propagation through the combined effects of surface integrity features [11]. Therefore, it is essential to examine how different MSTs affect the surface integrity properties of high-strength titanium alloys. Technological advances in nanocoatings have improved functionality by allowing precise control of the coating structure.

Ultrasonic surface rolling is a new MST technology that has recently attracted a lot of research interest. Static rolling and dynamic ultrasonic vibration are combined in the USR process to produce a smooth surface with an increased micro-hardness and compressive residual stresses on the target surface [22, 23]. This could allow the functionalization of microtextures under specific working conditions. An example proposal incorporating these characteristics is presented in Figure 1.26 as a multilayer ANN structure to illustrate the various dependencies covered by this critical assessment. Surface modifications and thin coatings are important components of microelectromechanical systems and electrical devices in the semiconductor and electronics industries. To further improve metal surface processing technology, several obstacles need to be addressed and future trends identified. Figure 1.26 shows the heatmaps of same cells obtained when building

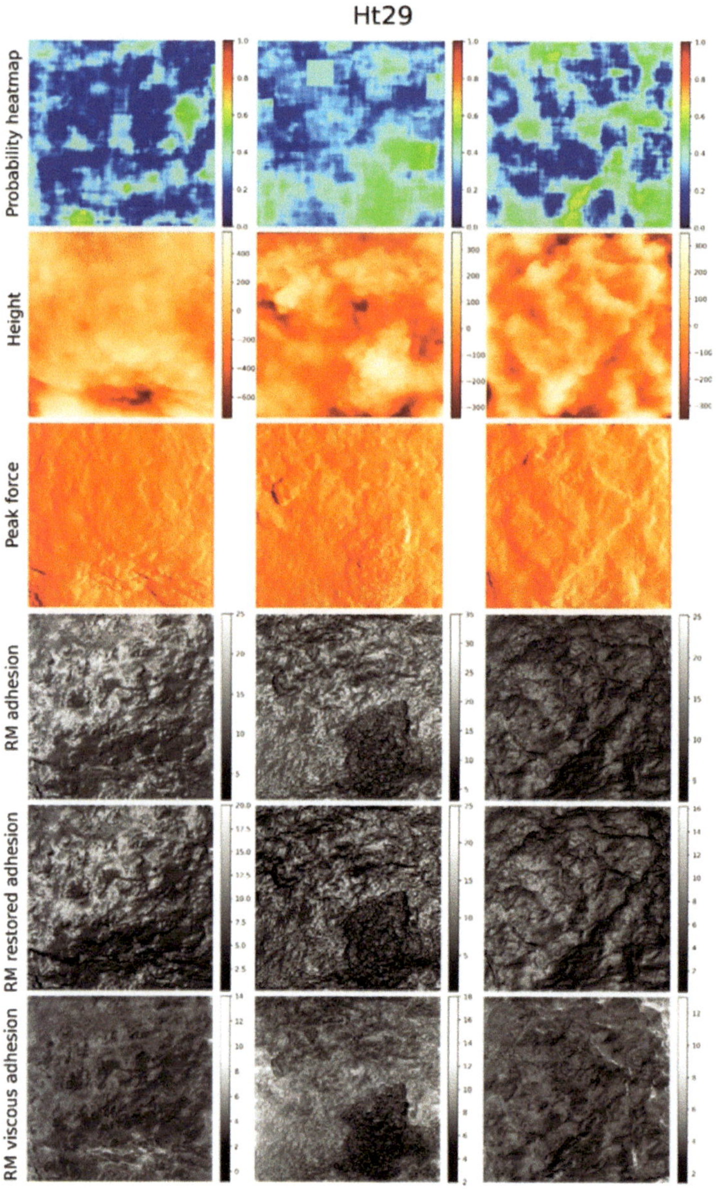

FIGURE 1.26 Cells with different aggressiveness displayed in heat maps created by combining the height channel and the three RM imaging channels. Three representative heat maps are shown for each class. At magnification of ¼ 10 × 10 μm, they are captured in 512 × 512 pixels, each image consisting of 5 × 5 μm recorded in 256 × 256 pixels. The top row shows heat maps that are the probabilities that 32 × 32 region surrounding each pixel belongs to its class. The next row shows cell height images on vertical scales in nanometers. The third row shows the feedback error, PeakForce, in arbitrary units; and the bottom three rows show maps of RM cell adhesion, recovered RM adhesion and RM viscous adhesion (nN is the vertical scale unit) [24] [open access].

the classifier using the four combined bands as mentioned earlier. We present the heatmaps, height, PeakForce and three resonance mode images that are used to build the classifier, just as we did when using a single band in the resonance mode [24].

This method typically uses a rolling tool that can be mounted on the spindle of any CNC machining center. The piezoelectric transducer generates ultrasonic vibrations with a frequency of 28–30 kHz, and air pressure is used to create a static load. A rotating rolling ball, loaded with air and vibrated by ultrasound, rolls and impacts the surfaces of components throughout the entire process [23]. The development of coatings to improve energy storage and conversion, as well as surface modification to enhance catalytic activity, have been the main goals of surface functionalization for energy applications. The mechanical, microstructural and fatigue characteristics of titanium alloys after LSP procedures have been the focus of numerous studies in recent years. It has been observed that LSP process parameters laser pulse energy, spot size and laser impact path have a significant impact on improving fatigue and surface integrity performance. Research has been conducted into how LSP procedures affect TC17 alloys. Advanced modeling is needed as there is a lack of research on the best texture characteristics to use in specific machining applications to maximize texture performance. The use of artificial intelligence (AI) and optimization modeling would be beneficial to develop systems that can accurately identify suitable textures and predict the performance of textured tools (wear rate, COF, adhesion), similar to the proposal in laser machining. To create reliable and affordable processes, surface treatment methods must overcome their limitations, which include adhesion issues, process complexity and durability. New approaches including biomimetic surfaces, nanotechnology and multifunctional coatings offer promising opportunities for future research and development. These potential application areas can focus on surface development for harsh environments, biocompatible surfaces, advanced characterization methodologies, integration of AI and machine learning, sustainable surface development and research and development opportunities for the industrial future. In conclusion, by improving the performance, protection and utility of materials, metal surface treatment technologies continue to evolve and transform various industries. This chapter discusses best practices, characterization techniques and applications, and provides a detailed overview of the topic. Surface engineering will continue to play an important role in meeting the changing needs of the industry by solving problems and exploring proposed future paths.

1.7 PROBLEMS

1. Describe the geometric surface layer of brittle materials and change its constituent surfaces without changing the composition of the material.
2. List some important surfaces and interfaces that make up a surface from a tribal perspective. Describe the potential consequences, such as wear damage or lubrication failure, if a triple failure occurs at these points.

3. How would the relative wear rates of surface treatment materials change if the abrasive particles are silicon carbide and characterization methods?
4. Why there is no visible adhesion on surface technologies when squeezing two metal objects?
5. How are the mechanical and tribological properties of adhesive film affected when two components with different mechanical strengths come into contact?

REFERENCES

1. Brzezinka, T., Rao, J., Paiva, J., Kohlscheen, J., Fox-Rabinovich, G., Veldhuis, S., & Endrino, J. (2019). DLC and DLC-WS2 coatings for machining of aluminium alloys. *Coatings*, *9*, 192.
2. Li, H., Ramezani, M., Chen, Z., & Singamneni, S. (2019). Effects of process parameters on temperature and stress distributions during selective laser melting of Ti–6Al–4V. *Trans. Indian Inst. Metal*, *72*, 3201–3214.
3. Awang, M., Khalili, A.A., & Pedapati, S.R. (2020). A review: Thin protective coating for wear protection in high-temperature application. *Metals*, *10*, 42.
4. Riquelme, A., & Rodrigo, P. (2021). An introduction on the laser cladding coatings on magnesium alloys. *Metals*, *11*, 1993.
5. Fotovvati, B., Namdari, N., & Dehghanghadikolaei, A. (2019). On coating techniques for surface protection: A review. *J. Manuf. Mater. Process.*, *3*, 28.
6. Dai, M., Zhou, K., Yuan, Z., Ding, Q., & Fu, Z. (2000). The cutting performance of diamond and DLC-coated cutting tools. *Diam. Relat. Mater.*, *9*, 1753–1757.
7. Nas, E., & Kara, F. (2022). Optimization of EDM machinability of hastelloy C22 super alloys. *Machines*, *10*(12), 1131.
8. Paredes, Á, Acuña, S.M., & Toledo, P.G. (2019). AFM study of pyrite oxidation and xanthate adsorption in the presence of seawater salts. *Metals*, *9*(11), 1177.
9. Hutchings, I., & Shipway, P. (2017). Tribology: Friction and Wear of Engineering Materials, Butterworth-Heinemann.
10. Quan, B., Li, J., & Chen, C. (2021). Effect of corrosion time on the synergistic corrosion of Q235 steel in sodium aluminate solutions. *Metals*, *11*(5), 753.
11. Kumar, C.S., Majumder, H., Khan, A., & Patel, S.K. (2020). Applicability of DLC and WC/C low friction coatings on Al2O3/TiCN mixed ceramic cutting tools for dry machining of hardened 52100 steel. *Ceram. Int.*, *46*, 11889–11897.
12. Wang, L., Liu, Y., Chen, H., & Wang, M. (2022). Modification methods of diamond like carbon coating and the performance in machining applications: A review. *Coatings*, *12*(2), 224.
13. Tebianian, M., Aghaie, S., Razavi Jafari, N.S., Elmi Hosseini, S.R., Pereira, A.B., Fernandes, F.A., & Huo, Y. (2023). A review of the metal additive manufacturing processes. *Materials*, *16*(24), 7514.
14. Močilnik, V., Predan, J., & Gubeljak, N. (2023). Effect of residual stresses on fatigue strength on the smooth and splined parts of spring torsion bars. *Metals*, *13*(6), 1004.
15. Ramezani, M., Mohd Ripin, Z., Pasang, T., & Jiang, C.-P. (2023). "Surface engineering of metals: Techniques, characterizations and applications. *Metals*, *13*(7), 1299.
16. Sadeghi, B., Cavaliere, P., Perrone, A., & Silvello, A. (2022). Crack initiation and growth behavior of HVOF stellite-6 coatings under bending loading. *Powders*, *1*(2), 62–74.

17. D'Orazio, A., el Mehtedi, M., Forcellese, A., Nardinocchi, A., & Simoncini, M. (2017). Tool wear and hole quality in drilling of CFRP/AA7075 stacks with DLC and nanocomposite TiAlN coated tools. *J. Manuf. Process.*, *30*, 582–592.
18. Huang, L., Yuan, J., Li, C., & Hong, D. (2018). Microstructure, tribological and cutting performance of Ti-DLC/α-C:H multilayer film on cemented carbide. *Surf. Coat. Technol.*, *353*, 163–170.
19. Grigoriev, S.N., Volosova, M.A., Vereschaka, A.A., Sitnikov, N.N., Milovich, F., Bublikov, J.I., Fyodorov, S.V., & Seleznev, A.E. (2020). Properties of (Cr,Al,Si) N-(DLC-Si) composite coatings deposited on a cutting ceramic substrate. *Ceram. Int.*, *46*, 18241–18255.
20. Nikolova, M.P., & Apostolova, M.D. (2023). Advances in multifunctional bioactive coatings for metallic bone implants. *Materials*, *16*, 183.
21. Costa, R.C., Nagay, B.E., Dini, C., Borges, M.H., Miranda, L.F., Cordeiro, J.M., Souza, J.G., Sukotjo, C., Cruz, N.C., & Barão, V.A. (2023). The race for the optimal antimicrobial surface: Perspectives and challenges related to plasma electrolytic oxidation coating for titanium-based implants. *Adv. Colloid Interface Sci.*, *311*, 102805.
22. Wang, C., Jiu, H., Zhang, L., Xu, R., Che, S., Guo, Z., Han, Y., Ma, J., Li, H., & Guo, F. (2023). Carbon coating hollow NiS nanoparticles anchored on Ti3C2 nanosheet as high-performance anode for lithium-ion batteries. *J. Alloys Compd.*, *956*, 170332.
23. Su, Y.L., Kao, W.H., & Chang, G.Y. (2023). Tribological, anti-corrosion, and electrical conductivity properties of CrCx coatings deposited on stainless steel 316L and used as metal bipolar plates for fuel cells. *J. Mater. Eng. Perform.*, *32*, 3739–3754.
24. Petrov, M., & Sokolov, I. (2023). Identification of geometrical features of cell surface responsible for cancer aggressiveness: Machine learning analysis of atomic force microscopy images of human colorectal epithelial cells. *Biomedicines*, *11*(1), 191.

2 Surface Engineering

2.1 INTRODUCTION

The coatings are proven to have exceptional abrasion resistance, which reduces friction losses and improves internal combustion engine performance. In many different industries, surface engineering is essential to improve the function and performance of metallic materials. Metals can now have better mechanical strength, wear resistance, corrosion protection and biocompatibility due to their ability to change their surface characteristics [1–4]. This field has been further advanced by recent developments in surface engineering techniques and characterization approaches that have led to innovative results and inventive applications [5]. Various techniques have been explored to improve bond strength, such as surface pretreatment, use of intermediate materials, and optimized substrate and coating procedures [6]. In addition, more attention has been paid to single-crystal coatings [7]. Because single-crystalline coatings lack grain boundaries and have strong crystalline order, they have better mechanical properties than polycrystalline coatings. They are attractive for wear-resistant applications as they provide better adhesion, reduced friction and greater wear resistance. Researchers can optimize the performance of single-crystalline coatings under challenging operating conditions and tune their properties to suit specific applications by precisely controlling their crystallographic orientation [8–12]. Corrosion is a major issue affecting the durability and attractiveness of vehicles in the automotive sector. Automotive parts now have effective corrosion protection that ensures longevity in a variety of situations, thanks in large part to surface treatment procedures. Automotive component industries are increasingly interested in zinc-nickel alloy coatings as an anti-corrosion solution [7]. As shown in Figure 2.1, the hybrid texture also reduced adhesion and wear scar more than individual textures. Chips passing through different tool textures changed the contact length and pressure distribution, so the hybrid texture performed better [7].

As a result of recent developments, surface technology has improved significantly. Through precise control of microstructure and composition at the nanoscale, improved mechanical and tribological properties can be achieved, leading to the emergence of nanostructure coatings [10–12] as a key area of research. These coatings are very attractive for various industrial applications due to their remarkable hardness, wear resistance and friction reduction. Moreover, surface modification techniques have been developed to overcome the difficulties associated with corrosion prevention. For metal components exposed to harsh environments, corrosion-resistant coatings [7, 13] and surface passivation techniques [14, 15]

 DOI: 10.1201/9781003584735-2

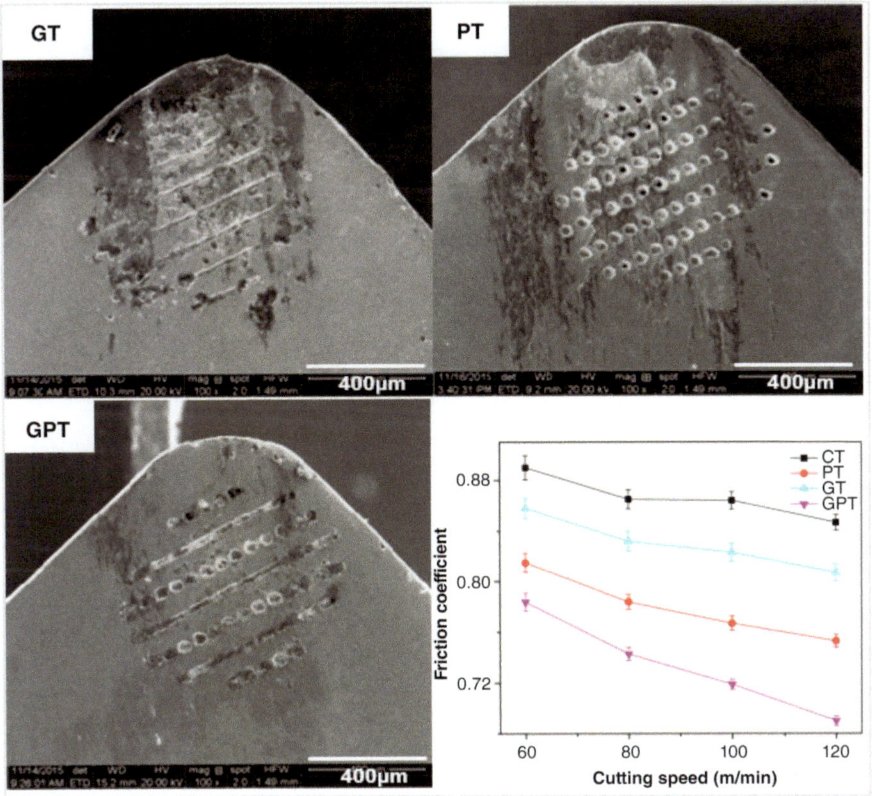

FIGURE 2.1 Wear trace on surface of WC rake after 150 seconds of cutting at 120 m/minute with different texture geometries: GPT-Hybrid, PT-Dimples and GT-Grooves. COF analysis at tool–chip interface [7] [open access].

provide increased durability and service life. These developments hold great promise for the automotive, aerospace and marine industries, where corrosion-related failures can have catastrophic consequences [16–20]. This particular bath, which uses trivalent chromium ions, is an environmentally friendly replacement for the conventional hexavalent chromium bath [21, 22]. Trivalent chromium coatings reduce the environmental impact of standard chrome plating methods while providing greater corrosion resistance, adhesion and visual appeal. For further improving the corrosion protection of automotive parts, researchers have explored nanocomposite coatings incorporating organic inhibitors or nanoparticles [23, 24]. The interdisciplinary field of surface engineering came into focus in the early 1980s when it was recognized that the vast majority of manufactured components could subject to catastrophic failure or degradation as a result of various surface phenomena such as wear, corrosion or fatigue during use. For machining hardened steels, Ze et al. [8] have used a femtosecond laser to cut

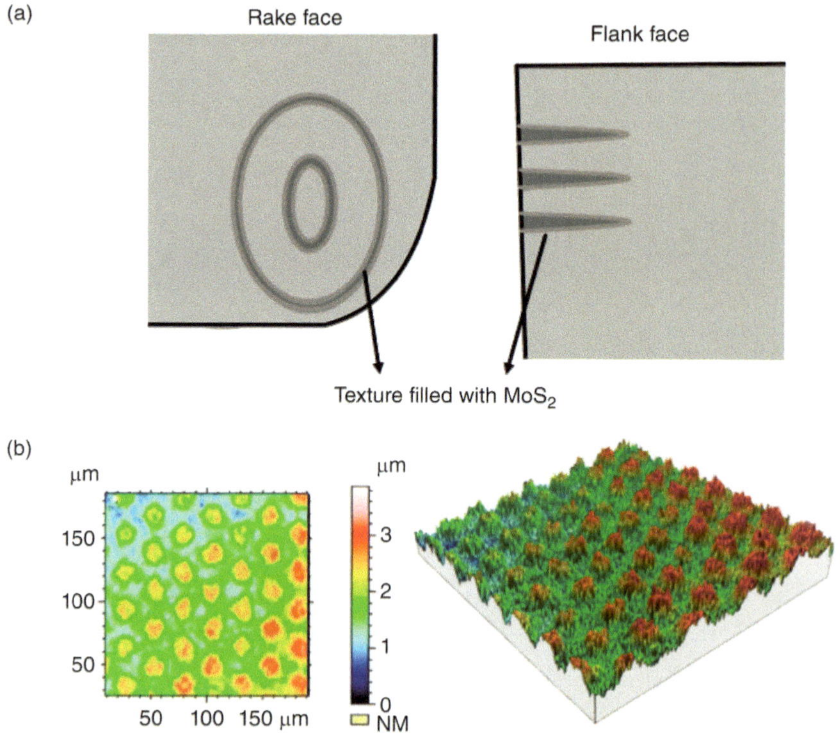

FIGURE 2.2 (a) Schematic of elliptical groove textures filled with MoS_2 on rake and side surfaces. (b) Honeycomb geometry is created using H10S WC tool [7] [open access].

circular and linear grooves filled with molybdenum disulfide (MoS_2) on a tungsten carbide (WC) lubricant (Figure 2.2a). Due to the typical crater wear shape, an elliptical geometry is chosen. In an attempt to reduce overall tool wear, the textures were spaced 150 µm from MCE. The elliptical shape is 20% more effective in reducing forces than linear grooves. Compared to the linear design, this shape facilitated the release and smearing of lubricant on rake face, resulting in a reduction of 1 mm to 0.8 mm in contact length at tool interface [7].

The increasing use of various surface treatment technologies, such as ion implantation, laser and electron beam processes, plasma thermochemical methods, new engineered coatings (e.g., electroless nickel plating) and duplex surface modification techniques, have driven advances in these technologies. However, the roots and core ideas of surface technology are found in traditional surface heat treatment methods, including tempering, nitriding and carburizing. When the surface of an engineered component is unable to sufficiently withstand external influences, failure typically occurs in the environment in which it is located. Its functionality depends on the selection of surface material with suitable thermal, optical, magnetic and electrical properties, as well as sufficient resistance to wear, tear and corrosion. On Al_2O_3, elliptical grooves have demonstrated a gradual gain compared to TiC-coated linear grooves. Texturizing reduces coefficient of friction (COF) by 11.9% while increasing cutting temperature, cutting force. After processing, the surface relief of samples is displayed on three-dimensional (3D) maps in Figures 2.3 and 2.4. Depending on the processing settings and tool used, the surface relief of each layer included in the sandwich structure is displayed [21].

The surface relief of metal layer of milled specimen in Al/CFRP configuration using the uncoated tool showed an oriented arrangement of defects (Figure 2.3a). Regular and irregular cavities and protrusions are observed. Many visible ridges and grooves are observed, leading to surface roughness change. For the carbon fiber reinforced polymer (CFRP) composites (Figure 2.3b), unidirectional surface full of random deformations is obtained. The surface of aluminum alloy has Sa, Sz, Sp, and Sv values that are 46%, 71%, 78% and 76% lower than composite layer, respectively. After machining CFRP/Al configuration of aluminum alloy

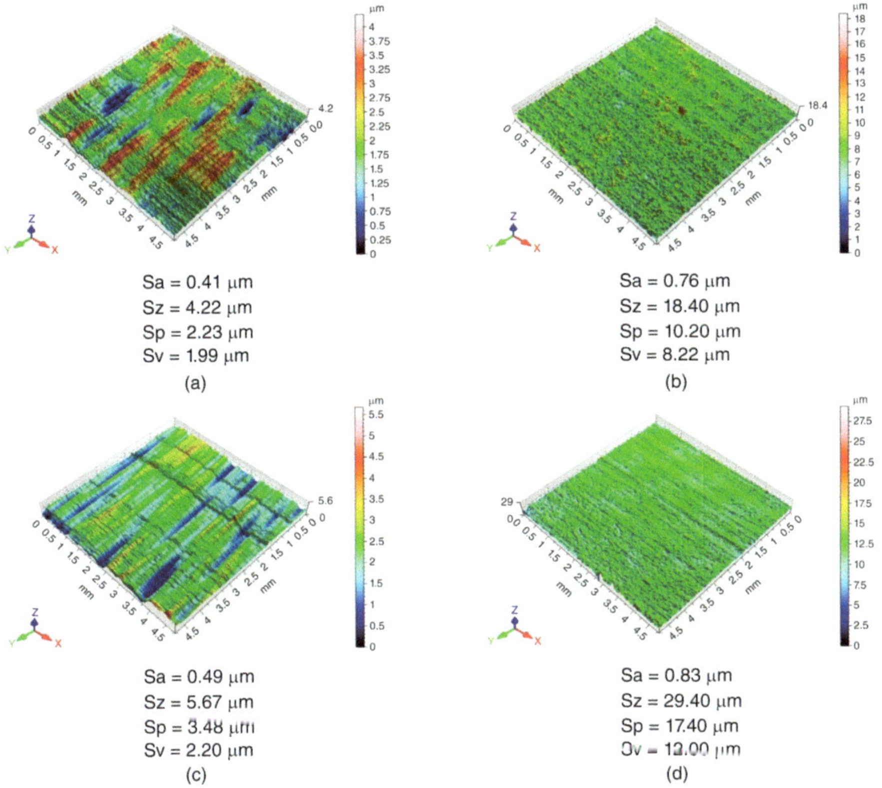

FIGURE 2.3 Surface topography after processing. Aluminum alloy is milled into Al/CFRP configuration using uncoated tool (a); the CFRP composite is milled into Al/CFRP configuration (b); aluminum alloy is milled into CFRP/Al configuration (c); and CFRP composite is milled into CFRP/Al configuration (d) using uncoated tool [21] [open access].

using the uncoated tool, an oriented surface with mixed structure is obtained (Figure 2.3c). A wide range of defects is present in surface relief. The defects are distributed on the surface of CFRP composite randomly and regularly, which lead to the formation of unidirectional surface (Figure 2.3d). The parameters recorded on the composite surface are approximately 82%, 81%, 80% and 82% higher than Sa, Sz, Sp and Sv parameters measured on metal surface. The aluminum alloy surface is milled using Al/CFRP configuration and TiAlN-coated tool to create oriented structure with a predetermined defect distribution (Figure 2.4a). The defects are uniformly distributed and scattered. For the CFRP composite (Figure 2.4b), unidirectional surface filled with random asperities is obtained. The microengineered structure surface has showed a slight change in the height of topographic features [21].

Using dual wavelength Nd:YVO4 picosecond laser, authors in reference [21] have studied honeycomb texture on WC and Sialon ceramics. Using an Al_2O_3 workpiece material, the tribological performance is evaluated using pin-and-disk tester.

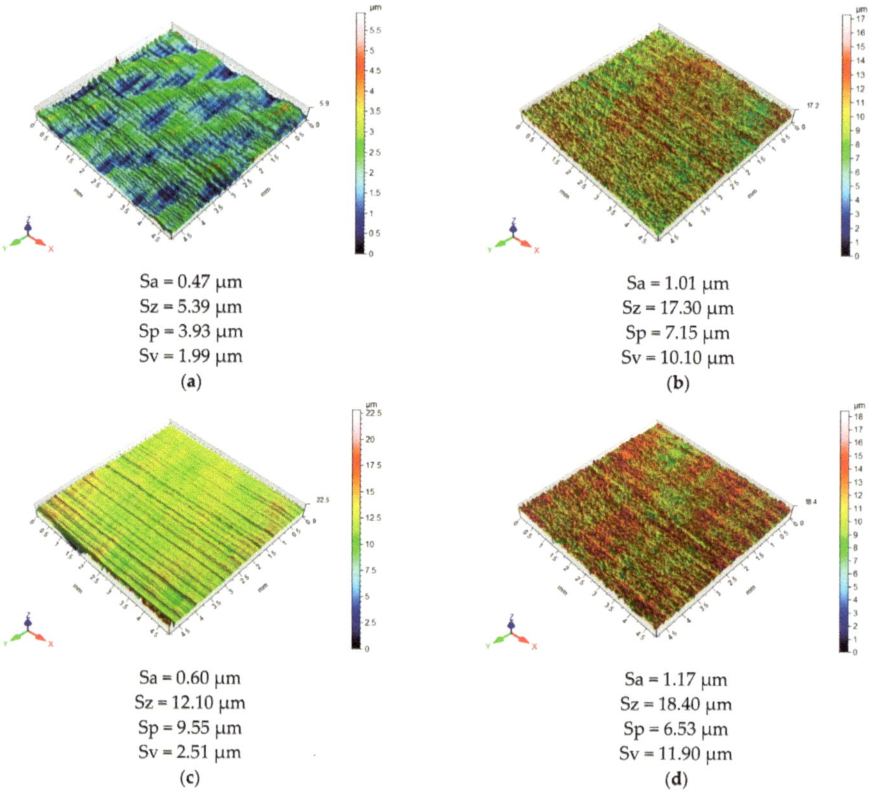

FIGURE 2.4 Surface topography after grinding process: (a) aluminum alloy in Al/CFRP configuration, (b) CFRP composite in Al/CFRP configuration, (c) aluminum alloy in CFRP/Al configuration, (d) CFRP composite in CFRP/Al configuration [21] [open access].

The textures are 30-μm wide and 650-μm deep (Figure 2.2b). The tribological performance is comparable to conventional textures; however, this texture design shows 20% less COF variability. To develop a composite material with qualities not inherent in base or surface material, surface engineering entails the application of both traditional and new surface technologies to components and structural materials. Existing engineered components are designed using different surface treatment techniques, but ideally, surface finishing should involve designing the part based on the surface treatment that will be used. Developments in surface engineering have also had an impact on the energy sector. To increase the catalytic activity of material and facilitate efficient energy conversion processes, surface modifications have been used. In addition, research has been conducted on coatings and surface functionalization to improve the durability and efficiency of energy storage systems such as fuel cells [17] and batteries [18]. Surface engineering techniques are used in the aerospace and defense industries to improve the performance and durability of vital components that are exposed to harsh environmental conditions [19]. To ensure engine components and automotive parts, which are reliable and safe, the automotive industry requires wear-resistant coatings and corrosion protection technologies [20, 21]. Meanwhile, the electronics and semiconductor industries require surface modification to produce high-performance devices [22].

These factors typically act together to cause aspiration system erosion and cavitation, steam turbine blade erosion due to thermal fatigue, slurry pump impeller galling and corrosion due to chloride ions. The corrosion process involves complex interactions between physicochemical variables and requires careful analysis due to variety of ways can occur, including vibration corrosion, stress corrosion and differential aeration. After identifying (diagnosing) the various types and forms of wear on component or equipment, a technical engineer uses materials science to figure out what coating or alloy metal, polymer, ceramic, or combination of these can extend its life. The use of alloys capable of resisting current wear patterns must also be determined by the engineer.

2.2 SELECTION OF SUITABLE MATERIALS

The selection of suitable materials for friction pair parts is often limited by non-tribological variables such as cost. Weight is an important element in addition to corrosion resistance. Strength, stiffness and other mechanical properties are also critical in engineering applications. These elements help define the range of potential solutions, even if they limit the variety of materials that can be used. Modification of a surface or coating to provide a combination of surface and sub-layer characteristics, which is a part of content of "surface engineering".

Wear is visible throughout the entire structure since it is a necessary component of engineering systems. However, due to the cost of replacing parts, lost production capacity, reduced energy efficiency and ultimately wear and tear on the device itself, this results in significant maintenance costs. Despite enormous progress, many challenges remain in the field of surface engineering. Further research and development are needed regarding the complexity of surface finishing processes, the need for precise control of deposition parameters and scalability of large-scale production methods. Moreover, the cost-effectiveness of these approaches and their ability to integrate surface technology with other manufacturing processes are critical factors for practical application. Looking to future, we see that new trends and technologies in the field of surface engineering have great potential. The design and optimization of geometric surfaces with tunable properties can benefit from advanced computer modeling and simulation techniques. In addition, research into new materials such as graphene and other 2D materials opens up the possibility of creating complex and practical coatings [7].

It is not until brown oxidized wear residues at the joint between the fixture and steel specimen of their fatigue testing equipment, which first revealed the phenomenon of fretting [25]. Fretting is currently defined as relative motion between contact surfaces under variable loads such as mechanical vibration, fatigue loading, electromagnetic vibration, fluid-induced vibration or thermal cycling that has a very small amplitude (displacement amplitude is typically on the order of microns) [2–4]. Typically, the contact surface is considered to be "fixed". In other words, abrasion occurs in mechanical components during clearance or "tight" fits. The design of tribological system must take into account many potential processes. An algorithm illustrating the stages of constructing a tribological system. One of the most popular materials for manufacturing mechanical parts is metal and its alloys. Due to their microstructure and composition are standardized (sometimes even worldwide), their mechanical properties are easier to predict.

The properties of non-metallic materials tend to change even with same composition because they are less controllable. However, the response of materials to tribological applications cannot be summarized in a single number, even when their mechanical and physical properties are equivalent. The basic mechanisms and specific wear modes largely determine the choice of materials and technologies for manufacturing technical surfaces in tribology. Therefore, the selection of wear-resistant materials will be considered depending on the type of wear being considered.

As a result, the applied load and design parameters will determine how much the effective pressure at interaction surface will decrease. However, the actual contact area will determine the pressure and surface quality of both triboelements will determine this. All of these factors must be taken into account as changes in driving speed or pressure can alter the wear process. For this reason, understanding the significance of subsequent wear levels is critical to the design or improvement of a friction pair [4–7]. If friction is a form of wear, the force and movement between surfaces must be considered. In addition, it is extremely important to regulate the supply of oxygen to surrounding air. Potential temperature fluctuations and thermal expansion between the two components are additional factors to consider. Components with torque and potential sources of vibration. In situations where misalignment occurs between sliding surfaces, such as in bearings, this is unavoidable since misalignment is a normal result of torque. Because surface tension can transfer torque from one part to another, it is an important factor in this situation. In this case, analysis of reduction in generated torque is necessary. Three factors are critical when it comes to in-service wear mechanisms involving contact stress, such as gears, cam followers and bearings; contact force, number of effective load cycles and constant number of cycles. They should be considered not only because of their possible inclusion among these three elements but also because degree of corrosion resistance that the pair of materials exhibits [7]. The steps involved in building a tribological system are shown in Figure 2.5.

The magnitude of the load and its geometry will be critical to reducing the forces involved removal of particles from the system will be a criterion to consider if wear is erosive or abrasive, and caused by solid particles. For example, wear particles or contamination may become present in lubricants. Removing these particles by filtration or inertial separation will be critical since the size of large particles has a greater impact on wear than small ones. However, the ratio (Hm/Ha $\geq$ 0.85) between material hardness (Hm) and abrasive hardness (Ha) should be greater than 0.85. One important aspect of erosion is the speed at which particles come into contact with the surface. Equally important is the density of impact material and same angle of incidence. Preventing sudden changes in fluid angle is something to consider when managing wear caused by water erosion.

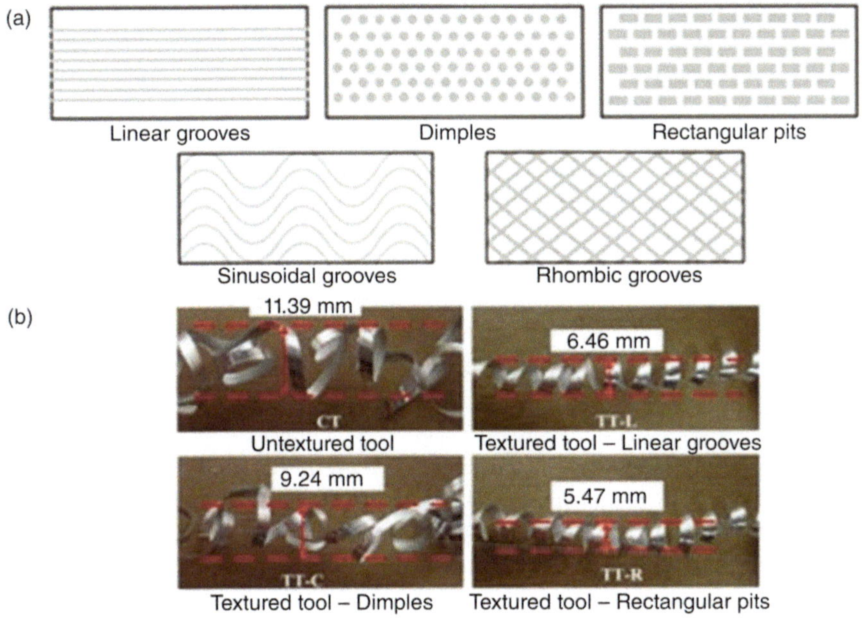

FIGURE 2.5 (a) Different types of microtextures include rhombic, sinusoidal, rectangular pits, depressions and linear grooves. (b) Different texture geometries form a chip (right) [7] [open access].

TABLE 2.1
Typical Coefficient K Values for Wear Lubricated by Sliding [22] [Open Access].

Type of Lubrication	K
Hydrodynamic	$< 10^{-13}$
Elasto-hydrodynamics	10^{-13}–10^{-9}
Limit	10^{-10}–10^{-6}
Solid lubrication	$\approx 10^{-6}$
Without lubrication (severe wear)	10^{-4}–10^{-2}

Lubricating bearings and other friction pairs are one of the best ways to minimize their wear. Under lubricant sliding conditions, the wear coefficient is represented by the constant K, value of which can be significantly reduced without compromising the hydrodynamic conditions of the lubricant. However, they are not always repairable, and K value with insufficient lubrication can approach 10^{-6}, depending on the characteristics of lubricant used. K stands for constant. About sliding wear in the Arshard equation, this means:

$$K = QH/W \tag{2.1}$$

where W is applied normal load, Q is amount of wear determined by the contact of all protrusions and P is contact pressure, which can be replaced by hardness of material being worn. Table 2.1 presents the permissible values of K according to A.S.M. standards.

Hydrodynamically lubricated sliding wear is the optimal condition, and designers should do everything possible to promote its use in real-world operating conditions. The minimum thickness of lubricating layer is the most important element controlling lubrication regime, in addition to surface roughness. Specialized monograms can be used for calculation which includes all relevant parameters and takes into account the additional component λ [26].

2.3 PROCEDURES

An algorithm has been developed that can be used to evaluate different wear and tear calculation formulas based on the many types of wear and tear that occur in the workplace [26]. For metal-to-metal sliding, K usually has a maximum value that is lower than for non-metal-to-metal and non-metal-to-non-metal sliding. With the same properties B, value of K when sliding metal on metal is much higher. Tribological compatibility, or ease with high values of molecular component of friction between the two metals can be established, which is a key factor in determining how much K is 64 drops if the two metals in a pair are in different states. This possibility has a close relationship with the characteristics of equilibrium diagram resulting from the interaction of two metals, as well as with corresponding crystalline and molecular structures of elements and the degree of solubility in solid state. These two combinations are completely insoluble, exhibit extremely poor solubility in the solid state, and, when appearing as two coexisting phases in liquid state, form tribologically compatible pairs. Identical pairs of metals are, of course, completely and mutually soluble, and also to some extent compatible. As shown on the map, other pairs have different solubility ratios. To obtain low K values, good tribological compatibility is required due to low mutual solubility. In general, low tribological compatibility and therefore relatively high K values are demonstrated by slip pairs with high mutual solubility. Compatibility is influenced by many factors, including mutual solubility and properties of surface coating (usually oxides) within slip pairs. In noble metals such as gold, platinum, silver and rhodium, low K values are usually associated with lack of basic oxide coatings, highlighting the importance of oxidative processes [27].

Certain metals with compact hexagonal structure show abnormal behavior because of their confinement. Plasticity is about cubic-shaped metals and chemical properties. For instance, their K values are significantly lower when operating together without oil than when greased with any hydrocarbon lubricant. The hardness of steel and other metals that develop oxide layers when sliding determines the stability of this layer and primary wear process. If the metal is hard enough to provide sufficient mechanical support to the oxide layer, then at low K values, moderate wear will occur through an oxidation mechanism. As a result, hardness can be greatly affected by hardness. Hardness can influence the wear rate of particular alloy's particles, although it is not a reliable indicator of the adhesive wear resistance of some metals. Wear resistance of various alloys. Sometimes other factors are more important, especially if microstructural components are present, such as carbides in steel and graphite in cast iron [28].

Using a microsecond fiber laser, microtextures in the form of pits, parallel grooves and perpendicular grooves with width 35–60 μm; depth 10–25 μm and spacing 70–150 μm are created. Each texture exhibits signs of crater wear. The resistance to further deformation is not enhanced by textures close to crater wear (Figure 2.6a). The study has found that the workpiece surface quality and chip formation are not affected by texture type. The workpiece material used and low cBN concentration in the tool (45 wt%) are identified as the reasons for the unstable performance. The extent to which a metal wears well at lower loads is not necessarily directly related to its resistance to severe adhesive wear and surface damage under high and abrupt loads. Some factors that influence a material's resistance to damage when sliding over a surface are the strength of bond formation, adhesion of oxide coating after rupture, and effectiveness of surface layer in maintaining adhesion. The strength of exchange bond is determined by solubility, and metals with strong bonds are more prone to surface damage when sliding [7].

Due to hexagonal metals, with fewer sliding surfaces, they are less ductile and less susceptible to damage than cubic metals. Material crushing, plowing marks

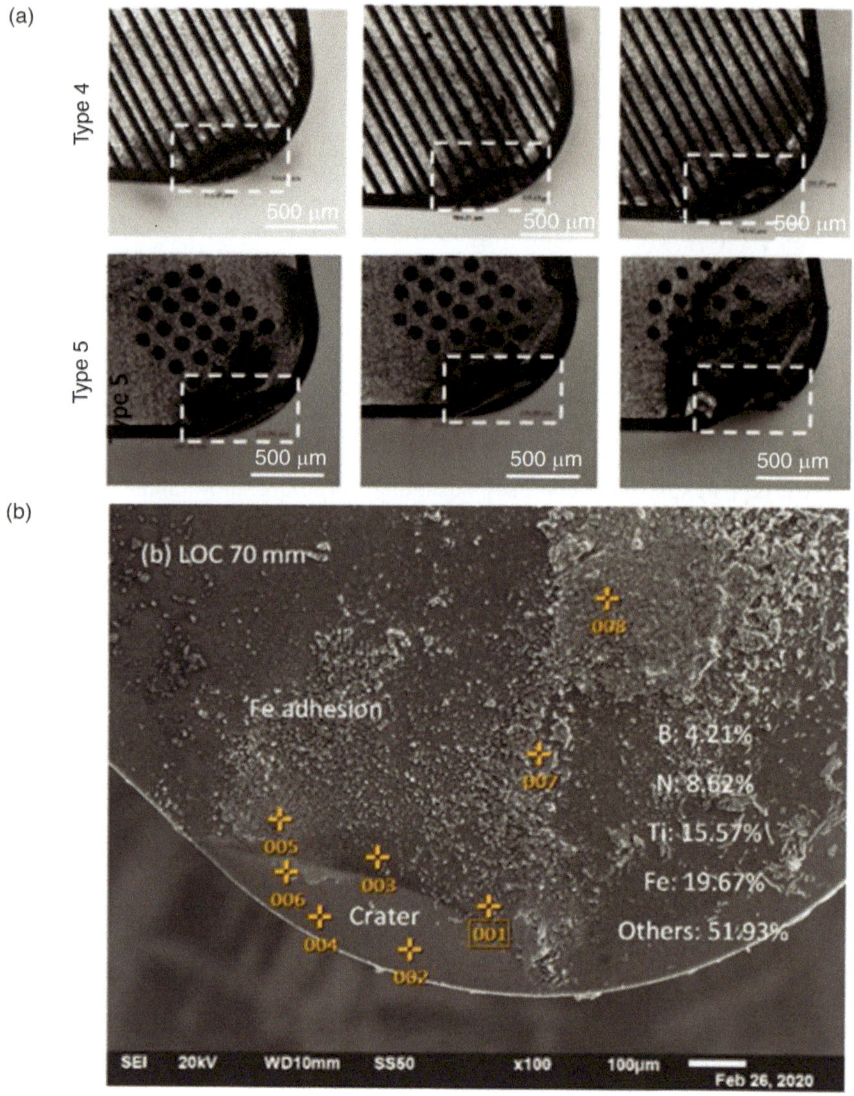

FIGURE 2.6 (a) Two of the obtained textures show the evolution of crater wear. Type 4: depth 10 μm, width 35 μm and distance 70 μm. Type 5: width 60 μm, distance 150 μm and depth 15 μm. (b) Chip adhesion and crater wear on the rake face of a PcBN tool [7] [open access].

and texture damage are the main wear defects of textured tools (Figure 2.7) [7]. However, microtextures continue to significantly reduce craters and surface wear. According to some studies, highly strain-hardened metals and alloys are less prone to surface damage during sliding.

However, the ability of this factor to predict outcomes is not perfect. For example, austenitic steels undergo a process that results in significant surface

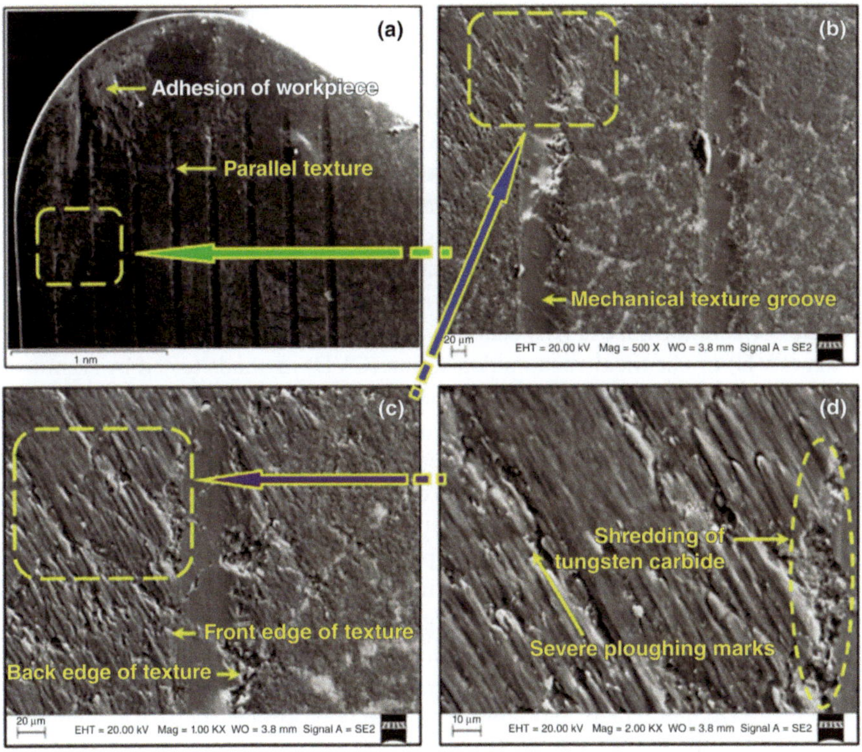

FIGURE 2.7 After 900 seconds of machining an H-13 steel workpiece, (a) the rake face of parallel microtextured cutting tool on workpiece, (b) each unique mechanical microtextured groove, (c) leading and trailing edges of textured groove, and (d) severe plowing marks [7] [open access].

damage when their structure changes to martensitic under high-strain hardening. Hardness itself does not always mean resistance to damage of sliding surface. For example, high concentrations of carbides or nitrides in steel exhibit high resistance to damage from sliding surfaces, greater than that of equal hardness, but with less brittle and hard particles in the mixture. A comparison graph of typical wear coefficient values K for different materials under sliding conditions and with different forms of lubrication [7].

Hard coatings or diffusion layers are suitable barriers against this type of attack since they have very low ductility. Resistance to damage is generally greater on rough surfaces, ideally with a random pattern (such as those resulting from sandblasting), most likely due to the restriction of joint growth and increased likelihood of damage. The wear rate of ceramic materials subjected to soft sliding may be comparable to or lower than that of dissimilar metals. This discovery indicates that ceramic materials may have much lower wear rates than metals, especially when combined with their high hardness. However, tribological applications are limited to a certain extent by the widespread use of ceramic materials. Their mechanical qualities, especially fracture toughness, could not be at proper level. For example, they will need to be manufactured using suitable powder metallurgy technology, which is typically very expensive and can result in both severe wear and light surface damage, so must be designed with extreme care. However, ceramic components such as silicon nitride valve parts, aluminum oxide hip tips and aluminum water pump bushings and seals can generally withstand severe abrasion during the tribological process [29].

Some of disadvantages of widespread use of ceramic materials in friction of vapor components can be avoided by using deposited materials on metal substrates or ceramic deposits deposited by plasma powder, physical vacuum deposition (PVD) or chemical vacuum deposition (CVD) methods tested by significant group of experts in the field of surface coating engineering. In all tribological applications of ceramic materials, the using of lubricant is particularly advantageous because it reduces surface adhesion, which reduces the likelihood of localized cracks causing severe wear. It is a critical to consider the potential chemical reaction that may occur between an unsuitable lubricant and the surface. They are most often used as simple bearings in polymeric materials and not often used as wear-resistant materials, but are sometimes found in dry or boundary slip conditions. However, some sufficiently strong polymeric materials can be used as bulk components in tribology. Among them, nylon (polyamide) and polyester sulfides play a significant role. In addition, they are usually used in the form of composites made from polymers and reinforced with appropriate fillers. Although some carbon fiber-reinforced polymers are used in race car gears, these materials are used as low-stress gears [30].

In other cases, where the contact pressure penetrates deep into the part, towards the entire surface layer, or even below it (negative gradients), methods for creating thicker surface layers are required. For example, the surface material of highly loaded gears must have high elastic strength to maintain the conditions of elastic interaction during operation. High contact forces are applied when sliding contact occurs under elastic conditions. On another hand, the high fracture toughness and resistance to fatigue cracks are necessary for tooth core and rest of gear, which are subjected to strong cyclic and sometimes shock loads during operation. In this case, it is recommended to use metal components that have undergone surface chemical and thermal treatment to combine these properties [23]. Microgrooves on WC-Co are produced using a laser at three laser speeds (11.8 J/cm^2 fluence) (Figure 2.8a). Although the highest speed only reached the uppermost layers of the texture, it is still sufficient to cause material spattering along the texture. The groove depth increases and its textural shape became more distinct as the speed decreased [31].

Unlike comparable forged steel components, these materials have good friction and are lightweight. The designer has the opportunity to choose from the vast array of surface materials available to engineers, rather than using materials of same volume as surface. The wide range of layer depth and hardness combinations that can be achieved on surfaces using these methods. It is clear

FIGURE 2.8 (a–c) Low and (d–f) high magnification SEM micrographs on worn surfaces of WC-Co-Cr coating under different applied loads [31] [open access].

that different methods offer different combinations of hardness and depth of surface layer. It is known that methods such as nickel plating, nickel chemistry and chrome plating do not exist. When wear expansion is limited and force acting on the surface is sufficient, surface deposition techniques such as PVD, CVD, or ion implants can be used to produce very thin layers, and high hardness acting on surface is quickly reduced, maintaining thin surface layer. This is due to the rapid achievement of stage of elastic interaction. When it comes to cutting precision engineering components such as dies and some cutters, these techniques can be of great benefit. In fact, the shelf life of cutting elements can be significantly extended by using PM carbide grades and NTi vacuum coating [22].

2.4 SEMICONDUCTOR AND ELECTRONICS SECTOR

The experts of industry utilize some of the smallest and most precise fabrication and fabrication services of any modern industry in electronics and semiconductor sectors. In addition to maintaining a spotlessly clean work area and staff of competent team members, they must ensure that uncontaminated raw materials such as pure silicon are handled with care when producing these products [24]. Considerable interest has been shown in plasma processing for the surface design of electronic devices [23]. Using plasma etching and deposition processes, surface properties can be precisely controlled, including chemical composition, wettability and surface roughness. Contaminants can be removed, surfaces can be activated to increase adhesion, or surface energy can be altered to improve the wetting properties of adhesives or coatings [32].

Because of silicon that the main raw material used to make semiconductors, it is not as conductive as other materials that are often machined, such as copper and silver. However, semiconductors can operate in a wide range of conditions and temperatures, unlike parts made from other raw materials. Due to their extreme portability, they are highly prized as chips for various electronic equipment and gadgets, including GPS systems, iPhones, PCs and laptop computers. Semiconductors, which convert absorbed light into electrical current, are as important to creating solar cell cells as microchips. Semiconductors can be used to create light-emitting diodes (LEDs), which are devices to emit light of specific colors when different materials are used during the manufacturing process. The semiconductor is responsible for the ability of some lasers to release energy and function [25]. Another important component of electronic surface technology is chemical processing. Chemical processes such as passivation, cleaning and etching are used to increase the chemical stability of components, remove impurities and improve surface uniformity. The focus of recent research has been to develop chemical treatments that are both environmentally friendly and use the least amount of hazardous ingredients to provide highly effective surface modification. New approaches based on water-based chemicals or environmentally friendly etchants have been explored to replace traditional chemical treatments, which use hazardous chemicals.

Effective inspection, quality control (QC), failure analysis, and research and development (R&D), including cross-sectional and purity analysis, are required by both electronics manufacturers and suppliers. This goal covers product innovation as well as the manufacturing of wafers, semiconductors, integrated circuit (IC) chips, voltaic components, printed circuit boards (PCBs) and assemblies (PCBAs), and battery systems. Solutions that help suppliers and manufacturers prepare cross-sections of components and conduct reliable and cost-effective defect analysis are beneficial. A diagram of polycrystalline diamond (PCD) and polycrystalline cubic boron nitride (PcBN) is shown in Figure 2.9a [7]. The ceramic composite known as WC is not found in nature. The covalent bond makes it hard, and binder gives it strength properties to counteract the brittleness of ceramics. Due to its beneficial metallic and ceramic properties, WC is strong and durable [7].

Microelectromechanical systems (MEMS) use thin film coatings. The production of MEMS, which are used in electronics and sensors, depends heavily on surface technology. MEMS devices are coated with thin films that provide mechanical stability, insulation and environmental protection that allowing for reliable and extremely sensitive MEMS devices [4]. These advances contribute to the sustainable production of electronic products [17]. Surface passivation is one of the most important surface modification methods to protect electronic components from moisture and other environmental factors [18]. Electronic equipment has thin films or coatings called passivation layers to prevent moisture absorption or corrosion from reducing their electrical performance. Organosilanes and conformal atomic layer coatings are two examples of passivation materials with enhanced barrier properties that have been developed recently [21]. These materials create a durable moisture barrier that prevents moisture vapor from entering delicate electronic parts and ensures long-term reliability.

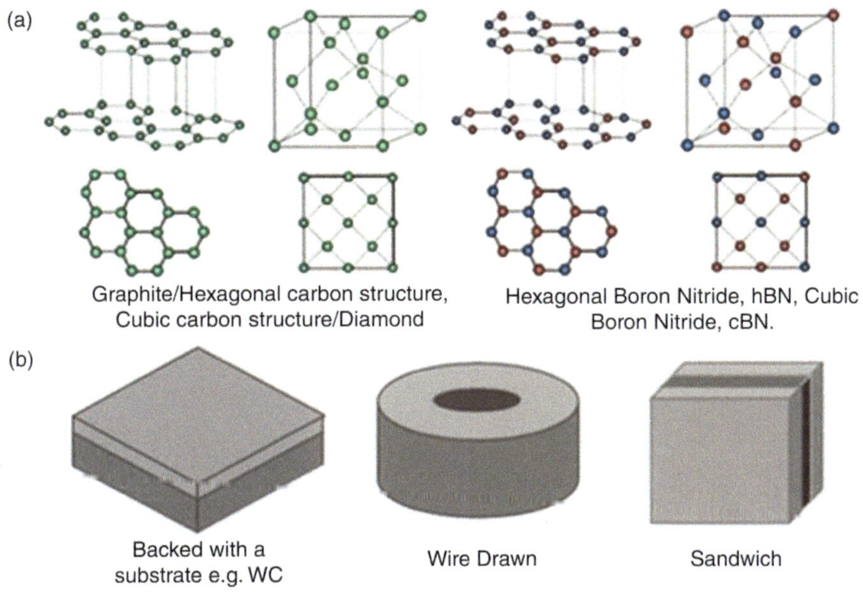

FIGURE 2.9 (a) Schematic diagram of atomic organization; (b) potential forms of polycrystalline materials taken from [7] [open access].

To stay ahead of competition, you can develop cutting-edge technology and gadget parts that work better than before at a competitive price [21]. There have also been many studies on metal films as thin film coatings for MEMS devices. In MEMS, the metals including titanium, gold and aluminum are deposited to provide optical reflectivity, electrical conductivity or junction points. The adhesion, uniformity and reliability of metal coatings have recently improved through research using complex deposition processes such as sputtering, evaporation or electroplating [22]. In addition, to facilitate the integration of MEMS devices used in biomedical applications with biological systems, research has been conducted to develop biocompatible metal coatings for these devices [23].

Different studies have also explored the integration of functional materials and surface modification in electrical devices. For example, research has focused on developing self-cleaning or antifouling coatings for electronic surfaces to improve the usefulness and reliability of electronics in harsh environments. By imparting hydrophobic or oleophobic properties to surface coatings, researchers hope to prevent the accumulation of dust, oil and other contaminants on surfaces and reduce the need for frequent cleaning and maintenance [24]. In addition, surface modifications have been explored as a potential way to incorporate new features into electronic devices. For example, nanoparticle self-assembly and nanoimprint lithography have been used to create micro- or nanostructures using surface patterning techniques with specific optical, electrical or mechanical properties.

These surface designs can be used in devices such as LEDs, optical filters or microfluidic devices, where precise control of surface characteristics is important to device performance [20]. Tiny powder grains ranging in size from 8 to 25 μm are sintered using metallic or ceramic binders to create polycrystalline composites [15]. The most popular binder for PCD and WC typically is cobalt [8]. Silicon, metallic and ceramic compounds such as aluminum (Al) or titanium (TiN) are used for PcBN [33, 34] (Figure 2.10). During sintering, the binder ingredient is a catalyst to create conglomerate masses of polycrystalline structures [7].

It is necessary to have high pressure, high temperature (HPHT) conditions, which should be between 5 and 20 GPa and 1300 and 2500°C, respectively [4]. The phases of sintering process can be varied to create structures with different grain sizes, bond strengths and binder percentages that are suitable for specific purposes. Sintering is an energy-intensive and expensive process if carried out on

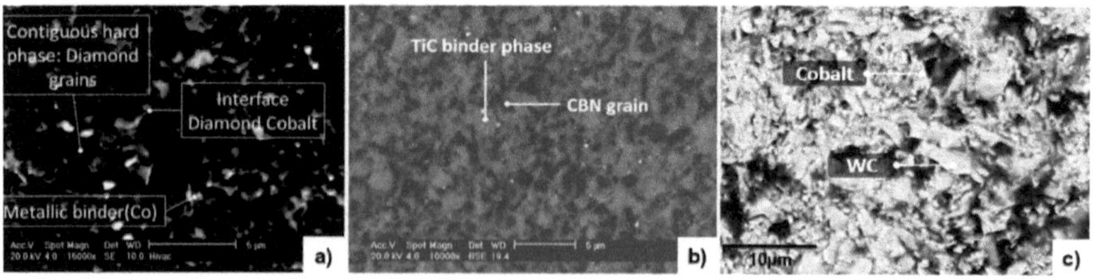

FIGURE 2.10 Examples of material and binder mixtures: (a) PCD-Co, (b) PcBN-TiC, and (c) WC-Co [7] [open access].

an industrial scale. Configurations for ultrahard polycrystalline materials that are commonly used are shown in Figure 2.10b. To limit unwanted phase changes, it is important to control the sintering parameters and procedure [7].

For example, in PCD, there is the possibility of forming graphitic structures [17], amorphous carbon regions [33] and nanoscopic graphite regions in cobalt sections [18]. By using a master template to print resist material onto a substrate, nanoimprint lithography creates nanoscale patterns. Using this method, it is possible to create nanostructures of specific sizes, shapes and locations, which allows the optical and electrical characteristics of surfaces to be controlled [21]. Self-assembly of nanoparticles is an additional method for creating nanostructured coatings [4]. During the self-assembly process, nanoparticles can form distinct structures or patterns, imparting special mechanical, electrical or optical properties to surfaces. Hard and superhard materials are difficult to manufacture using conventional technologies due to their mechanical properties. Turning inserts, drills, end mills, milling and finishing tools are some of the applications of these materials in the cutting tool industry [32].

Through nanoscale modification of optical, electrical and mechanical surface properties, researchers can develop functional coatings that provide increased efficiency, better device performance and new opportunities for innovation. The field of medical implants has evolved significantly with surface engineering approaches that meet the critical requirements of biocompatibility and successful integration into the human body [20]. An overview with justified reasons why laser micromachining is useful for the cutting tool industry can be found in Figure 2.11. Manufacturers still use casting processes, pressing small grains

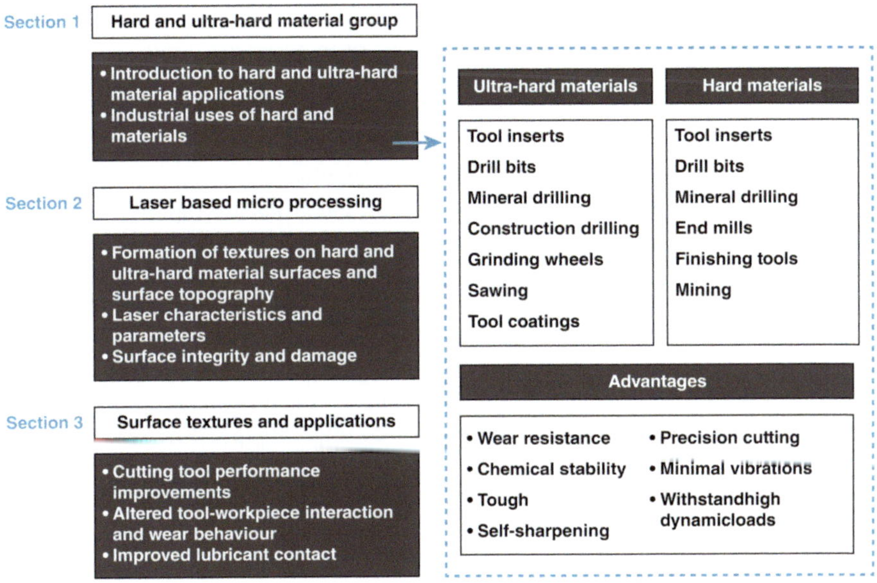

FIGURE 2.11 Common industrial applications of WC, PcBN and PCD [7] [open access].

into molds under HPHT conditions and processing each insert for an average of 10 minutes, to produce cutting tools [7].

In terms of surface development for medical implants, bioactive coatings are at forefront. To promote bone integration, hydroxyapatite (HA), a biocompatible substance that closely mimics the mineral component of real bone, is often used as a coating. By offering the surface that promotes bone cell attachment and proliferation, hydroxyapatite coatings help the implant and surrounding bone tissue establish a strong bond [22]. Varied studies have explored the application of new methods to enhance the bioactivity, uniformity and adhesion of HA coatings to implant surfaces, such as electrochemical deposition or biomimetic strategies [23]. Numerous studies have been conducted in the semiconductor electronics sector in the United States, and the scope of research on this topic is still expanding [11]. The industry's track record of employment and manufacturing development, as well as its inventive success, make it the most attractive sector for investment in the postwar American economy. Since the federal government has been active in the semiconductor sector since its inception, government programs have aimed to improve the innovation performance of other industries that are looking to the growth of semiconductor industry for guidance [24]. The contention that the military market is a critical to innovation in the US economy, where the market demands drives invention to be supported by the semiconductor industry in academic discussions on these issues [32]. However, these and other statements should be interpreted with extreme caution. What sets the semiconductor business apart from other high-tech industries, such as commercial and military aviation, is the strength and closeness of this connection. According to analysis of significant innovations throughout the history of this business, process innovation has always been critical to successful product innovation.

The absorption behavior and mechanisms of material are controlled by wavelength (Figure 2.12a). Due to its maximum absorption, the wavelength of 1064 nm is often used for the ablation of hard and ultrahard materials [7]. However, even in insulating materials such as PCD and PcBN, the absorptivity and optical breakdown efficiency are improved at shorter wavelengths (532 nm) due to multiphoton absorption [6]. The excitation and mobility of electrons are rapidly triggered by multiphoton absorption. The advantages are higher ablation rates with improved precision. PCD-Co10% and WC-Co4.6% are ablated by Eberle et al. [11] at two different wavelengths; 532 nm and 1064 nm. When using 532 nm, the ablation rate is consistently higher at different fluences and pulse durations (Figure 2.12b, c). Moreover, immobilization of functional groups can influence cellular responses and promote certain tissue integration. Examples include the addition of growth factors or peptide sequences to the surface of implants [33, 34]. Medical equipment, devices and surfaces that are regularly handled can now be treated with antimicrobial coatings to prevent the growth and spread of bacteria, viruses, and fungi [7].

Much research has been done on silver-based coatings due to their powerful antibacterial properties. The silver ions emitted by these coatings damage the membranes of microbial cells and inhibit their growth. The goal of more

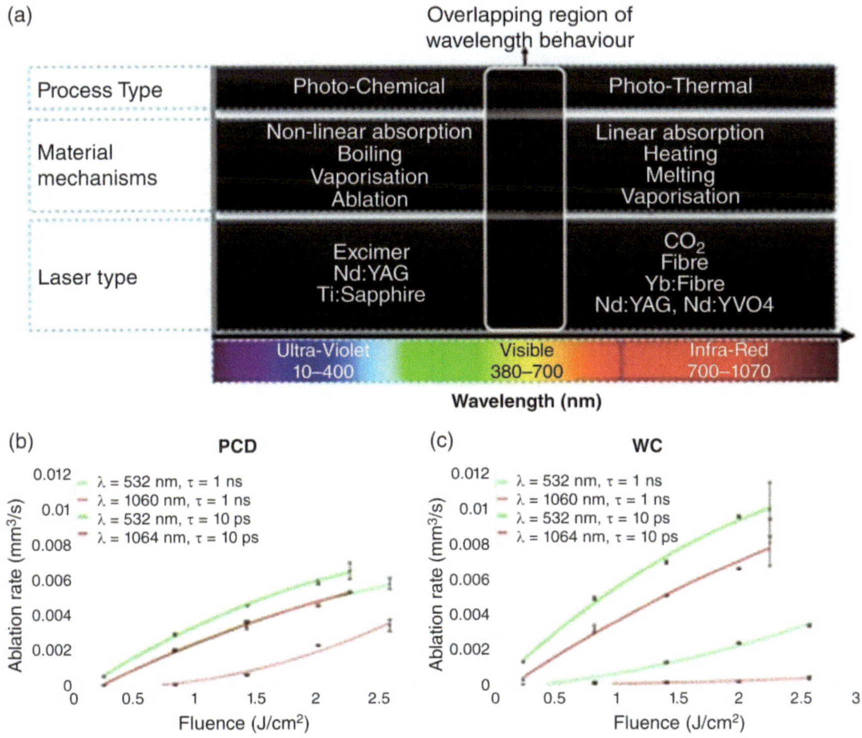

FIGURE 2.12 (a) Brief overview of material behavior at different wavelengths; (b) PCD ablation rate. (c) WC ablation rate [7] [open access].

recent research is to create silver-based coatings with controlled release mechanisms, long-lasting antibacterial activity and minimal risk of damage to human cells [35]. Researchers have explored combinations of broad-spectrum antibacterial metals, such as copper alloys, which promise to reduce surface bacterial contamination [24]. In addition, antibacterial polymers have been developed for surface treatment for medical purposes. These polymers provide powerful and durable antibacterial properties. Quaternary ammonium compounds and polymer nanoparticles with antibacterial agents are two examples of this type of polymer. Touch surfaces, medical devices and even textiles can be coated to reduce the risk of healthcare-related illnesses and improve worker and patient safety [29].

Additionally, advances in surface technology have made it easier to create self-cleaning surfaces in healthcare settings. These surfaces use advanced technologies such as photocatalytic coatings or super hydrophobic materials that repel water, prevent contaminants from sticking and make cleaning easier. These self-cleaning surfaces help improve hygiene and reduce the risk of cross-contamination in healthcare settings by reducing the adhesion of bacteria and organic waste [23]. The goal of current research is to create multifunctional coatings that have more than just antibacterial activity. For example, coatings with photocatalytic properties that can generate reactive oxygen species have been explored using materials such as titanium dioxide (TiO_2) exposed to light [29].

These coatings have been demonstrated to effectively decompose organic contaminants and kill bacteria from surfaces when exposed to UV light. Likewise, techniques such as nanoparticle encapsulation or integration into biodegradable polymers have also been explored to reduce the adhesion of proteins, biofilms and other contaminants to surfaces [30]. In healthcare settings, antimicrobial coatings are used on more than just devices and medical equipment. High-touch surfaces in hospitals, such as door handles, bed rails and handrails are designed to prevent the transmission of infections [36].

2.5 ENERGY USE

Over the past few decades, global urbanization has grown significantly, both in an organized and unorganized manner. The result of planned urbanization, which is encouraged by governments to improve the economic conditions of countries and regions, is organized urbanization. However, in most cases, urbanization in the world is an inevitable phenomenon, driven, among other things, by the desire of families to gain access to health, educational services and employment opportunities. Urban concentration occurs as a result of people moving from rural areas to metropolitan areas for certain reasons. The urbanization of industry and population have led to sustainable development problems associated with environmental degradation and depletion of economic resources. Surface modification to enhance catalytic activity. Many energy-related applications have developed as a result of surface engineering techniques used to increase the catalytic activity of metal surfaces. Other study has focused on changing the surface properties of catalysts to increase their selectivity and catalytic efficiency [30]. Monitoring the surface roughness of catalysts is one approach. Changing surface morphology at the nanoscale or introducing hierarchical structures can have a significant impact on the catalytic activity.

The pulse duration or irradiation pulse width affects how much energy is distributed and deposited onto target material (Figure 2.13a) [22]. It affects the absorption behavior like wavelength. There is a linear absorption in the nanosecond (ns) regime, which gradually changes to a nonlinear absorption in the femtosecond (fs) regime [23]. A higher level of surface roughness leads to higher catalytic efficiency because it increases the active surface area, improves reactant adsorption and promotes dispersion of reactant molecule. Furthermore, surface functionalization, which adds specific chemical groups or functional bonds to change the catalytic capabilities of the catalyst surface, can modify the electronic structure of catalyst surface. The question that needs to be addressed is whether the increasing environmental emissions are due to partly concentrated urban areas [7].

A well-known mechanism states that when an economy or region shifts from an agricultural to an industrial structure, the demand for fossil fuel energy for domestic and industrial purposes increases. This increasing in demand could

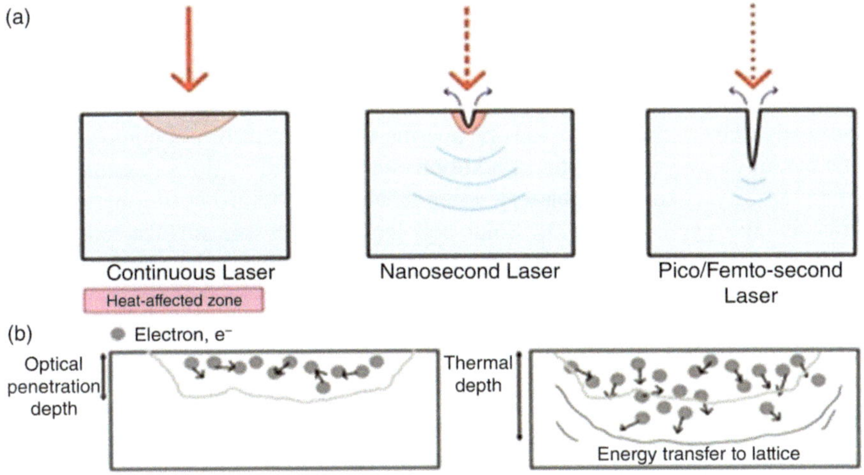

FIGURE 2.13 (a) Modified diagram comparing the effect of pulse duration on target surface. Blue dashed lines represent shock waves. Material expulsion and electron energy transfer at optical penetration depth compared to thermal depth (11 mm for PCD, 5.40 mm for PcBN and 1.2 mm for WC) are demonstrated by shorter pulse durations [7] [open access].

result in both gaseous and aqueous emissions. The situation is aggravated by using traditional industrial devices with a high energy consumption in production processes. The idea that megacities and industrial areas serve as the center of economic activity for entire economies and regions is true. Given this, it makes sense to explore the links between urban concentration, nonrenewable energy intensity, economic development and environmental emissions to gain a deep understanding of both environmental and economic sustainability [29, 36].

The development of bimetallic or doped structures is an important component of surface engineering in catalysis. Researchers can fine-tune the surface composition and electrical structure to improve catalytic performance by mixing two or more metals with different catalytic abilities. When two metals are combined, bimetallic catalysts often exhibit a synergistic effect that improves catalytic activity, selectivity and stability. Numerous studies have been carried out on these catalysts in processes such as selective oxidation reactions [28], carbon dioxide conversion [29] and hydrogen synthesis [30]. These surface changes and catalytic advances have a significant impact on creating sustainable energy solutions [37–41]. Researchers want to make hydrogen production processes more efficient, less energy-intensive and overall more sustainable by optimizing the surface characteristics of catalysts [42]. The coatings are to improve energy conversion and storage [43]. Surface engineering approaches have led to advances in areas such as solar cells and batteries [44].

In the world of lithium-ion batteries, surface coatings on electrode materials have been extensively studied to improve their performance. These coatings have many benefits, including improving the movement of ions and electrons within the battery, reducing unexpected negative interactions and increasing the stability of electrode materials during cycling. For example, it is possible to reduce negative reactions and improve battery stability by coating electrode materials with protective layers, such as oxides or polymers, to prevent direct contact between the electrode and electrolyte. In addition, surface modifications can improve the compatibility of different battery components, thereby promoting better charge transfer and cycling [35, 43]. Solar cell technologies use surface technologies to improve light absorption, charge transport and device performance. Surface mounting techniques such as surface mount can effectively diffuse and trap light within solar cells using a pattern or nanostructure, which widens the photon path and increases absorption [45].

Moreover, light reflection from the cell surface can be reduced by modifying the surface with AR coatings, which increases light absorption and overall device performance [35]. To improve charge extraction, minimize surface recombination losses and increase the efficiency of solar cells, surface passivation techniques have been applied [45]. In addition, new energy storage and conversion technologies pose challenges are explored using surface engineering methodologies. For example, surface modifications are being explored to improve the efficiency of electrochemical energy storage devices such as supercapacitors [44]. Challenges and potential paths significant advances have been made in surface finishing technology to improve the functionality and performance of metals in a variety of applications. But there are still problems to be solved, and there is a chance for more progress in this area in the future.

2.5.1 Limitations of Existing Surface Engineering Methods

Cost and complexity of the procedure. Many surfaces treatment methods are expensive to process because they require complex procedures, specialized tools and experienced operators. Because these methods require specialized tools and knowledge, their practical application in industrial settings can be difficult. It is critical to create scalable and affordable surface treatment methods that can be quickly put into practice in industrial sector service life and reliability. The performance of metals can be improved through surface coatings and modifications, but their service life and durability may still be an issue. Over time, coatings can deteriorate, providing less protection against environmental factors such as wear and corrosion. The service life of surface treatment materials must be extended [6].

The success of surface treatment methods depends on the coating and substrate having sufficient adhesion. However, when using certain combinations of materials, it may be difficult to achieve strong and long-lasting adhesion. Establishing methods to enhance coating adhesion and compatibility with substrates is essential for reliable, durable surface treatment solutions and complex geometric pattern. Processing limitations may arise when surface engineering techniques are used on complex structures or geometries. In such circumstances, achieving uniform coverage and hiding power can be challenging, resulting in variations in performance across the entire surface. To meet the need of many industries and applications, it is vital to find effective solutions for the surface treatment of complex shapes and structures [7].

The performance and quality of surface-engineered products can be affected by changes in material properties, environmental factors or process parameters. To ensure consistency and reliability of surface design, process control must be strengthened and strict quality assurance procedures must be established integration with currently used production processes. It may be difficult to integrate surface treatment techniques into current production processes, and it may be necessary to replace production lines or equipment. For surface finishing procedures to become widespread and ensure efficient production, it is necessary to integrate seamlessly with modern manufacturing systems. Continued research and development in surface engineering will be required to overcome these limitations [7].

Certain surface treatment procedures may require the use of hazardous chemicals, consume large amounts of energy or produce waste, all of which can have adverse effects on the environment. To ensure sustainable and ethical manufacturing practices, surface treatment methods must have the least possible negative impact on the environment. Important considerations include creating environmentally friendly substitutes and increasing process efficiency. Some surface treatment methods may be limited in the types of materials on which they can be used. To serve a wider range of applications and industries, there is a need to expand the range of materials to which surface engineering techniques can be applied.

Obtaining repeatable and consistent surface finishing results can be difficult, especially in large-scale production [28–30, 36–38]. Hexagonal boron nitride (hBN), which is created by the laser process, acts as a solid lubricant, helping to dissipate heat in the cutting zone. The behavior of the chip in deformation zone is modified by chamfer texturing. The wear resistance is reduced by parallel grooves. On the side surface, cross-hatching and perpendicular hatching have shown an excellent wear resistance due to different processes (Figure 2.14).

2.5.2 Physical Properties

- *Flowability of materials*: This is a property of a material that indicates how easily it flows when in a liquid state. It is the reciprocal of viscosity of liquid material.
- *Weldability of materials*: This is a property of material that indicates how easily it is possible to weld two pieces of material together using heat, pressure or both.
- *Elasticity of materials*: This is a property of material that allows it to return to its original size when a force or load is removed. Porosity of materials contains certain dissolved gases when it melts. These gases dissipate and leave voids when the substance solidifies. The number of voids in a substance is indicated by its porosity. solids.

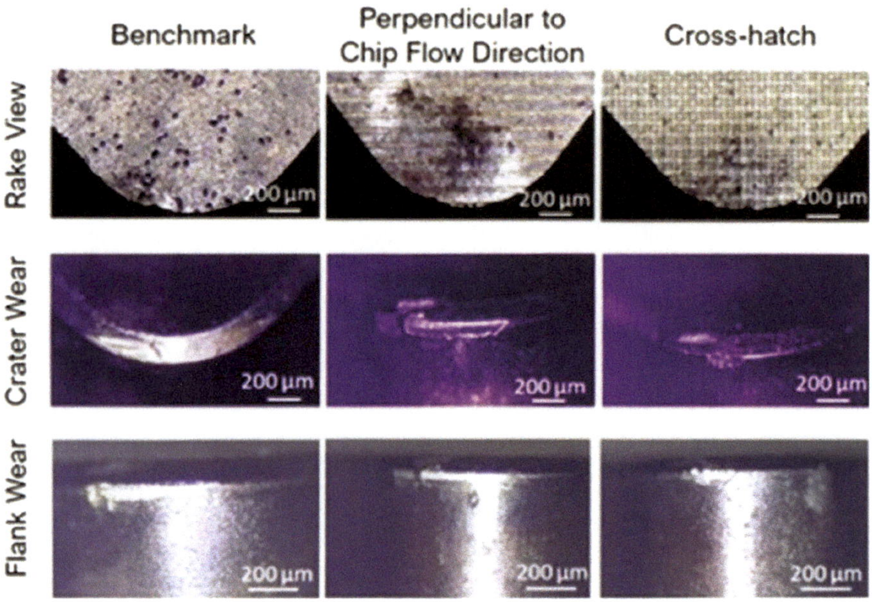

FIGURE 2.14 The wear mechanism of textured PcBN tools [7] [open access].

- *Thermal conductivity of materials*: This is a property of a substance that indicates how easily heat can be transferred through it. "The amount of heat transferred by unit thickness of material normal to the surface of unit area per unit time when the temperature gradient across the piece of material is unity in the steady state" is the definition of thermal conductivity of material. Watt per meter per K is its SI unit.
- *Electrical conductivity of materials*: The ease with which material can conduct electricity is shown by the property of material. It is denoted by π. This is the specific resistance of material multiplied by itself [35].

Biomimetic surfaces taking inspiration from the natural world, biomimetic surfaces seek to mimic the special qualities and features found in natural systems. An intriguing avenue for surface engineering research is creating surfaces replicating the self-cleaning characteristics of lotus leaves, sticky gecko feet or drag-reducing shark skin. Biomimetic surfaces can lead to new solutions in industries such as healthcare, automotive and aerospace [15]. The topographic images obtained for Figure 2.15c, d, e, f, and highly oriented pyrolytic graphite (HOPG (Figure 2.15a, b panels) electrodes are shown in Figure 2.15. The sample after one in HOPG sulfuric acid electrolyte (see Figure 2.15a) is characterized by flat terraces and clear step edges. Panels a, c and e correspond to the clean electrodes [46].

The other panels refer to the sample after one CV. After the intercalation procedure [45], the situation changed dramatically: transparent bubbles (Figure 2.15b), a result of deformation of trapped basal plane gas, are evident near

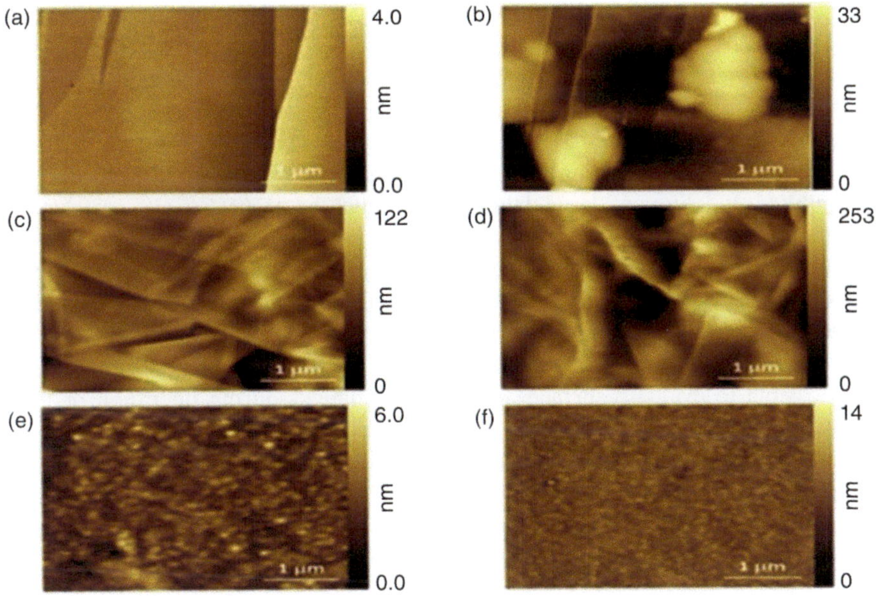

FIGURE 2.15 Before HOPG (a); after HOPG (b); GF (c); GF (d); GC (e) and GC (f) electrochemical treatment, AFM images of electrodes used were obtained [46] [open access].

or at step edges. The latter represented intercalation regions of solvated ions, which are preferred [46]. Figure 2.15c shows GF. The surface morphology is found to be a result of stacking of multiple thin graphite layers. The figure shows numerous stages, defects and pores. Thus, as mentioned earlier, the expected gas evolution during an anodic sweep in acidic electrolytes may find additional pathways for its outflow. The atomic force microscopy (AFM) image obtained after CV treatment provides strong evidence for this theory (see Figure 2.15d). The application of 3D printing or additive manufacturing as a surface finishing method is receiving increasing attention. By carefully applying ingredients layer by layer, this technique produces coatings with unique characteristics and complex surface structures. Surfaces with integrated characteristics, roughness profiles, or graded compositions can be produced using additive manufacturing processes, creating new opportunities for customized surface engineering solutions [14]. Intelligent surface design is one of the newest developments in the field of surface engineering. These surfaces can dynamically adapt to changes in the environment. For example, in response to external stimuli such as pH, light or temperature, smart surfaces can change their optical, wetting or adhesive properties.

The ability of molecules to self-assemble is another important criterion for surface coverage. Mechanical strength, adhesion and wear resistance will be enhanced through a combination of self-assembly and healing [24]. Once the self-assembled molecules are on surface of substrate, they polymerize to form a coating. The ability of coating technologies to self-heal from chemical or physical corrosion will be improved in the future [21, 22]. The application of nanotechnology in the field of surface engineering creates new opportunities for tuning characteristics, improving functions and creating complex applications. There is growing interest in applying artificial intelligence (AI) and machine learning techniques to surface design. These technological advances can help the process of designing and optimizing surface structures, coatings, and modifications using computer modeling and large data sets to predict the properties and performance of materials. In addition, using machine learning algorithms, new materials with customized surface properties can be developed, thereby simplifying surface design procedures [14]. Real-time monitoring and characterization of surfaces are becoming increasingly important in the field of surface engineering. Using in situ techniques such as spectroscopy, microscopy and sensor approaches, it is possible to evaluate surface and coating changes during operation or under equivalent conditions. Using this information, preventative maintenance plans can be created, which can also be used to optimize surface finishing operations and ensure the quality and reliability of surface finishing [14]. Table 2.2 lists the latest developments and methods of metal surface treatment that are previously discussed, along with their advantages and applications.

2.5.3 Promising Areas for Future Research

One of the most significant advances in surface technology has been smart coatings. The potential applications for them are numerous and have the potential to completely transform various industries. In this section, we will look at smart coatings, their potential applications and their future prospects [17–21].

The surface characteristics of materials often have a major impact on process efficiency. Important factors include corrosion resistance, selectivity and reaction speed. (You can learn more about the critical role that surface technology plays as a cross-sectional technology and how it impacts the development of renewable energy technologies in the Table 2.2). New coatings that react on their own are called smart coatings. They respond to external stimuli and therefore offer benefits beyond the basic passive purposes of a coating (e.g., decoration, protection). Heat, light irradiation, mechanical induction, temperature changes, pressure changes, pH changes, strong corrosive ions and more can act as stimuli. For this reason, "stimuli-responsive" or "environmentally responsive" coatings are other terms for smart coatings [22–24].

In theory, smart coatings always consist of active and passive parts that must be studied together to achieve their intended goals. Various technical strategies are used to achieve these "smart" qualities, including mixing organic and inorganic phases and using micro- and nanoparticle formulations. The functional element of these smart coatings can be either a resin or any of a wide range of potential additives (e.g., pigments, bioactives, enzymes, RFID devices, nanomaterials, inhibitors, microelectromechanical devices, microencapsulated actives, etc.). Table 2.2 presents metal surface finishing technologies and trends and recent advancements and tendencies[22].

TABLE 2.2
New Developments in Metal Surface Finishing Technologies and Trends [22] [Open Access].

Emerging Trend/ Technology	Potential Benefits	Potential Applications
Multifunctional coatings	• Enhanced performance and functionality • Reduced need for additional treatments or maintenance	• Automotive coatings • Aerospace coatings • Antifouling marine coatings
Nanotechnology and nanomaterials	• Improved surface properties • Tailored functionalities • Enhanced material performance	• Wear-resistant coatings • Corrosion-resistant coatings • Biomedical implants • Electronics
Biomimetic surfaces	• Self-cleaning properties • Adhesive properties • Drag reduction properties	• Aerospace surfaces • Automotive surfaces • Marine antifouling surfaces
Additive manufacturing for surface engineering	• Tailored surface structures • Graded compositions • Embedded functionalities	• Customized surface engineering • Aerospace components • Biomedical implants
Smart surfaces	• Adaptive response to external stimuli • Dynamic control over surface properties	• Adaptive optics • Microfluidics • Sensors
Machine learning and artificial intelligence	• Design optimization • Predictive modeling • Faster and more efficient processes • Real-time assessment	• Coating development • Surface modification • Material discovery
In situ characterization and monitoring	• Quality assurance • Proactive maintenance strategies	• Coating performance monitoring • Surface modification evaluation • Wear and corrosion monitoring

The development of smart coatings has seen a significant breakthrough with the introduction of nanoparticles. Thanks to nanotechnology, surfaces can now be precisely tailored at the microscopic level to perform specific tasks, such as better corrosion protection, hardness and abrasion resistance. Metal and metal oxide nanoparticles with various functional qualities and improved performance are increasingly being used in smart coatings [26–28]. Their potential applications are many and varied, as are the potential environmental signals that influence smart coatings. Some of the most promising application areas are listed in the subsequent paragraph.

The automotive sector uses smart coatings as self-healing coatings to protect surfaces from corrosion and abrasion. Dirt-repellent coatings are available, while it is also possible to manufacture transparent or color-changing adaptive coatings. The service life of components is extended by self-healing coatings. Multifunctional smart coatings have the potential to further improve the efficiency of solar cells. Self-cleaning ability, photoactivity and high transparency are the most sought-after qualities in solar cell applications, although self-healing and antimicrobial activities have also gained attention recently. Thus, "intelligent" coatings consisting many components have been developed to produce films that serve multiple purposes.

Surface engineering approaches must be scalable and easy to implement to be of significant value across a variety of industries. The goal should be to develop surface treatment methods that are easy to use by industrial manufacturers. This means addressing issues such as robust quality control methods, affordability, process scalability and ease of interoperability with existing production procedures. The goal of future research should be to create surfaces whose characteristics can dynamically change in response to external stimuli [36, 37]. This includes surfaces that can switch between different states of adhesion, wettability, or optical properties, opening up new possibilities and applications.

In particular, research into materials with stimulus-responsive properties and development of surface engineering techniques to control their dynamic behavior will be important. Nature serves as a source of inspiration for surface engineering. Subsequent research should focus on understanding and replicating the unique properties and functions found in natural systems, such as self-cleaning, anti-fog and anti-reflective surfaces. Table 2.2 displays the latest developments and trends in metal surface finishing technologies. By modeling nature, researchers can develop new surface engineering solutions that provide improved stability, durability and performance. Many applications at different scales, from the nanoscale to the microscale, require surface engineering. The main goal of research should be the development of methods that allow precise control of surface properties and structure over a wide range of length scales. Among these are surface engineering methods that are hierarchical and use surface changes in many dimensions to optimize performance and provide synergistic effects [38–41].

As additive manufacturing matures, it becomes possible to integrate surface engineering techniques directly into the 3D printing process. This includes changing the surface during printing or producing new printing materials with integrated surface features. A wide range of techniques are called surface engineering and aim to design and modify the surface properties of components. To maximize material volume and surface properties, two main types of surface treatment methods are available [41]. These are surface modification and surface coating. Surface shape design is used in other processes as well. Applying a layer of molten or semi-molten material or chemicals to a substrate is called surface coating. Changing and increasing the functionality of surface, rather than changing the composition of bulk materials, is one of the main goals of surface coating. Physical vapor deposition (PVD), chemical vapor deposition (CVD), sol–gel, cladding, electroplating, plasma and thermal spray are some of the few types of surface coating methods [11]. Surface modification procedures may include diffusion treatments (e.g., nitriding and carburization) and high-energy treatments (e.g., ion implantation), as well as flame, induction, laser or electron beam hardening. Surface modification techniques can be used to change the mechanical or physical properties of components, reduce friction and improve surface resistance to wear and corrosion. Surface treatments, including laser cladding, can also be used in conjunction with coating procedures. Combining the benefits of surface modification and coating meets the specific needs and ensures the product's suitability. Surface design should theoretically involve designing components based on the surface treatment applied [35, 42–45]. However, surface-based methods are often used in the design of existing engineered components.

- No universal correlation has been found between wear, despite the potential presence of a lubricant as a third body or key component of one of elements of pair (for example, graphite in molten iron or molybdenum sulfide in some nylon-based polymers) and destroy priceless items. Interaction. On another hand, lubrication usually reduces wear and friction. It is better to use a minimal amount of lubricant than not to use any lubricant to reduce wear.
- It is best to avoid using the same slip-wear materials. Highly compatible mineral materials are preferred because they exhibit very low or no solid solubility on equilibrium diagrams.
- Many surfaces treatment methods, including PVD, CVD, thermal and thermochemical surface treatments, can be used to produce high surface hardness, which is useful in a variety of situations.
- Carbides or nitrides in the outer layer have an advantage even if the surface hardness of steel is significantly reduced. The high-hardness surfaces resulting from deformation are resistant to adhesives, etching and severe abrasion. A relationship can be established between surface quality and etch resistance.
- The rough surface created by surface bombardment is more resistant to etching. Sliding wear is resistant to surface layers created by PVD, CVD, ion implantation, nitrate, or carburization processes. This resistance is determined by the speed of impact, angle of impact and density of the impacting substance [45].
- The advantage in these circumstances is low ductility and high hardness.

Surface engineering is an important and dynamic field that improves the performance and the quality of materials, significantly impacting many different sectors. Continuous technological advances coupled with international collaborations will drive the development of surface engineering and make it a fundamental part of modern manufacturing and materials science. Surface engineering will be essential to shaping a sustainable and technologically advanced future by addressing current challenges and seizing future opportunities.

2.6 PROBLEMS

1. How would you rate the wear of the mechanisms and the surface roughness of used coins?
2. Important factors are high corrosion resistance and the method of manufacturing the coin (including stamping of small parts of the coin surface). When coins are in circulation, it is also important that they retain their surface detail over a long period as this means longevity. Explain how the wear rate of a newly proposed coinage alloy can be calculated.
3. Explain the surface roughness measurement techniques available for the following tribological applications. DLC coatings are sometimes used on automotive engine camshafts in place of electroplated chrome plating.
4. Explain the physical properties and new developments and technologies.
5. Should we expect that two moving objects will have more actual points of contact than two stationary objects? How can wear and particle transport affect the number of contacts points?

REFERENCES

1. Ramezani, M., Pasang, T., Chen, Z., Neitzert, T., & Au, D. (2015). Evaluation of carbon diffusion in heat treatment of H13 tool steel under different atmospheric conditions. *J. Mater. Res. Technol*, *4*, 114–125.
2. Li, H., Ramezani, M., Chen, Z., & Singamneni, S. (2019). Effects of process parameters on temperature and stress distributions during selective laser melting of Ti–6Al–4V. *Trans. Indian Inst. Met*, *72*, 3201–3214.
3. Awang, M., Khalili, A.A., & Pedapati, S.R. (2020). A review: Thin protective coating for wear protection in high-temperature application. *Metals*, *10*, 42.
4. Riquelme, A., & Rodrigo, P. (2021). An introduction on the laser cladding coatings on magnesium alloys. *Metals*, *11*, 1993.
5. Fotovvati, B., Namdari, N., & Dehghanghadikolaei, A. (2019). On coating techniques for surface protection: A review. *J. Manuf. Mater. Process*, *3*, 28.
6. Vázquez-Martínez, J.M., Del Sol, I., Batista, M., Vidal, S.R.F., Hernandez, P.M., & Salguero, J. (2019). Experimental study of laser texturing processes on the lubricant retention of carbide (WC-Co) surfaces. *Key Eng. Mater*, *813*, 55–61.
7. Hazzan, K.E., Pacella, M., & See, T.L. (2021). Laser processing of hard and ultra-hard materials for micro-machining and surface engineering applications. *Micromachines*, *12*(8), 895.
8. Ze, W., Jianxin, D., Yang, C., Youqiang, X., & Jun, Z. (2012). Performance of the self-lubricating textured tools in dry cutting of ti-6Al-4V. *Int. J. Adv. Manuf. Technol*, *62*, 943–951.
9. Hutchings, I., & Shipway, P. (2017). Tribology: Friction and Wear of Engineering Materials, Butterworth-Heinemann.
10. Oppong Boakye, G., Ormsdóttir, A.M., Gunnarsson, B.G., Irukuvarghula, S., Khan, R., & Karlsdóttir, S.N. (2021). The effect of polytetrafluoroethylene (PTFE) particles on microstructural and tribological properties of electroless Ni-P+ PTFE duplex coatings developed for geothermal applications. *Coatings*, *11*(6), 670.
11. Akhtar, A.N., Murtaza, G., Shafique, M.A., & Haidyrah, A.S. (2021). Effect of Cu ions implantation on structural, electronic, optical and dielectric properties of polymethyl methacrylate (PMMA). *Polymers*, *13*(6), 973.

12. Nsilani Kouediatouka, A., Ma, Q., Liu, Q., Mawignon, F.J., Rafique, F., & Dong, G. (2022). Design methodology and application of surface texture: A review. *Coatings*, *12*(7), 1015.
13. Močilnik, V., Predan, J., & Gubeljak, N. (2023). Effect of residual stresses on fatigue strength on the smooth and splined parts of spring torsion bars. *Metals*, *13*(6), 1004.
14. Ramezani, M., Mohd Ripin, Z., Pasang, T., & Jiang, C.-P. (2023). Surface engineering of metals: Techniques, characterizations and applications. *Metals*, *13*(7), 1299.
15. Nazarzade, K., Naderi, R., & Mahdavian, M. (2023). Fabrication of corrosion-resistant conversion coating based on ZIF-8 on electro-galvanized steel utilizing pulse electrodeposition technique. *Surf. Coat. Technol*, *457*, 129292.
16. Katić, J., Krivačić, S., Petrović, Ž, Mikić, D., & Marciuš, M. (2023). Titanium implant alloy modified by electrochemically deposited functional bioactive calcium phosphate coatings. *Coatings*, *13*, 640.
17. Nikolova, M.P., & Apostolova, M.D. (2023). Advances in multifunctional bioactive coatings for metallic bone implants. *Materials*, *16*, 183.
18. Costa, R.C., Nagay, B.E., Dini, C., Borges, M.H., Miranda, L.F., Cordeiro, J.M., Souza, J.G., Sukotjo, C., Cruz, N.C., & Barão, V.A. (2023). The race for the optimal antimicrobial surface: Perspectives and challenges related to plasma electrolytic oxidation coating for titanium-based implants. *Adv. Colloid Interface Sci*, *311*, 102805.
19. Wang, C., Jiu, H., Zhang, L., Xu, R., Che, S., Guo, Z., Han, Y., Ma, J., Li, H., & Guo, F. (2023). Carbon coating hollow NiS nanoparticles anchored on Ti3C2 nanosheet as high-performance anode for lithium-ion batteries. *J. Alloys Compd.*, *956*, 170332.
20. Su, Y.L., Kao, W.H., & Chang, G.Y. (2023). Tribological, anti-corrosion, and electrical conductivity properties of CrCx coatings deposited on stainless steel 316L and used as metal bipolar plates for fuel cells. *J. Mater. Eng. Perform*, *32*, 3739–3754.
21. Doluk, E., Rudawska, A., & Miturska-Barańska, I. (2022). Investigation of the surface roughness and surface uniformity of a hybrid sandwich structure after machining. *Materials*, *15*(20), 7299.9.
22. Chowdhury, M.A. (2019). Friction, Lubrication and Wear, BoD-Books on Demand.
23. Hutchings, I., & Shipway, P. (2017). Tribology: Friction and Wear of Engineering Materials, Butterworth-Heinemann, Department of Materials Science and Metallurgy, University of Cambridge, United Kingdom.
24. Martínez-Pérez, F. (2022). Surface engineering. Application on wear. *Revista Ciencias Técnicas Agropecuarias*, *31*(4).
25. Martínez, P.F. (2017) Industrial Maintenance. Concepts and Applications, Ed. Editorial Cuba Azúcar, La Habana, Cuba.
26. Liu, Y., Deng, J., Wang, W., Duan, R., Meng, R., Ge, D., & Li, X. (2018). Effect of texture parameters on cutting performance of flank-faced textured carbide tools in dry cutting of green Al2O3 ceramics. *Ceram. Int.*, *44*, 13205–13217.
27. Sugihara, T., & Enomoto, T. (2013). Crater and flank wear resistance of cutting tools having micro textured surfaces. *Precis. Eng.*, *37*, 888–896.
28. Wang, Q., Yang, Y., Yao, P., Zhang, Z., Yu, S., Zhu, H., & Huang, C. (2021). Friction and cutting characteristics of micro-textured diamond tools fabricated with femtosecond laser. *Tribol. Int.*, *154*, 106720.
29. Pang, K., & Wang, D. (2020). Study on the performances of the drilling process of nickel-based superalloy Inconel 718 with differently micro-textured drilling tools. *Int. J. Mech. Sci.*, *180*, 105658.
30. Kümmel, J., Braun, D., Gibmeier, J., Schneider, J., Greiner, C., Schulze, V., & Wanner, A. (2015). Study on micro texturing of uncoated cemented carbide cutting tools for wear improvement and built-up edge stabilisation. *J. Mater. Process. Technol.*, *215*, 62–70.

31. Mazaheri, Y., Khodaveysi, E., Roknian, M., Sheikhi, M., & Heidarpour, A. (2022). 75Cr3C2-25NiCr and 86WC-10Co-4Cr high wear-and corrosion-resistant cermet coatings deposited on A356 substrate by high-velocity oxy-fuel technique. *Coatings*, 12(10), 1408.
32. Ludema, K.C., & Ajayi, O.O. (2018). Friction, Wear, Lubrication: A Textbook in Tribology, CRC Press.
33. `Leyendecker, T., Lemmer, O., Jurgens, A., Esser, S., & Ebberink, J. (1991). Industrial application of crystalline diamond-coated tools. *Surf. Coat. Technol*, *48*, 253–260.
34. Kozaderov, O., Swiatowska, J., Dragoe, D., Burliaev, D., & Volovitch, P. (2021). Effect of Cr(III) passivation layer on surface modifications of zinc-nickel coatings in chloride solutions. *J. Solid State Electrochem*, *25*, 1161–1173.
35. Hao, X., Cui, W., Li, L., Li, H., Khan, A.M., & He, N. (2018). Cutting performance of textured polycrystalline diamond tools with composite lyophilic/lyophobic wettabilities. *J. Mater. Process. Technol.*, *260*, 1–8.
36. Guimarães, B., Fernandes, C.M., Figueiredo, D., Carvalho, O., Silva, F.S., & Miranda, G. (2020). Effect of laser surface texturing on the wettability of WC-co cutting tools. *Int. J. Adv. Manuf. Technol*, *111*, 1991–1999.
37. Niketh, S., & Samuel, G. (2017). Surface texturing for tribology enhancement and its application on drill tool for the sustainable machining of titanium alloy. *J. Clean. Prod*, *167*, 253–270.
38. Wakuda, M., Yamauchi, Y., Kanzaki, S., & Yasuda, Y. (2003). Effect of surface texturing on friction reduction between ceramic and steel materials under lubricated sliding contact. *Wear*, *254*, 356–363.
39. Rao, C.M., Rao, S.S., & Herbert, M.A. (2018). Development of novel cutting tool with a micro-hole pattern on PCD insert in machining of titanium alloy. *J. Manuf. Process*, *36*, 93–103.
40. Wenlong, S., Jianxin, D., Hui, Z., Pei, Y., Jun, Z., & Xing, A. (2011). Performance of a cemented carbide self-lubricating tool embedded with solid lubricants in dry machining. *J. Manuf. Process*, *13*, 8–15.
41. Jianxin, D., Wenlong, S., & Hui, Z. (2009). Design, fabrication and properties of a self-lubricated tool in dry cutting. *Int. J. Mach. Tools Manuf.*, *49*, 66–72.
42. Xing, Y., Deng, J., Feng, X., & Yu, S. (2013). Effect of laser surface texturing on Si3N4/TiC ceramic sliding against steel under dry friction. *Mater. Des.*, *52*, 234–245.
43. Moravčíková, J., Moravčík, R., Kusý, M., & Necpal, M. (2018). Influence of laser surface texturing on tribological performance of tool steels. *J. Mater. Eng. Perform*, *27*, 5417–5426.
44. Xing, Y., Zhang, K., Huang, P., Liu, L., & Wu, Z. (2020). Assessment machining of micro-channel textures on PCD by laser-induced plasma and ultra-short pulsed laser ablation. *Opt. Laser Technol.*, *125*, 106057.
45. Meng, R., Deng, J., Liu, Y., Duan, R., & Zhang, G. (2018). Improving tribological performance of cemented carbides by combining laser surface texturing and W-S-C solid lubricant coating. *Int. J. Refract. Met. Hard Mater.*, *72*, 163–171.
46. Filoni, C., Shirzadi, B., Menegazzo, M., Martinelli, E., Di Natale, C., Li Bassi, A., & Bussetti, G. (2021). Compared EC-AFM analysis of laser-induced graphene and graphite electrodes in sulfuric acid electrolyte. *Molecules*, *26*(23), 7333.

3 Global Perspective of Surface Engineering

3.1 INTRODUCTION

Wear of moving parts is a common long-term problem that often occurs in operating environments and results in a permanent decline in regular productivity, operational efficiency and other performance-dependent components. Then, it becomes necessary to immediately stop the machine or the entire study process to replace the damaged components with new ones [1]. Surface technology is a concept that is continually used in industry to improve efficiency, component life and efficiency of high-tech manufacturing processes. To ensure and validate optimal product reliability at minimal capital cost, the industries are constantly inventing new materials, coatings, manufacturing technologies and innovative designs that can perform successfully in any operating environment. The broad and open-ended discipline of surface engineering is dedicated to improving the performance, durability, sustainability and material quality of engineered components.

Surface properties such as wettability, visual, optical fastness, corrosion resistance, wear resistance, tear resistance, or tribological properties can be independently changed or improved by surface treatment technology. Surface engineering, which can be achieved by applying various surface treatment technologies to create an environmentally sustainable surface, entails modifying the phases and properties of the surface to minimize surface deprivation over time. The goal of surface engineering is to improve component wear resistance and adjust friction characteristics such as surface hardening, carburization and coating, with or without composition modifications [2].

By applying surface layers to substrate materials, the coating is a surface engineering approach that improves the tribological performance of materials; the techniques include physical vapor deposition (PVD), thermal chemical spraying, high-velocity oxygen deposition (HVOF) and vapor deposition. Surface treatment is the most effective means of preventing corrosion, wear, erosion and other related problems. Corrosion, wear and erosion lead to the gradual breakdown of metal surfaces and components, which can lead to cost savings, reduced plant efficiency and shutdown consequences apparatus.

From numerous studies conducted by ASM [3, 4], it is clear that the US economy spends $300 billion annually – or 4.2% of GDP – on protecting metals from corrosion. The combination of wear and corrosion can have a dangerous effect on components, resulting in material wastage, reduced material life, and much more than just one process affecting individual parts. The research has shown

DOI: 10.1201/9781003584735-3

that corrosion can only occur in the absence of mechanical wear, although the reverse is not possible. Many industries, including pulp, mining, paper, chemicals and energy, are negatively impacted by a combination of corrosion and wear production. Due to the relatively small area that absorbs the load, the two surfaces come into contact, relative motion occurs and contact stresses increase. This will cause wear and friction and may even cause failure. There are high-stress areas in a modern engine or transmission where metal-on-metal pressure forces all the lubricant out and causes heat to build up. This heat is enough to instantly fuse two parts under extreme pressure and friction, just before their movement tears them apart. The ongoing process of welding and breaking down on an atomic scale causes wear and resistance that is harmful. If gears are components of a spacecraft, the entire multimillion-dollar spacecraft would fail if just one component failed. Therefore, it is very useful to have surfaces that reduce wear and combat friction.

3.2 JOURNEY IN SURFACE ENGINEERING

The history of surface technology dates back to the prehistoric period when people of Greek and Chinese origin used solid organic materials for heat treating, carburizing and tempering. With the advent of significant research into modern surface technology in the 20th century, a new chapter in the history of surface technology is written. Composite materials are created using laser, plasma and ion beam processes [1–3, 5, 6]. Some 20 years have passed the founding of the first surface research institute, which is established at the Birmingham Research Institute in the UK in 1983. At the end of 1985, the first worldwide journal on surface engineering is published. In 1987, China has published the country's first surface engineering journal. With the fields of electromagnetic energy, thermal spraying techniques, laser melting techniques, nanosurface engineering, etc. having undergone extensive research, the surface engineering research is currently expanding extremely rapidly. Over the past 20 years, the surface engineering has evolved into three distinct phases: thermal sputtering, physical vapor deposition (PVD) and chemical vapor deposition (CVD) (laser, ion implantation, and multiple surface treatments). AFM images provided more precise data on the shape and surface structure of the protein [7]. AFM images of human serum albumin (HSA), HGG and HF before and after their interaction with fatty acid derivatives (FCD) are shown in Figure 3.1. AFM image of HSA has revealed nanoparticles with a height of about 4–8 nm (Figure 3.1a), which was consistent with the flattened ellipsoidal shape described by the heart-shaped structure and side length of 80 Å, which was represented at 2.8 Å resolution by crystallographic methods [7, 8]. The height increases to 10–15 nm and HSA–FCD dispersion typically refers to a system involving HSA and a fatty compound, often represented as FCD is transformed into aggregation by adding FCD (Figure 3.1d). This indicated that HSA formed the complex and FCD adsorbed on each other. The spherical shape is also visible in AFM image of HGG typically refers to "high-grade glioma," a type of aggressive brain tumor that includes glioblastoma multiforme (GBM) and

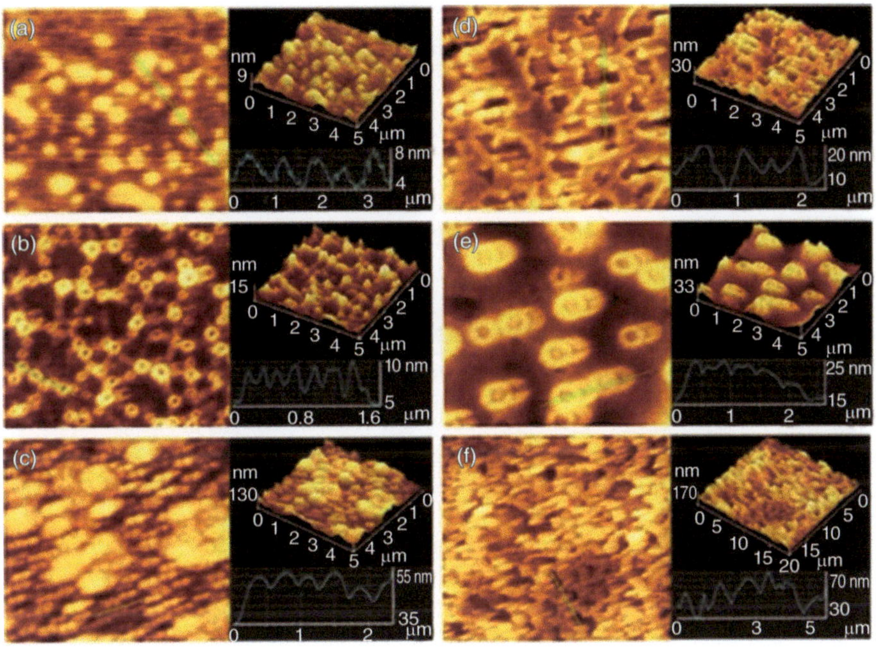

FIGURE 3.1 2D and 3D AFM images of corresponding volume and height analyses of (a) HSA, (b) HGG, (c) HF, (d) FCD-HSA, (e) FCD-HGG, and (f) FCD-HF [7] [open access].

anaplastic astrocytoma. The difference is that with an average height of 3–6 nm, the center of particle surface is uneven and hollow (Figure 3.1b). Its height is approximately equal to that of immunoglobulin G (4 nm) [7] and comparable to the height of HGG measured by ellipsometry, which has an average height of 45 Å [7]. The height of HGG increased to 9–11 nm after interaction with FCD and obvious adhesion phenomena are visible (Figure 3.1e). In addition, the morphology of HF is a thin ribbon with several nodules (Figure 3.1c), which closely matched the three-node scheme reported in other studies. The dense particles resulted in an uneven height of the substrate, which limited the accuracy of the study. The height of HF was 5–15 nm (Figure 3.1c) [7].

The second group consists of several methods for applying surface coatings. The use of nanotechnology in materials based on nanostructures ranks third. Engineering methods for surface improvement are shown in Figure 3.1.

3.3 TECHNICAL CHARACTERISTICS OF SURFACE

By improving surface properties, including wear, erosion, corrosion, adhesion, abrasiveness and oxidation, the surface engineering strives to provide the best possible modifications to a component's surface, microstructure/composition and grain structure [2]. In surface modification, surfaces, microstructures, or

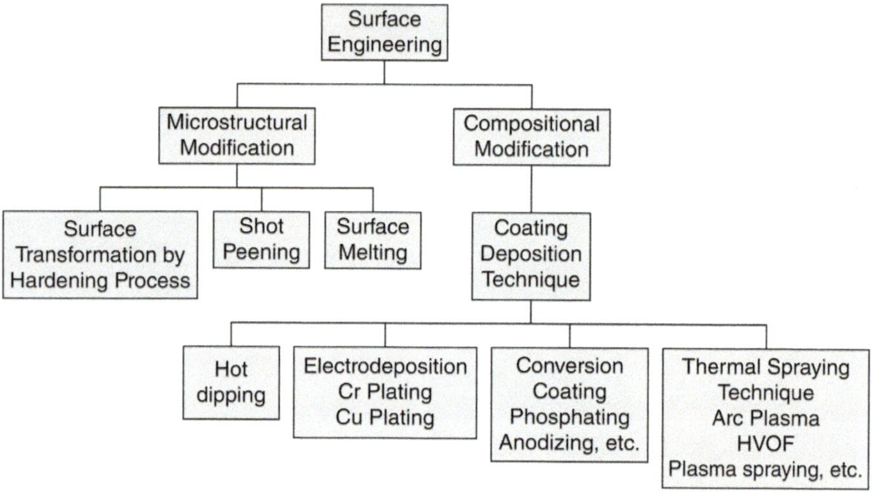

FIGURE 3.2 The primary taxonomy of manufacturing/metallurgical industry relating to surface construction [2] [open access].

compositions are often modified simultaneously for any segment of a component using both traditional and complex processing techniques. The main classification of the metallurgy and surface engineering industries is shown in Figure 3.2. One of the characteristics of surface technology is wear, which is the gradual removal of material from working surface of an object during its movement [2]. Alternatively, tribological interaction between the component surfaces and atmosphere tends to cause material loss at the component surface. Wear is the process of losing material from surface of part. The wear process is significantly influenced by the macrostructure, microstructure and surface roughness.

Another type of wear, in which material loss begins when a hard surface slides over a soft one is called abrasive wear [2]. Solid particles in liquid or water can cause erosive wear by rolling or sliding on the surface of the part [3]. Erosion tends to occur more often in caustic environments. Loss of material and important properties in an aquatic environment as a result of corrosion-related chemical and electrochemical processes. When a material is exposed to an oxidizing environment (rich in oxygen), depletion of the material occurs.

3.4 TYPES OF SURFACE GEOMETRY

Different antibacterial properties are introduced depending on the intended use. For example, a bone scaffold is created using a silver laser melting process that is found to be able to eliminate 99.9% of S. aureus in just 14 hours [2]. At the macroscopic level, the PHA samples were transparent to visible light when deposited as thin coatings. The surface morphology of the PHA samples was examined to separate the uniform material fractions for AFM tip spacing testing. After spin coating on a glass substrate with arithmetic mean height Sa and root mean square

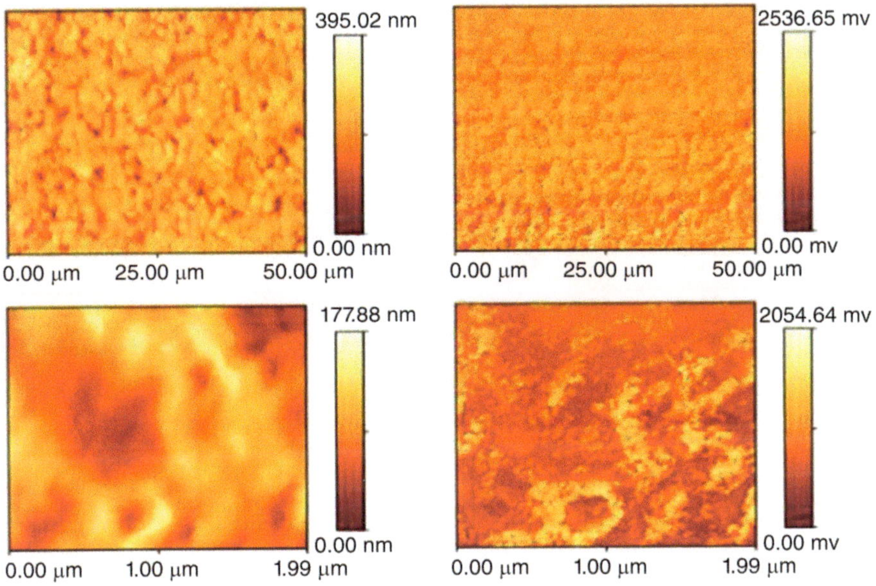

FIGURE 3.3 Images of surface morphology acquired by AFM in tapping mode for PHA thin films. Top: height (left) and demodulation (right) 50 × 50 (µm) images; bottom: height (left) and demodulation (right) 2 × 2 (µm) images [9] [open access].

height Sq of 16.7 nm and 22.8 nm, respectively, the morphology of the PHA was examined using tapping mode AFM. Note that Figure 3.3 shows the 50 × 50 (µm) and 2 × 2 (µm) detuning and height images.

The relationship between surface charge and wettability and surface chemistry affects cell adhesion, morphology, proliferation and differentiation [10]. The biocompatibility and immunogenicity of materials are largely dependent on surface chemistry. Atomic energy microscopy images of the surface of chalcocite in 1×10^{-4} M PAX solution at pH 6 for 10 minutes are shown in Figure 3.4. When the metal surface is in contact with 1×10^{-4} M PAX solution for 10 minutes, a large amount of adsorbed material appears on the chalcocite, as can be seen by comparing Figure 3.4a, the height image, with confirm the typical roughness exhibited by PHA samples, a line plot is attached to a 50 × 50 (µm) height image (Figure 3.2). The roughness parameters include defining a reference plane with area A and specifying Z(x,y) as the interval between points and the reference plane.

The amount of heat supplied to the surface layer during the hardening process, the duration of heating, and heat loss during conductive cooling are the main determining factors of surface hardening. Beam hardening, flame hardening and induction hardening are common methods of surface hardening, as well as laser hardening. When a substance is heated to its melting point and then allowed to solidify, it is said to have undergone surface melting [2, 3, 5, 6]. In the second category, compositional qualities (microstructure and mechanical properties) are modified to transform the surface information of the component.

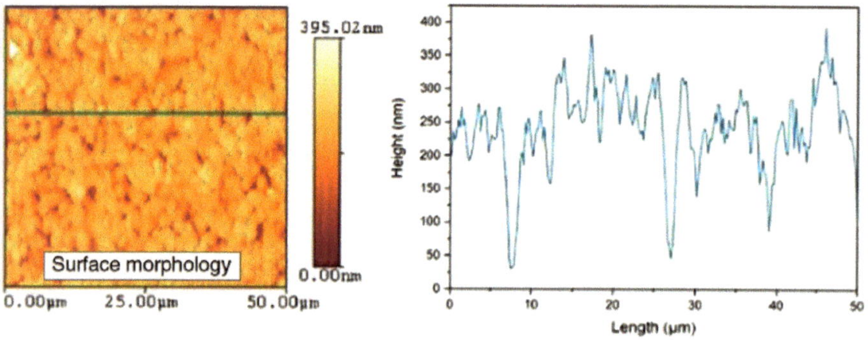

FIGURE 3.4 Height 50 × 50 (µm) image of surface morphology acquired by AFM in tapping mode for PHA thin films sample (left) and its line profile (right).

These procedures include carbonitriding, boriding, nitriding, chrome plating and carburizing (cementation). The cementation process involves the diffusion of carbon particles onto the surface of the substrate (steel). At the same time, at the hardening stage, an austenitic and martensitic structure developed during the carburization process. This structure provides hardness and serves as the basis for material distortions resulting from changes in the structure and volume of the lattice. The third category of methods are those in which materials such as paints, coating techniques, electroplating, PVD, CVD and fused deposition are applied to the surface of components. Coating is a surface treatment process that involves applying surface-treated layers to base materials to improve their mechanical properties, corrosion resistance, corrosion resistance, reduce the coefficient of friction and improve thermal insulation and aesthetic appearance. By providing surface treatment procedures including metallurgical, chemical and mechanical treatment, the above contents can be improved [2, 11, 12].

3.5 SURFACE MODELING

As surface treatment technologies rapidly evolve to meet the needs of industry, surface treatment technology plays a critical role in combating material degradation problems. There are many different surface modeling solutions available today that can provide a range of opportunities to enhance and optimize the performance of materials and components in challenging operational or hostile situations.

Process modeling improves a range of technical activities and provides important technical data as well as fundamental modifications to chemical, physical, mechanical, biological and other processes. Twenty years ago, system designers have created sophisticated models of cementing process, a surface treatment procedure. It is based on outdated diffusion laws [13]. These intelligent models have been ingeniously implemented in industry to control processes and have been combined with them to create new designs such as bearing and gear designs [2, 12]. Research and development (R&D) in the current environment are progressing rapidly, despite the outdated statistical design of experiments and the fact that process modeling can only reduce and support experiment-based research, but not eliminate it. The experimental design model can be tested; However, the opposite cannot be done. Programs that use simulation models can greatly benefit experimental research; For example, in the field of thermodynamics, the feasibility of applicable research is used to improve efficiency. The surface design system must be executed with care to achieve optimal results. performance and proper design of the surface treatment system is necessary. As shown in Figure 3.5, the coupled phase field modeling has been used in models at macroscale [14], microscale [15] and multiscale [16] levels. Most of the phase field models that are presented in these experiments are able to estimate the fatigue life and construct the fracture morphology accurately. At the same time, several approaches are proposed to improve the performance of the phase field models in an attempt to further

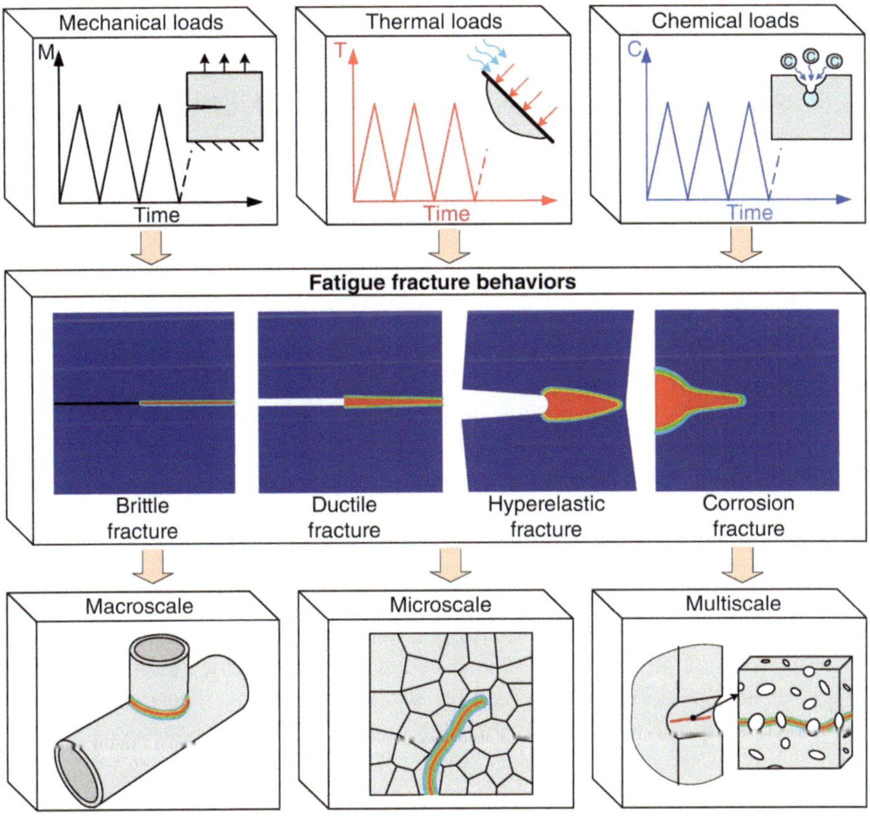

FIGURE 3.5 Phase field methods are used to simulate fatigue failure behavior [13] [open access].

improve the convergence, accuracy and efficiency. By using these strategies, the performance of the phase field models has been improved to a satisfactory degree. The purpose of this study is to review the developments in these areas [13].

Fundamental laboratory studies of vapor and liquid in the early stages. On the other hand, direct experiments with vapors and liquids can be calibrated without data by combining a simulator with thermodynamic computational models. Simulations provide a new insight into solutions that are difficult to calculate experimentally. For example, for process integration, a simulation process can solve prototype computational fluid dynamics models [2, 17]. The mechanical and tribological properties of surface geometries can be difficult to simulate using simulation models. Because in fact, different materials cause the properties of the substrate or component (hardness, tribology, structural gradation, chemical and mechanical surface roughness) to change depending on environmental conditions. It is difficult to attempt to determine precise correlations between various material properties and their associated components without knowing these interactions, which mimic the geometric properties of ceramic, metallic, non-metallic and composite material surfaces [2, 4]. For this reason, it is critical to understand how a material's properties relate to each other before creating simulation models of it.

3.6 ENVIRONMENTAL AND SURFACE ENGINEERING

Surface engineering deals with the mechanical, physical, or chemical manipulation of the microstructure and/or composition of a component's surface, which may involve the addition of material to alter the surface quality of the component. One of the most noticeable effects is the significant extension of the life of various components in many industries. It also helps save energy by increasing efficiency through the use of lighter components, which significantly reduces friction and increases combustion temperatures. This study provides an introduction to surface engineering, including its various technologies, industrial uses and positive and negative environmental impacts. Some high-stress areas in a modern engine or transmission allow heat to build up as pressure on the metal expels all the lubricating oil. This heat is enough to instantly fuse two parts together under intense pressure and friction before them movement tears them apart. Resistance and corrosion result from the continuous welding process and atomic scale destruction, including damage. If the spaceship's gears are separate parts and one of them broke, the multimillion-dollar ship as a whole would be lost. Therefore, it is highly desirable to have specially designed surfaces to reduce wear and combat friction. Many materials are designed with specific properties in mind, although they are not specifically optimized for surface properties. These problems can be solved using surface engineering. The surface modification methods include:

- Implanting alloy atoms at different depths increases impact strength and fatigue properties;
- Apply thick or thin surface layers (Surface coating)
- Reshape the component surface to distribute pressure evenly.

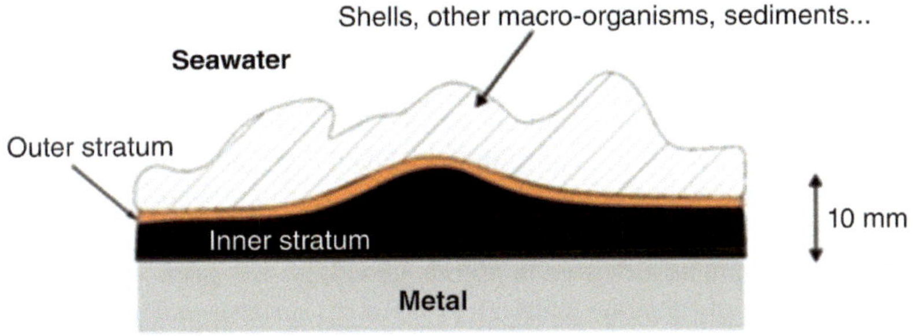

FIGURE 3.6 Schematic illustration (cross-section) of layer covering a carbon steel coupon after it had been immersed in natural seawater for six to eleven years [14] [open access].

3.7 WEAR, FRICTION AND CORROSION

Problems of wear, corrosion and friction can be solved using surface engineering techniques. Wear, corrosion and friction are the three most common causes of technical failures. In developed countries, wear, corrosion and friction account for 7% of GDP; however, through the implementation of efficient tribological systems, this percentage could be as low as 1%. Globally, Europe consumes 34% of all lubricants produced; 50% of this amount is recycled, and another 2.5 million tons are released into the environment. Figure 3.6 illustrates damage to an uncoated shaft due to seawater corrosion. The main cause of corrosion is the electrochemical reaction between metals and the materials in their environment, it shows damage to the shaft due to rust without a corresponding layer. The annual cost of corrosion is estimated to be approximately £30 billion in the UK and $276 billion in the USA [14–16, 18–20].

Wear is the progressive loss of material from a surface. Adhesive and abrasive wear are the most common types of wear (see Figure 3.7). Abrasive wear is caused by hard particles or hard projections pressing against and along a hard surface. Local bonding between contacting hard surfaces results in material loss from one surface or transfer between two, causing adhesive wear [14]. Adhesive wear is one example of how a softer material can destroy a harder one. Figure 3.7 shows both the thickness and shape of the TiN coating.

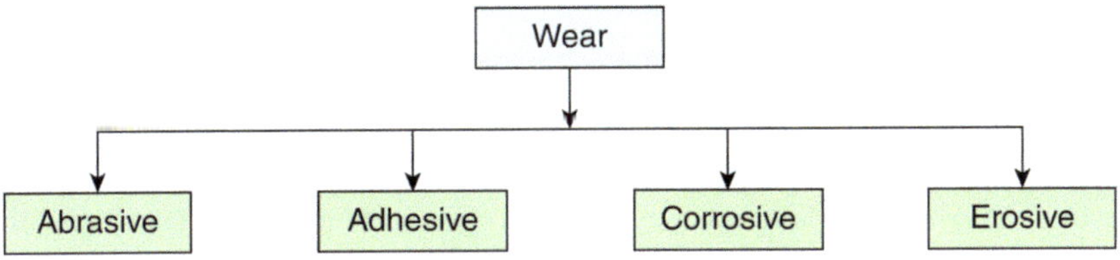

FIGURE 3.7 Different kinds of wear [14] [open access].

The film shows a uniform surface morphology with obvious circular features at the end of the columns (see Figure 3.8a). The film has a columnar morphology where the columns grow as the substrate rises to the top of the film, consistent with the cross-sectional morphology (see Figure 3.8b, c). The layer thickness is 3 μm. Notably, the deposition procedure involved two bias levels, namely −105 V and −90 V. As seen in Figure 3.8c, this resulted in the creation of two distinct zones on the film. The presence of Ti and N is confirmed by EDS analysis performed on the film, as indicated in Figure 3.8d [21].

The structure of substrate material, grouping of surfaces and subsurface materials are the main focus of surface engineering. Under operating conditions, these qualities affect system performance. When creating a surface engineering system, the principles of modeling, property and operation performance modeling, R&D and simulation are used. The basis of surface engineering design is developed using all these interrelated design principles [3, 11]. The modeling method involves the use of virtual models.

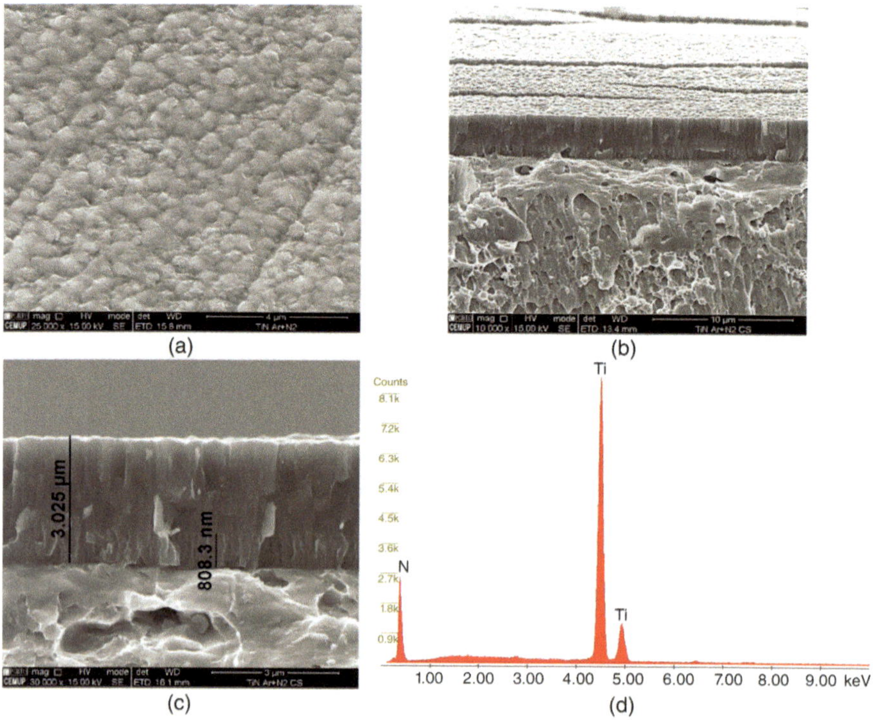

FIGURE 3.8 Morphological characteristics of TiN coating: coating surface (a), coating surface and thin film cross-section (b), thin film cross-section (c), and EDS spectra of TiN (d) [21] [open access].

3.8 APPLICATION OF SURFACES

This chapter examines the use of thin films on engineered surfaces to protect machine and device parts from mechanical and environmental damage, extend their service life, and/or improve their aesthetic appeal. The various applications in the field of surface engineering can be easily divided into three main groups. Applications requiring protection from environmental factors are classified as (1) surface passivation; (2) They provide protection against mechanical damage. And (3) decorative. Let's give a quick overview of surfaces before diving into the details of these applications. The destructive effect on a material caused by interaction with its environment is known as corrosion [6]. Metals degrade through chemical and electrochemical processes. The elements that determine a metal's susceptibility to corrosion are its purity, temperature, aqueous pH, environmental conditions, presence of a passive layer and potential electrochemical processes [5]. Eh-pH diagrams facilitate the display of the theoretical susceptibility of metals to corrosion and equilibrium state of a system, both of which can be predicted using thermodynamic data [3]. Redox and pH diagrams can be used to predict when corrosion will occur and how passive and resistant metal is to corrosion. The rate of change is determined by a kinetic property such as exchange current at equilibrium. Many factors including the composition of the metallic material, environment, mechanical stress, geometry, temperature and time can cause different forms of corrosion, and there is generally more than one type of corrosion [6]. When every metal surface is subject to corrosion, it is called general (universal) corrosion. Galvanic corrosion usually occurs in aqueous environments when two different metals with different electrical properties come together. The material with the higher electrical potential becomes the cathode, while the material with lower electrical potential melts and becomes anode. The term localized corrosion refers to many important corrosion conditions. Pitting, crevice corrosion, stress corrosion cracking, scaling, filiform corrosion and intergranular corrosion are some examples of such conditions. Regional corrosion is not the same as general corrosion, in which the entire metal surface is active; instead, it is usually limited to a few areas. In addition, surfaces usually contain various adsorbed contaminants and are not clean. It is shown that pitting corrosion started and developed in the absence of p-Cl-BHA. The Cu[0] crystals synthesized in slow crystallization experiments at low concentrations of ascorbic acid exhibited a variety of morphologies under the scanning electron microscope, including tetrahedral or truncated tetrahedral, trapezoidal polyhedral, elongated needle-like, semi-prismatic, and isotropic (such as cubes, truncated cubes, and octahedral cubes) (Figure 3.9).

Prevention of copper corrosion in 0.5 M NaCl and 0.1 M NaSO is achieved by dibenzyl sulfoxide (DBSO) and parachloro benzo hydroxamic acid (p-Cl-BHA), respectively. The theoretical distribution of copper ion species in solution, as inferred from PHREEQC calculations and water/mineral equilibrium calculations, is shown in Figure 3.7, and will be briefly addressed in this section. The dissolution trend of copper differs significantly from that of other metals for pH

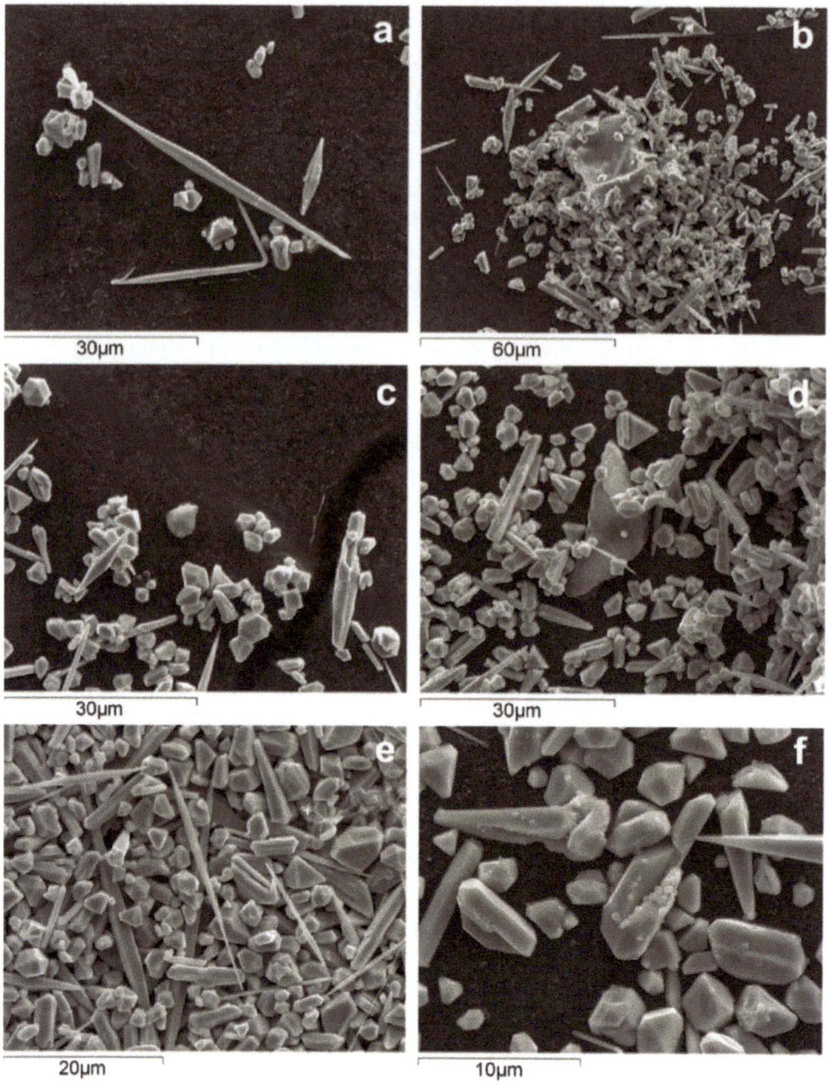

FIGURE 3.9 Scanning electron microscopy (SEM) images of the Cu crystals obtained in batch reduction experiments with Cu(II)-rich acidic solutions (pH 3.95, 25°C) and ascorbic acid 10 mM. Panels (a–f) display different textural and crystallographic features in a single sample (same sample as in Figure 3.9 b, c) [22] [open access].

dependence under experimental settings. The saturation state of metallic copper is pH-independent, according to our calculations for T = 25°C and pe = 1, but other metals show a strong pH dependence (Figure 3.10a). As a result, throughout the pH range studied (0–7), copper-rich mine waters will be saturated for metallic copper, exhibiting a constant saturation index (SI) of about 4.0, indicating a large

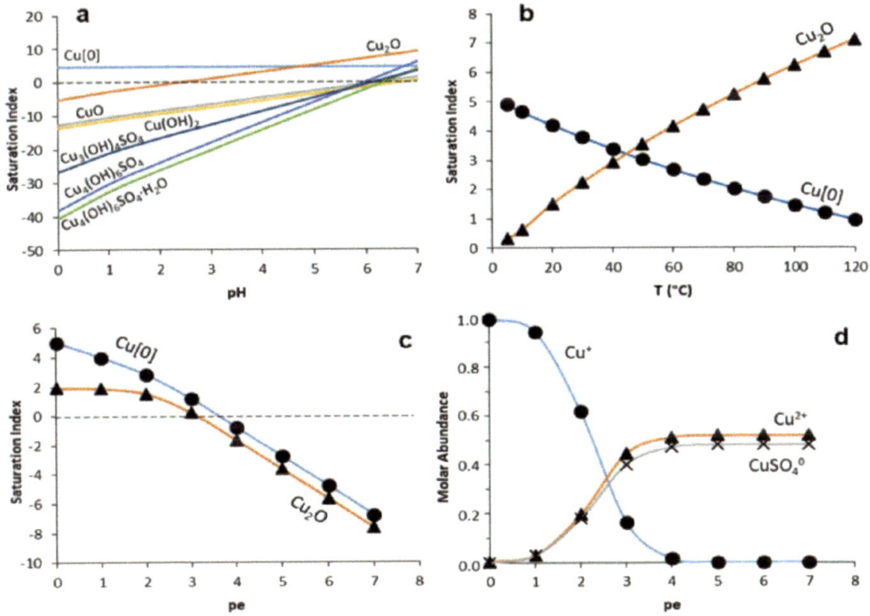

FIGURE 3.10 Water/mineral equilibrium trends in the system Cu-O-S at 1 bar: (a) Saturation index of different Cu-containing phases with respect to the Cu(II)-rich acidic solution as a function of pH (pe = 1); (b) evolution of the saturation index of metallic copper (Cu[0]) and cuprite (Cu_2O) as a function of temperature (T) at conditions of pH = 4 and pe = 1; (c) evolution of the saturation index of Cu[0] and Cu_2O as a function of electronic potential (pe = Eh/59.2) at conditions of T = 25°C and pH = 4; (d) molar proportion of major ionic copper species in solution (Cu^+, Cu^{2+}, $CuSO_4^0$) as a function of pe. The computations considered in all cases Cu = 3×10^{-3} M, S = 0.2 M and O_2 = 2×10^{-4} M [23] [open access].

thermodynamic suitability for metallic copper crystallization (Figure 3.7a). On the other hand, copper hydroxide ($Cu(OH)_2$), copper oxide (tenorite, CuO), and other copper sulfates ($Cu_3(OH)_4SO_4$, $Cu_4(OH)_6SO_4$, and $Cu_4(OH)_6SO_4 \cdot H_2O$) are all highly undersaturated at low pH and only reach saturation at nearly neutral pH. Among the later minerals, cuprite (Cu_2O) is an anomaly in that it reaches saturation at pH values above 3.0 and shows even greater saturation than metallic copper at pH > 4.5 (Figure 3.10a). Therefore, during reduction studies at pH values ranging from 3 to 5, the dissolution behavior of cuprite may be of interest to understand the partitioning between dissolved and solid copper in acidic solutions.

Therefore, the authors have suggested that by converting copper into a less soluble and more stable sulfide molecule, dibenzyl sulfoxide prevented copper corrosion in sulfate solutions. Typically, the coatings formed on the surface of tool are several microns thick. At the cutting edge, they improve wear resistance while reducing friction and dispersion. Land management plays an important role in the transport sector. Coating technologies account for about 6% of the production costs of engines and gearboxes [2, 3, 5, 6]. In the transportation sector, surface coatings are often used in three main areas: powertrains, vehicle components and permanent structures. For example, the surface of an ordinary metal consists of thin, often translucent (-0.01-0.1) layer equivalent oxide with pores and cracks. Water molecules, oxygen molecules and some, both organic and inorganic polar compounds, have weak bonds with oxide layer. Various interactions can occur at the boundary between two objects. These characteristics of surface interaction are neither intrinsic nor fundamental.

Similar to thermal conductivity or Young's modulus, but based on several external factors including cleanliness of environment, surfaces and whether surfaces move relative to each other and surface characteristics, of course. Attention should be paid to the surface characteristics, which are critical in interfacial interactions. The following factors determine interfacial interactions: (1) chemical activity of surface or the tendency of surface to form a surface film, the composition of which differs from primary one; (2) surface adsorption capacity; (3) surface energy, which quantifies the effort required to create a new surface; and (4) interfacial energy, which evaluates the compatibility of two interacting surfaces. As the surfaces move relative to each other, the mechanical characteristics become more apparent. Numerous temperature-dependent properties also become important when surfaces move at very high speeds relative to each other. Devices such as plumbing, mechanical parts, cutting tools and other objects can be subjected to surface damage from a variety of surface contacts, either individually or in combination. A better way to address the damage problem would be to use a single material to create devices that must also be resistant to the surrounding chemical environment to ensure the required mechanical, electrical and thermal properties. Unfortunately, due to incompatible needs in some situations and cost or weight considerations in others, such a solution is usually not possible.

Parts subject to air corrosion are covered with thermally sprayed polymer coatings. In addition, MoS_2 can be applied by magnetron PVD to spacecraft parts such as gears and ball bearings. As a result, the transmission system generates less heat and the gears are protected from rising temperatures. The performance of a component can be improved by surface modification. For example, surface modification is necessary to improve the wear resistance of the slat track. Conventional hard chrome plating can cause problems during machining. For this purpose, a surface modification method using plasma nitriding has recently been developed and applied. This does not require additional post-machining procedures and, unlike chrome plating, does not reduce fatigue life.

3.8.1 Applications for the Sports Industry

Titanium oxide surface treatment technology has proven to be a successful advancement in improving the performance of engine parts and therefore improving the performance of racing cars. Any engine will perform better if its inertial mass is reduced because it will accelerate faster, use less fuel, and have better acceleration. Titanium alloys have a wide range of qualities, including high ones. In chemically reactive situations, motors with low friction coefficients and high-temperature wear contacts are preferred. An elegant solution to the problem of preventing surface damage is provided by the use of thin surface films and surface treatments that increase the reliability and serviceability of devices while reducing material losses. Tensile stresses (red contours) are generated in the valleys of surface texture for the intact coating, and some of these stresses propagate to the interfaces. Compressive stresses are found at the top of surface. The diamond-like carbon (DLC) coating surface, interface between the DLC and binder layer, and the contact between binder layer and substrate are most likely sites for coating failure at defect nucleation stage. It is shown that the scenario illustrated in Figure 3.11, where the friction coefficient is m = 0.2, results in the appearance of surface roughness irregularities ("peaks"), as well as interfaces between the DLC film and binder layer/substrate, which show signs of damage accumulation and crack development [24].

The advantage of thin film protection and decoration is that engineered components can be made from inexpensive and readily available materials. For example, a small layer of corrosion-resistant material applied to the inside surface of a regular water pipe will prevent corrosion as well as make the entire pipe from special, expensive corrosion-resistant alloys. Numerous technical systems, including automobiles, aircraft engines, turbine blades, ball bearings, machine tools,

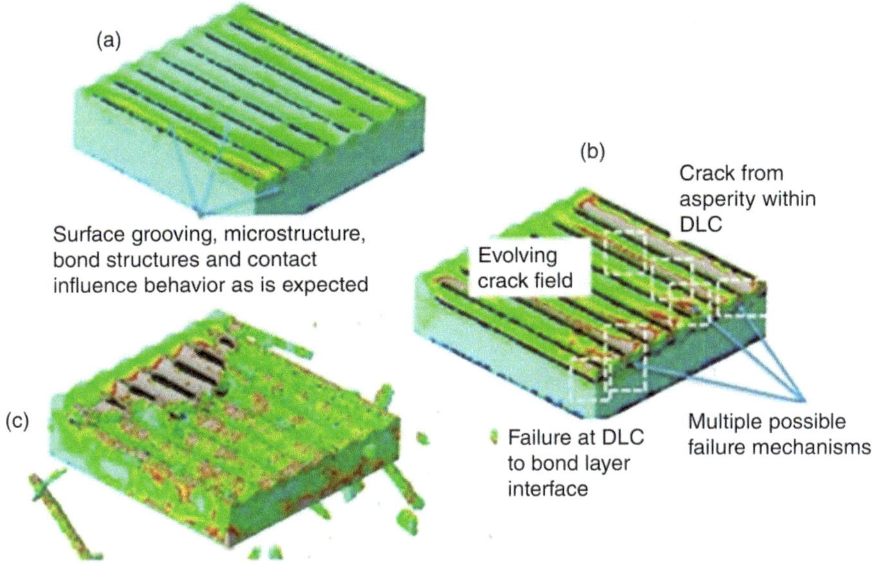

FIGURE 3.11 Evolution of first primary stress state at different stages of damage accumulation of coating system: (a) sliding on a smooth DLC surface before damage initiation, (b) damage initiation with several possible failure mechanisms, and (c) final DLC coating that cracked and failed completely [20] [open access].

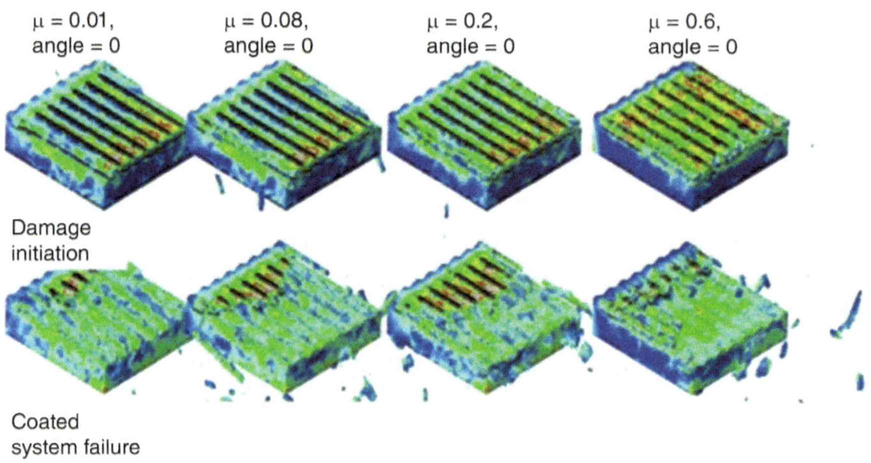

FIGURE 3.12 DLC failure behavior with slip surface topography on flat DLC. Large-scale crack formation (top) and coating system failure (bottom) [20].

ultra-high vacuum and high-temperature nuclear reactor components, plumbing, bridges, furnaces, ship components, etc. [11]. This behavior appears to be typical in situations where the friction coefficient is significant, highlighting both damages caused by the surface topography and various contacts. The final failure in this scenario occurs at substrate-binder layer interface and through the coating, as partially visible in Figure 3.10c. Figure 3.12 shows the behavior of different micromechanical coating systems modeled by the finite element method as a function of the friction levels indicated by the MD results. The first set of numbers represented the circumstances under which damage initiation occurred, and the second set represented the circumstances under which the coating failure occurred. It can be seen that the failure process of a particular system is friction-dependent, meaning that catastrophic failure first occurred in the DLC, then in the bond layer, and finally in the substrate bond layer [20].

MD modeling can be used to relate the microstructure-scale mechanisms of DLC coatings to friction, and cause and effect inferences can be made from this relationship. Due to the altered surface chemistry of DLC coatings, friction is lower in lubricated contacts, and micromechanical stress and strain are reduced accordingly, which improves the coating's resistance to fracture and increases its load-bearing capacity. Friction control of golf clubs is a sporting industry [13]. The spin of a golf ball is affected by the friction between the club face and the ball during impact. The classification of materials applied in the tribological system in Figure 3.13. The distance, spin, and flow direction of a golf ball are determined by controlling the friction between the club and the ball. Figure 3.14 illustrates how surface technologies can be used to combat this. Introduced a new variational fatigue phase field model in the field of macroscale brittle fracture research using the phenomenological method to study a one-dimensional example. The model

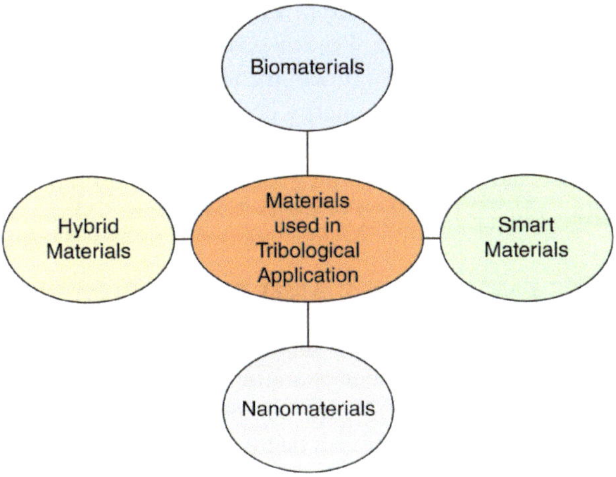

FIGURE 3.13 Classification of materials applied in the tribological system [open access].

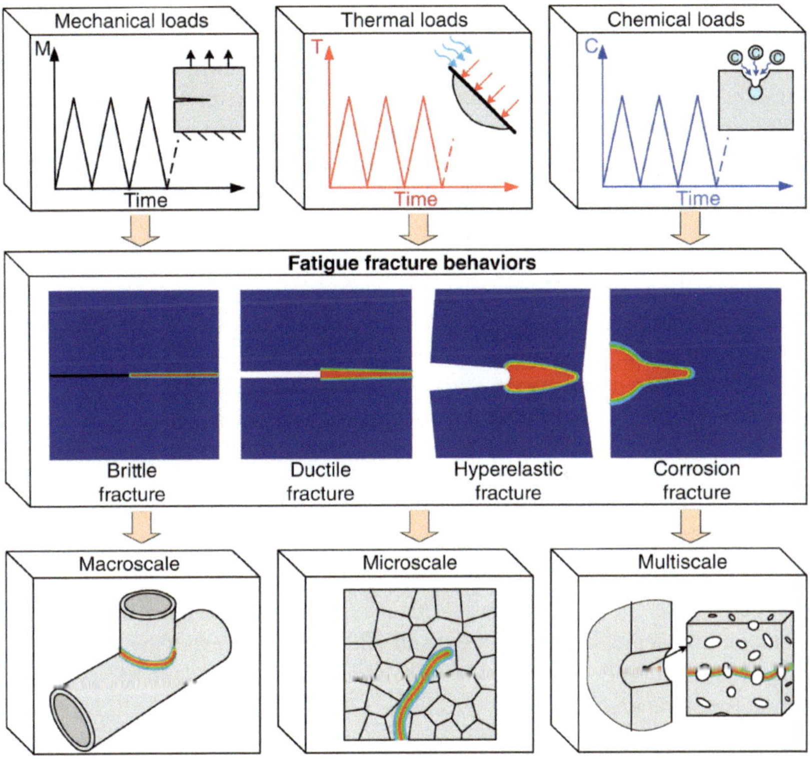

FIGURE 3.14 Brittle fracture on a large scale. 2D model of the plate with holes and 3D model of the steering arm [13] [open access].

can extract the established patterns in describing the mean stress effect and describe the typical Wöhler curve. The fracture toughness was effectively reduced by adding the fatigue degradation function and unique fatigue history variable to the conventional regularized free energy equation [13]. The Paris law and crack propagation velocity curve are naturally recovered, as shown by the numerical results, which also show that the framework can handle complex geometries and loading conditions in 2D (Figure 3.14a) and 3D (Figure 3.14b).

3.9 SURFACE ENGINEERING AND TRIBOLOGY

This is a natural phenomenon when moving parts and components of tools or machines gradually wear out under operating conditions. This is a natural phenomenon where moving parts and other components of tools or machines gradually wear out during use, resulting in a decreasing in functionality and efficiency of item. The need to replace worn machine parts or tools with new ones, which usually leads to the need to temporarily stop the machine or entire technological process. Meanwhile, global industry is constantly working to improve the efficiency and effectiveness of technical processes, as well as machines and vehicles. As a result, it becomes more difficult to study with tools and machine parts. As a result, the industry is constantly seeking advanced design and tooling materials and manufacturing technologies that, under certain operating conditions, guarantee maximum durability at a reasonable unit cost. As a result, the industry is constantly searching for innovative and advanced structural and tooling materials and manufacturing processes that will guarantee maximum durability at a reasonable level under specific operating conditions and unit costs. The development of composite materials and their integration into industry are two of the most important and exciting developments in the field of materials science and industrial processes. Many research centers around the world are conducting research and development of processes for the production of metal matrix composites (MCCs) [2, 12].

These types of composite materials combine the advantages of a reinforcing material with the high ductility of a metal matrix. As a result, these composite materials have low specific gravity, high tensile strength, and outstanding fatigue strength. Dispersed particles or high-hardness phases are usually responsible for the increased wear resistance of composite coatings. The invention and subsequent widespread use of nanoparticles and amorphous materials is another major advance in materials science and manufacturing technology in recent years [11, 12]. However, due to the great potential of this class of materials with grain sizes on the nanometer scale, typically in the sub-100 nm range, extensive research is still being conducted and scientists from around the world are concentrating on production methods and uses.

For the known and traditional structural materials, even by fine-tuning the grain and forming a nanostructure, the mechanical, chemical and physical properties can be significantly improved. For example, increasing the ductility of brittle materials, increasing the wear resistance and hardness of contact surfaces, etc.

For example, when grinding copper to a grain size of about 8 nm, its tensile strength can increase five times to 2500 MPa. However, when copper is ground to a grain size of about 50 microns, the strength does not exceed 500 MPa. The most advanced coating and surface treatment technologies can be used to create micro- or nanoscale coatings from advanced composite and metal materials, as well as surface layers with unique characteristics. These types of composite materials combine the advantages of a reinforcing material with the high flexibility of a metal matrix. As a result, these composite materials simultaneously have low specific gravity, high tensile strength and outstanding fatigue resistance. The corrosion resistance of composite coatings is usually increased by dispersed particles or phases of high hardness. The advent and current widespread use of nanoparticles and amorphous materials is another notable advance in materials science and manufacturing processes in recent years [2, 4, 11, 12, 17].

However, due to the great potential of this class of materials, which are characterized by grain sizes on the nanometer scale, typically in the sub-100 nm range, extensive research is still ongoing and scientists from around the world are focusing on the production of nanomaterials. Both metallic and nanostructure composite materials have beneficial properties (e.g., mechanical strength) that are often far superior to those of traditional materials. Through cleaning physical, chemical, and mechanical properties can be significantly improved, such as increased hardness and strength, increased wear resistance of contact surfaces, increased flexibility of brittle materials, etc., even in the case of traditional and already-known structural materials. For example, the tensile strength of copper when refined to a grain size of about 8 nm can increase fivefold to 2500 MPa, although grains down to 50 microns in size do not reach 500 MPa.

Another challenge is to structure surface layers to improve their tribological characteristics and wear resistance using various surface treatment methods such as heat treatment and thermochemical treatment. Many methods of using surface heat treatment to improve the tribological properties of metals and alloys have been studied and documented in detail in scientific papers [4, 11, 12, 17]. The most advanced and interesting are ion implantation, laser and plasma surface treatment. The basic idea of surface thermal treatment methods is to use extremely high rates of heating and cooling to change the microstructure of the surface layer without changing its chemical composition. The previously mentioned methods can also be used depending on the parts and processing characteristics of thermochemical processing. By enriching and thermochemically treating the surface layer with various materials, it is possible to significantly reduce the coefficient of friction, increase wear resistance and corrosion resistance. The most common and widely used thermochemical surface treatments of metals and alloys, especially titanium and its alloys are oxidation, carburization and nitriding.

All factors are taken into account, the type and intensity of the load, as well as environmental elements such as temperature, strong liquid or gaseous substances, etc. Therefore, an individual approach to each working surface, selection of suitable material and design are required. An additional and invaluable source of information on the phenomena and mechanisms of surface wear, as well as strategies for improving the tribological performance of working surfaces, is found in tribology and surface technology [12]. Gear materials, surface topography, lubricant characteristics and operating conditions are some of the important aspects they discovered and highlighted as influencing micro-pitting behavior. In addition, they discussed wear processes and offered suggestions to improve the resistance of gears to micropitting.

The knowledge offered in this review can be very useful to those who design heavy-duty and high-speed equipment in the modern world. Conversely, in a very comparable problem, a study is conducted and described gear wear. They have noted that the durability and reliability of transmission are greatly affected by type of used oil [11]. As a result, the scientists studied how test samples with a new type of W-DLC/CrN coating reacted to three currently available industrial gear oils. The study used one mineral oil and two synthetic oils based on polyalkylene glycol (PAG) and polyalphaolefin (PAO). They have demonstrated that a PAO type synthetic oil provided the best operating conditions and maximum durability as demonstrated by abrasion, scratch and pitting tests.

For example, the tensile strength of copper with granules up to 50 microns in size does not exceed 500 MPa; but once the grains are refined to 8 nm, the strength can increase fivefold to 2500 MPa. By using the most advanced coating and surface engineering technologies, it can be possible to create coatings from modern metallic and composite materials on a micro- or nanometer scale, as well as surface layers with unique characteristics. Currently, the following coating technologies are most widely used in industry: Arc and plasma surfacing technologies guarantee high melting efficiency of the consumable filler material. However, the disadvantage is that the relatively large amount of heat introduced into the support material makes it difficult to obtain fine-grained structures. For this reason, these technologies are usually used for cladding large and fairly large parts. Laser and electron beam cladding provides a tight heat-affected zone, minimal dilution and very low heat input. High-velocity oxygen fuel (HVOF) spraying, cold spray (CS) at supersonic gas jet speeds, and flame, arc and plasma spraying are examples of thermal spraying techniques that can create coatings with structures and chemistry that are not possible with coating technology. The disadvantage is the heterogeneity of coatings and often high porosity, which reduces their usefulness and narrows the range of practical applications. Coating technologies such as PVD and CVD enable the production of metallic and composite nanostructured coatings. However, in this case, the coatings are thin, the manufacturing method is labor-intensive, and the cost is prohibitive.

To obtain self-lubricating coatings, a vanadium-containing iron-based coating is proposed to be applied by electric arc spraying. Dry sliding tests are used to study wear characteristics, and electron microscopy and energy-dispersive X-ray spectroscopy (EDX) are used to analyze worn surfaces. They have found that because certain vanadium oxides are predominant and promote self-lubrication of coating, coatings containing vanadium exhibit a noticeable reduction in the coefficient of friction above 450°C. Thus, direct current reactive magnetron sputtering is used [11, 12], using reactive high-power pulsed magnetron sputtering, a top SiNx layer is deposited onto CoCrMo disk samples after the deposition of CrN layer.

The coatings under study are examined in terms of phase composition, wear rate, hardness distribution and surface roughness.

The influence of PTFE emulsion and TiO_2 particles on the characteristics of Cu–Sn composite coatings is studied in [11, 17] strength, hardness and tribological characteristics of the coatings under study. They have demonstrated that the reduced coefficient of friction of the sprayed coating (0.1) and increased resistance to wear and corrosion resulted from the inclusion of both PTFE and TiO_2. The surface of Ti_6Al_4V alloy is laser microtextured [4] to improve its wettability, wear resistance and sliding friction properties. Using water as a coupling fluid, they have measured contact angles to study the effect of processing parameters on the tribological characteristics of the titanium alloy, such as laser beam scanning speed and pulse energy density. Microscopic observations have been used to study and determine wear mechanisms. They found a significant correlation between laser pattern characteristics and wear behavior. Microtexturing of the surface compared to untreated Ti_6Al_4V alloy surfaces resulted in wear reduction of up to 70%.

Another example is in reference [11] that demonstration the using of laser beams in surface engineering to create surface layers and restore shell characteristics. The idea of coating old turbine blades is considered by the authors when studying laser cladding of the nickel-based superalloy K417G. In addition, crack sensitivity is controlled by laser remelting of the surface. In addition to the type and intensity of load, the type and degree of wear is also influenced by environmental variables such as temperature, aggressive liquid or gaseous agents, etc. Therefore, an individual approach to each working surface is necessary to ensure the durability and reliability of machine or tool components. This includes selecting the right material, designing its microstructure and using the best manufacturing or processing technology.

3.10 SURFACE COATING PROCESSES AND METHODS

There are many substrate coating methods and technologies available for purchase.

3.10.1 Coatings for Thin Films

Thin film technologies are used in many industries, including sports equipment, microelectronics, optics, magnetism, biomedicine, micromechanics, etc. We can use thin layers of paint every day. We could encounter them every day when we pick up our phones, open doors, or ish our hands. Without them, our touch screens will not work, and they protect products from premature wear due to corrosion. Since the 1960s, thin film technology has been used industrially to make reflective and semiconductor materials. Over the years, the deposition mechanisms and properties of thin films have become increasingly clear to us. As thin film coating technologies advance, their applications in aerospace, energy, semiconductor, optoelectronics and mobile medical devices are becoming more common. These innovations will be particularly important for the development of advanced application areas, including digitalization, smart technologies and data technologies.

While CVD is used for coatings thicker than 1 mm, PVD technologies are limited to the creation of thin films with thicknesses ranging from 0.1 μm to 0.1 mm [13, 20, 25–28]. Thin layers of TiN and TiC are often applied by CVD to components prone to corrosion and abrasion, as well as metalworking and forming tools. Films resembling diamonds are deposited using the CVD method. The fundamental idea of CVD process is that the feed gases in the chamber undergo chemical reactions with each other. As a result of this reaction, a solid-phase substance is formed, which condenses on the surface of the substrate. Any reaction associated with cardiovascular disease has three phases:

- Synthesis of a volatile carrier material such as nickel carbonyl;
- Sending the gas to the deposition site without destroying it;
- The chemical reaction required to create a coating on the substrate.

After 180 minutes of sputtering at 400°C, 100 nm, W-coated diamond particles are obtained. The sputtering current and voltage are adjusted to 0.9 A and 600 V, respectively. The sample holder is oscillated 35 times per minute to ensure the uniformity of the coating. Figure 3.15 shows the shape of diamond particles for both before and after sputtering with the W coating. The W coating produced using the magnetron sputtering method is complete and uniform compared to other coating methods. The thickness of the 100 nm W coating produced by the magnetron sputtering method is shown in Figure 3.15d. It should be noted that the coating is intentionally degraded to determine its thickness. The AFM results from Bruker Corporation; Karlsruhe, Germany) show that the thickness of the W coating (Figure 3.15e) is approximately 100 nm [28].

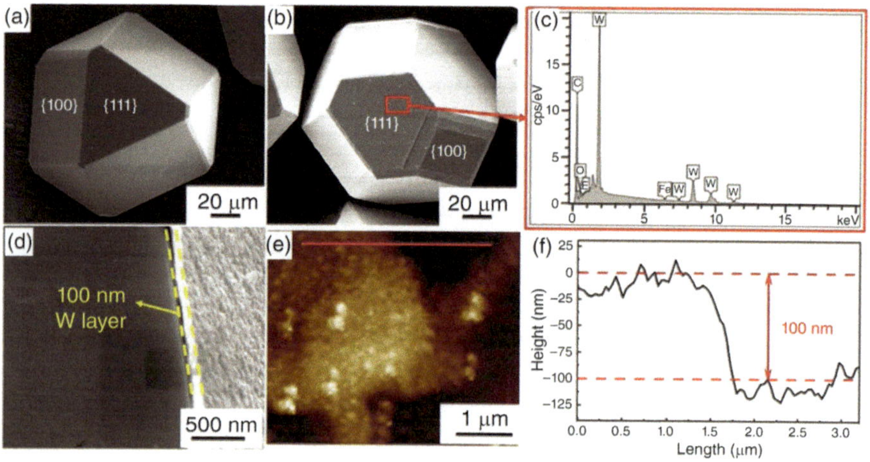

FIGURE 3.15 Surface morphology of (a) uncoated diamond; (b) W-coated diamond; (c) EDS analysis results of red areas in (b); (d) surface morphology of W-coated diamond particles after coating degradation; (e, f) diamond morphology characterized by AFM [28] [open access].

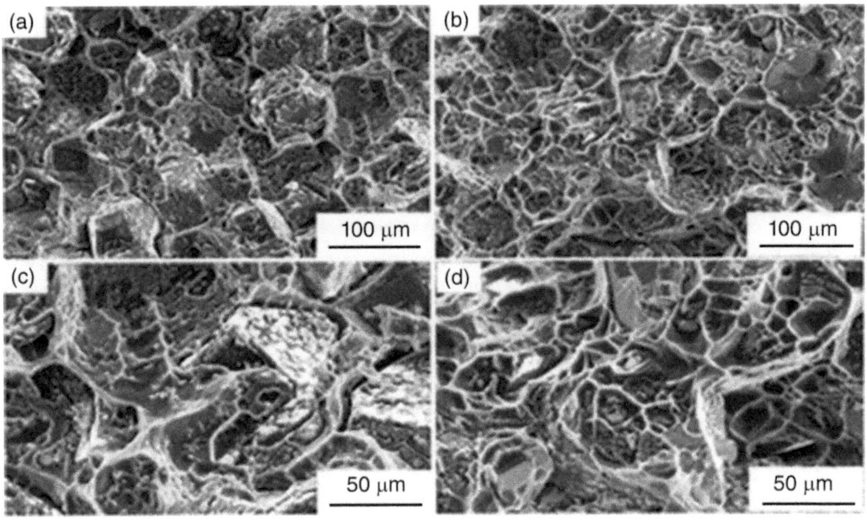

FIGURE 3.16 Fracture morphology of diamond -aluminum composites. (a) Uncoated, {100} interface with Al_4C_3 formation. (b) W-coated, $Al_{12}W$ formation limits Al -diamond reaction. (c) Uncoated, {111} interface with mechanical bonding. (d) W-coated, reduced interfacial reaction due to W layer [28] [open access].

In addition, the temperature is not high enough for the interaction of tungsten and diamond. In the composite preparation process, tungsten and aluminum react preferentially to form $Al_{12}W$, which shortens the contact period of diamond and aluminum. However, a small amount of Al_4C_3 is inevitably formed by aluminum and diamond due to the consumption of W coating. The fracture morphology of diamond/aluminum composites is shown in Figure 3.16. Figure 3.16a shows that there is clear interface selectivity in the bonding between different crystalline surfaces of diamond and aluminum matrix in uncoated diamond/aluminum composites. While, diamond {100} can react with aluminum to form blocky Al_4C_3, there is no chemical interaction between diamond {111} and aluminum due to short heat retention time, indicating mechanical bonding. The in situ-produced Al_4C_3 can achieve strong interface bonding by pinning the interface [28].

The surface morphology of composites after corrosion is shown in Figure 3.17. The interface between diamond and aluminum, or where the pitting mainly starts is the place where the etch pits first propagate for uncoated diamond/aluminum composite. $Al(OH)_3$ will be formed when water reacts with Al_4C_3, which has hygroscopic properties [28].

There is no Al_4C_3 interface product in the etch pit. The pit nucleation starts at the site of Al_4C_3 hydrolysis and pit becomes larger and larger as corrosion progresses. There are some pits at the edges of the diamond in W-coated diamond/aluminum composite, which are caused by traces of Al_4C_3. However, the corrosion of interface is less severe than uncoated diamond/aluminum composite, moreover, part of corrosion of aluminum matrix has exposed the hidden diamond particles. Figure 3.18 shows the surface morphology of

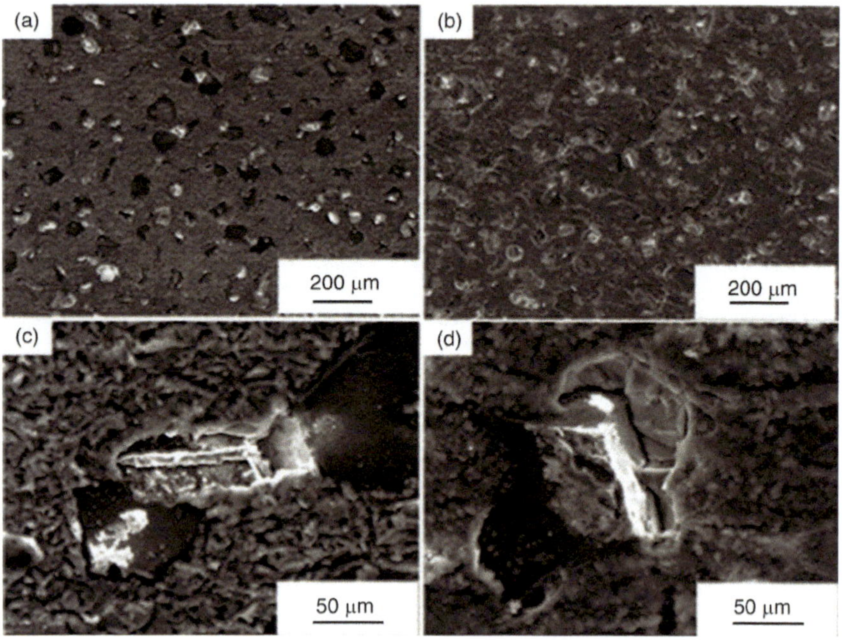

FIGURE 3.17 Surface morphology of corroded diamond-aluminum composite: (a, b) diamond-aluminum composite without coating; (c, d) with W-coating [28] [open access].

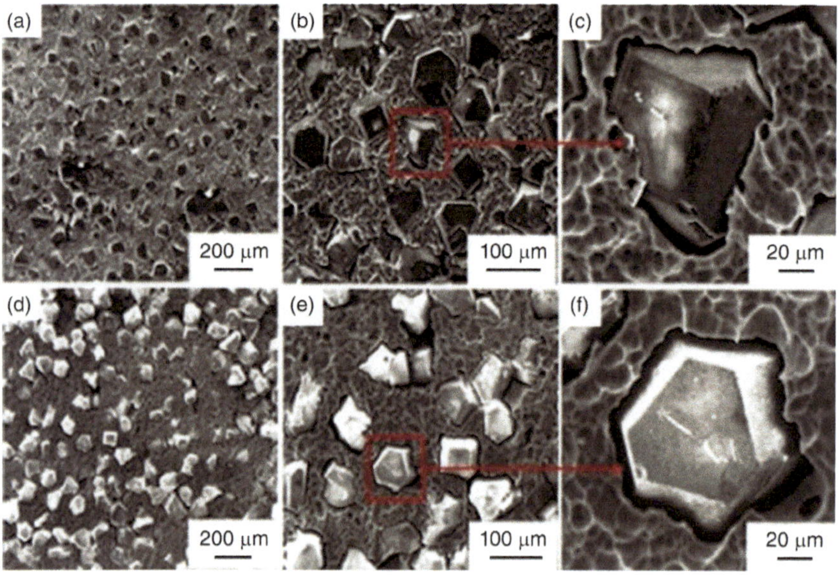

FIGURE 3.18 Surface morphology of diamond-aluminum composite after electrically accelerated corrosion. The diamond-aluminum composites in figures (a–c) are uncoated, while those in figures (d–f) are tungsten coated [28] [open access].

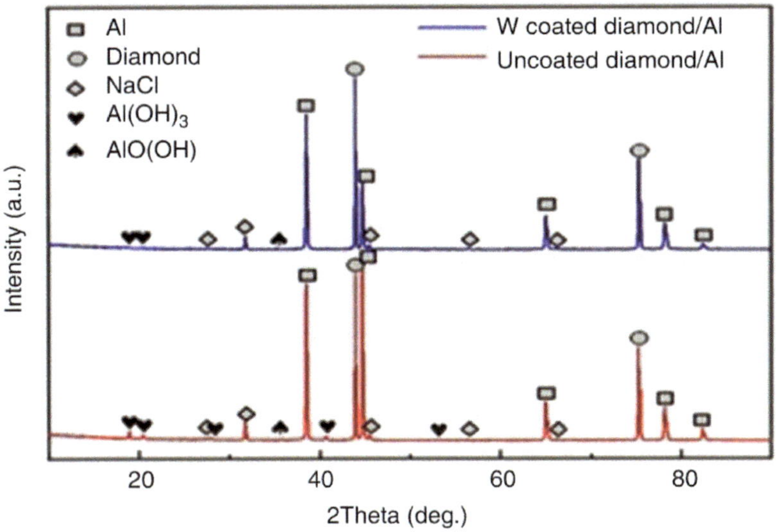

FIGURE 3.19 XRD pattern of a diamond-aluminum composite after corrosion [28] [open access].

diamond-aluminum composite during electrically accelerated corrosion. Before corrosion, most of the diamond particles in bare diamond-aluminum composite are embedded in the aluminum matrix. After corrosion, a significant number of exposed diamond particles are exposed as aluminum on the surface dissolves. Interface exposure is the primary form of corrosion, which mainly affects the aluminum matrix and the interface between diamond and aluminum [28].

The hydrolysis of Al_4C_3 is accelerated by electrification, and a noticeable gap is formed between aluminum and diamond. The XRD pattern of the corroded diamond-aluminum composite is shown in Figure 3.19. We have found that the composite contained NaCl, $Al(OH)_3$, and AlOOH in addition to the diffraction peaks of diamond and Al. Since the product on the surface of the sample waiss not cleaned during the XRD test, the corrosion environment of NaCl is found. The corrosion product of $Al(OH)_3$ in the bare diamond-aluminum composite is larger than that in the W-coated diamond-aluminum composite based on the intensity and number of the diffraction peaks [28]. The corrosion process of the aluminum-diamond composite is shown in Figure 3.20.

The composite was initially immersed in a NaCl solution. The potential difference between the aluminum and diamond electrodes leads to the formation of easily hydrolyzed Al_4C_3. At the interface of diamond-aluminum composites, where Al_4C_3 combines with water to form $Al(OH)_3$, corrosion predominantly occurs. The reduction in the number of diffraction peaks and the decrease in the strength of the Al_4C_3 diffraction peak in XRD indicate a lower corrosion product of $Al(OH)_3$ in the W-coated diamond-aluminum composite compared to

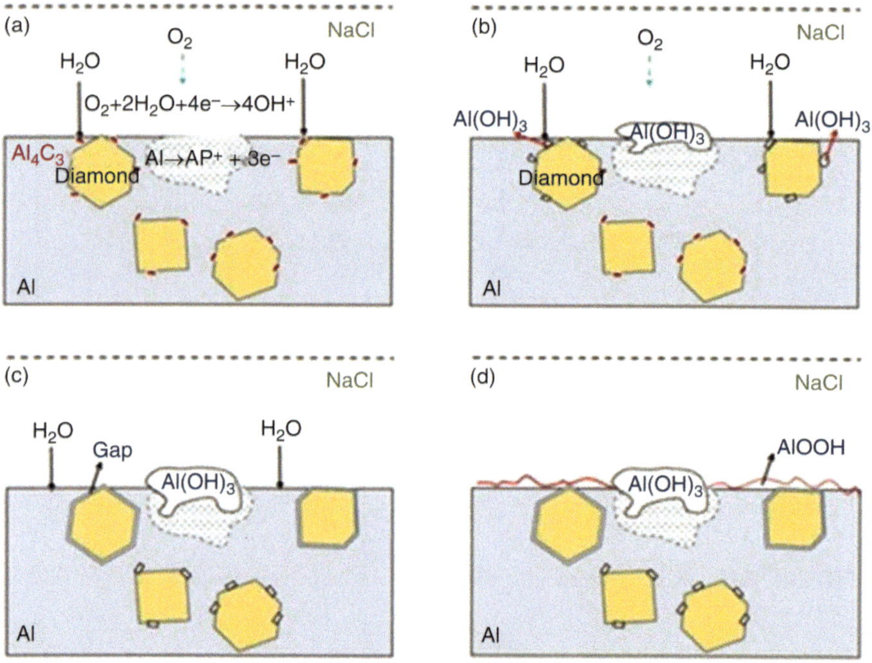

FIGURE 3.20 Schematic diagram showing the corrosion process of diamond/aluminum composite: (a) hydrolysis of Al_4 C_3 and corrosion of aluminum; (b) formation and accumulation of corrosion product $Al(OH)_3$; (c) corrosion leads to gaps between aluminum and diamond; (d) oxidation reaction of aluminum matrix produces AlOOH, preventing further corrosion [35] [open access].

the uncoated counterpart. Therefore, the fracture shape in Figure 3.21 further characterizes the composite after corrosion. The application of a nanoscale W layer on the surface of diamond improved the corrosion resistance of the diamond-aluminum composite. Compressed gas infiltration was used to prepare the nanoscale W coating and uncoated diamond-aluminum composites. The introduction of nanoscale W coating effectively improved the bonding between the diamond and aluminum matrix, suppressed the formation of the interface product Al_4C_3, and shortened the contact time between diamond and aluminum matrix. The total immersion experiment showed that the diamond-aluminum composite with nanoscale W coating had a smaller weight loss due to corrosion than the uncoated counterpart. Polarization curve measurement confirmed that the nanoscale W coating successfully reduced the corrosion rate of the composite. The main causes of corrosion in diamond-aluminum composites are Al_4C_3 hydrolysis [28].

Corrosion significantly reduces the mechanical properties of diamond-aluminum composites. After full immersion corrosion treatment, the flexural strength of uncoated diamond-aluminum composite dropped from 222 MPa to 180 MPa,

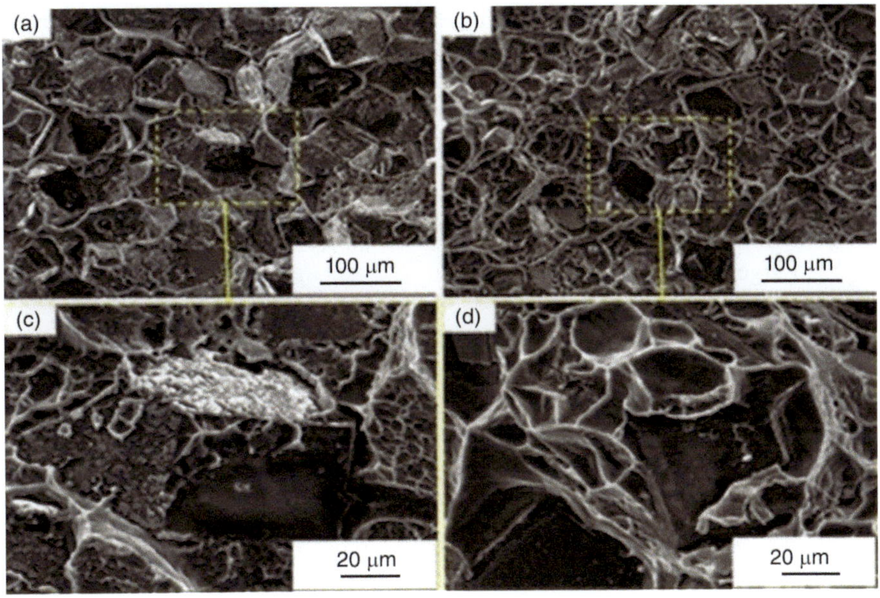

FIGURE 3.21 Morphology of fracture of diamond/aluminum composite after corrosion: (a, b) without coating with nanoscale W-coating (c, d); [28] [open access].

while the flexural strength of the W-coated composite dropped from 319 MPa to 195 MPa. The resulting W-coated diamond-aluminum composites showed higher flexural strength compared to the diamond-aluminum composites (188 MPa), Cr-coated diamond-aluminum composite (137 MPa), W-coated diamond-aluminum composite (304 MPa), and Ti-coated diamond-aluminum composite (313 MPa) described in the literature [4, 11, 12, 28]. The flexural strength of the W-coated diamond-aluminum composite is higher than that of the uncoated composite both before and after corrosion. The flexural strength of W-coated diamond/aluminum composite drops significantly, to 38.9%, after corrosion.

3.10.2 Spraying Procedure

In this process, using high-energy particle bombardment, the coating material is lifted off the solid surface during the spraying process. It is a spraying method in which coatings of nanoparticles are applied to nonwoven materials without immersing the substance in a liquid. The disadvantages of the classical spraying method [29–34] include its inapplicability to multilayer materials (e.g., filters) since it is impossible to guarantee the penetration of nanoparticles into deeper layers of the material, and when using a nanosol, the unevenness of resulting coating, since the sprayed suspension of nanoparticles when sprayed forms droplets of different size, which is sufficient to maintain the plasma discharge. Subsequently, the target is negatively biased (0.5–5 kV for a constant current device) to allow the bombardment of positive ions from the plasma. This voltage accelerates the ions which contact the target and cause the material to be ejected (sputtering). The limitation of glow discharge atomization method is that only the working gas pressure can be changed to independently control the given current density and voltage.

Magnetron sputtering is a process of sputtering by a glow discharge with magnetic acceleration. Three advantages of this method are:

1. Large deposition areas
2. Reduced substrate heating
3. Relatively high deposition rates

This greatly expands the range of applications by revolutionizing the deposition process. The magnetic field increases the efficiency of the available electrons in magnetron sputtering. For example, MoS_2, DLC, TiAlN, TiN, CrN and TiC. One of the surface application methods is magnetron sputtering.

3.10.3 Application by Thermal Spray

One of the most flexible methods for applying protective coatings is thermal spraying. Figure 3.22 shows that HVOF technology makes it easier to create denser bond coatings, especially when using hydrogen. A 200-kW gun can

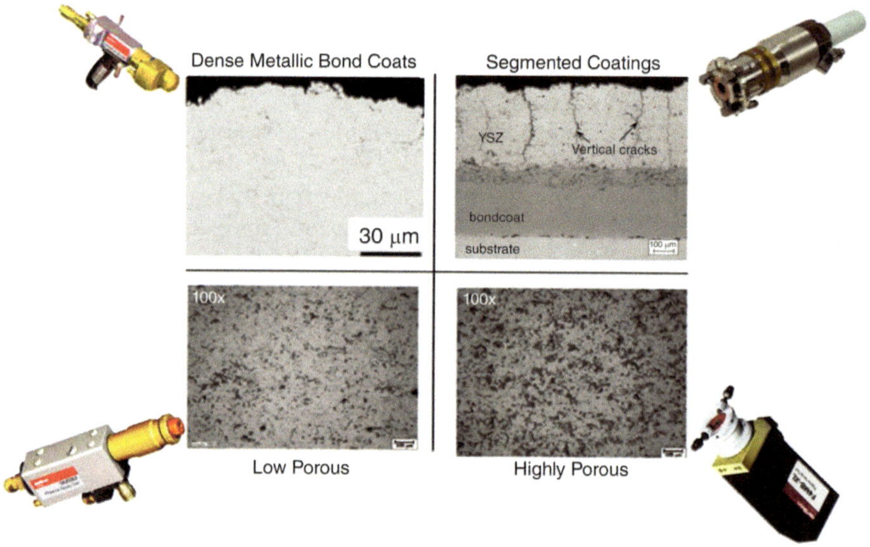

FIGURE 3.22 Gun arsenal and their capabilities for creating microstructures: Segmented coatings (Plazjet 200 kW gun), high-porosity TBC coatings (F4MB gun), low-porosity TBC coatings (9MB gun) and dense metal bond coating (HVOF DJ2600 gun) [39] [open access].

create segmented coatings more easily than a 50–100 kW gun can create porous coatings. Low-power guns can also produce dense segmented coatings, but the process is time-consuming and results in significant powder waste. It is wise to choose a gun that is short, capable of coating external components, and controllable for the user. Coating internal diameter parts using a long ID gun is also suitable. Commercially available Superjet Eutalloy burner from Eutectic Castolin, shown in Figure 3.21 is an example of flame powder atomizing equipment. The Superjet burner operates on the principle of separate supply of acetylene and oxygen through needle valves, which ensures precise flame control. As oxygen passes through the injector, powder is sucked into the system. Acetylene gas is simultaneously fed independently into the mixing unit. The powder is now mixed with both gases and fed through the system to the spray tip where it is ignited. Superjet technology has many applications, such as joining and straightening of metal components, surfacing or surfacing processes, errors in tool finishing, and shape changes and deaths on the job [24, 34–38].

It is useful to know what to expect from a jet section at a given standoff distance before performing a jet test. The particle density at the core versus the periphery, spot size, particle temperature and velocity can be determined by cutting the jet transversely to the powder fall direction. An overview of the jet cross-section is shown in Figure 3.23. Future lens technology should be able to monitor the jet axially and evaluate the composition of the jet cross-section.

A jet plate test involves moving the gun in front of a flat plate and focusing it on a spot for five to ten seconds to examine the spot size and determine if there

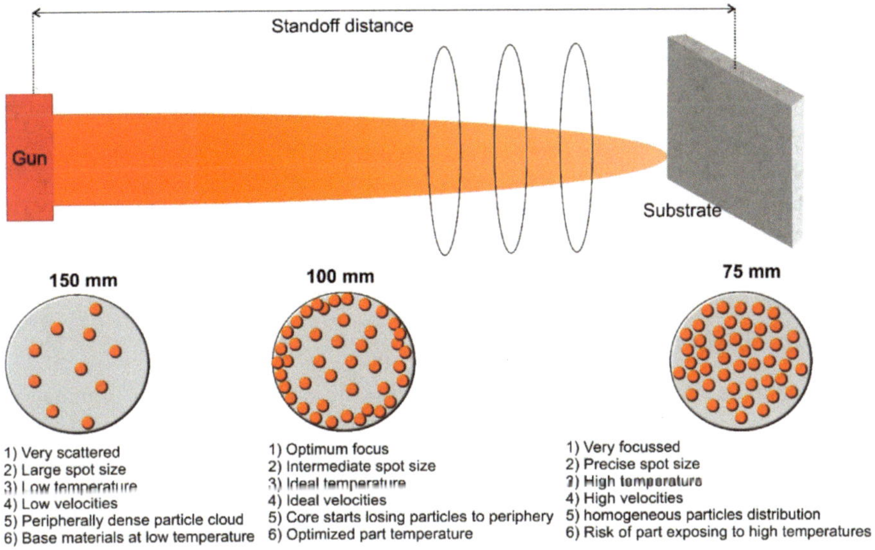

FIGURE 3.23 Illustration of a torch plate and how to ensure efficient deposition when using thermal spray (the term "standoff distance" refers to the front of the nozzle in the gun) [39] [open access].

are any problems with the equipment or powder injector [39]. There are several methods for coating metals and alloys, including diffusion bonding, soldering, welding, laser cladding and strain cladding. The anodizing process depends on the presence of a substrate on which the film grows. Anodized coatings are primarily used as oxide electrical insulators and protective coatings [8, 24, 40].

3.11 SURFACE MODIFICATION

3.11.1 Shot Blasting, Both Macro and Micro

One way to modify the cold surface finish of a part is shot peening. Demand for lightweight energy-absorbing structures in the transportation sector is growing as a result of regulations related to crashworthiness, pollution and fuel efficiency. Carbon fiber reinforced polymer (CFRP) composites are ideal for this type of application because they have good specific mechanical properties. The automotive and aerospace industries are increasingly interested in thermoplastic composites due to their short processing cycles, high fracture toughness, infinite shelf life, good solvent resistance, and inherent recyclability [2, 8, 41]. However, more than 75% of CFRP is produced using a thermoset matrix [14]. For most automotive applications, all-composite structures are considered too expensive and labor-intensive to manufacture, but multi-material designs provide a practical alternative. But this poses a problem for combining technologies.

Traditional mechanical fastening methods such as riveting or riveting are often used to join different materials, but they also introduce significant weight increases, the potential for galvanic corrosion and stress concentrations. An adhesive bond provides improved acoustic and vibration isolation, more uniform stress distribution, improved corrosion resistance, and an easier solution. However, to ensure a reliable connection, the bonded surfaces must be treated. Depending on the degree of hardening of spring, shot peening can extend its service life by 400–1200%. The fatigue life of gears can also be increased by more than 500% by shot peening. Exhaust megaphones, rudders, axles, torsion bars and marine propellers are among other items that are often shot blasted. Increasing operating loads or fatigue life is beneficial to everyone [14–16].

3.11.2 Implantation of Ion

Ion implantation is a technique used to change the surface properties of materials. It involves the introduction of ions from one solid material to another, changing the surface chemistry and physical properties of materials. The target is the place where the ions collide, the accelerator is the place where the ions are electrically accelerated to high energy, and the ion source is the place where ions of desired element can be created. The resulting functions and depth of penetration of ions into a solid are determined by the energy, type of ions and composition of target. Ion implantation is widely used in both the semiconductor and mechanical industries. For example, the most commonly used method is ion implantation of such impurities. Use of ion implantation in the semiconductor field. Steel materials can also be implanted with nitrogen ions or other materials. Structural and chemical changes to the implant can stop corrosion and crack development. This type of ion implantation typically uses a radio frequency antenna to excite the plasma used as the source. Typically, ion implantation procedures produce nitrides, borides, carbides, and other compounds [18, 19].

3.12 TECHNIQUES FOR TESTING SURFACE COATINGS

For surface-engineered products, surface testing is an important step in the quality assurance process. Surface quality can be verified using a wide range of testing methods. They are usually divided into destructive and nondestructive. If many homogeneous products are manufactured, then destructive testing is allowed on a limited area of the surface of the part. Adhesion/abrasion tests, chemical analyses, porosity measurements, hardness tests, flexural tests, and other destructive tests are commonly used with surface coatings. During the coating process, defects of all shapes and sizes can be added to the coating [20]. Applied Ni and Cr metallic interlayers to improve the adhesion between the DLC coating and silicon substrate [42]. Because they have formed stable carbides and utilized more sites for the nucleation and growth of carbon atoms, DLC coatings obtained on Cr layers had much smoother morphology (Figure 3.24a, c), more sp^3 bonds, and significantly lower residual stress than those on the Ni layers. In addition, as shown in Figure 3.24, 40 nm thick Cr interlayer not only improved the adhesion of DLC films but also provided the lowest CoF, excellent wear resistance, and highest hardness (27 GPa).

The most common nondestructive testing methods for searching for defects are capillary testing, radiography, eddy current, ultrasonic and optical methods [13, 25]. Eddy current has several applications: it can evaluate physical properties,

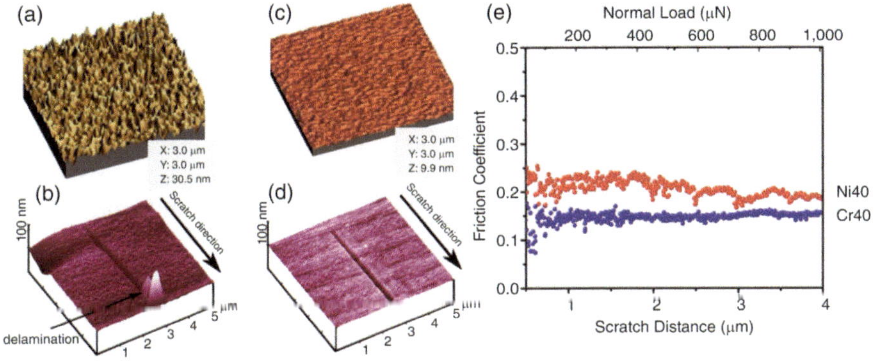

FIGURE 3.24 Ni and Cr metal interlayers used in DLC coating. AFM images of Ni and Cr nanolayers (a); AFM images of scratch test of DLC nanolayers on Ni (b), Ni and Cr nanolayers (c), Cr (d), and (e) coefficient of friction with increasing load [42] [open access].

identify surface and subsurface defects in components, determine the thickness of non-conductive surface coatings, and identify structural parts. Ultrasonic techniques are often used as a nondestructive testing method to detect tiny surface cracks and internal material defects. used to regularly inspect tracks, cars, trains and aircraft for early signs of fatigue cracks [26, 27].

Full-screen non-contact shearography eliminates the need for hydraulic coupling. Shearography also has a significant advantage over these two methods because it is faster than ultrasound testing and less sensitive than electron speckle interferometry (ESPI). Shearography is already used in the aerospace industry for the evaluation of composites in the rubber industry for nondestructive testing of tires [28, 39, 42–44]. This chapter provides a brief overview of innovative developments in the field of surface engineering and provides information on the various approaches, history of surface engineering, properties, methods, advantages, principles and applications of surface modeling. All manufacturing industries can be significantly impacted by surface treatment technologies that are generally considered advanced. According to this view, greater emphasis should be placed on research and development (R&D), which focuses on creative and flexible ideas for modifying surface structure to achieve optimal and cost-effective solutions and the effective use of non-standard surface treatment methods and tools.

The surface properties of structural materials cannot be ignored in design as they have a major impact on the functionality and life of the part. The engineering environment is complex. They tend to be complex and combine surface deterioration caused by chemicals and physical stress. To counteract these situations, surface treatments can improve the overall performance of components and extend their service life. In short, surface engineering technology offers practical solutions to complex situations. Future applications in the sports industry for surface engineering will be numerous. Compared to destructive testing, nondestructive testing has many advantages. Due to product reliability, there is a need for improved quality and more sophisticated nondestructive testing methods. Optical interferometric methods such as ESPI and ESPSI provide the advantages of non-contact and full-field methods. The industry is rapidly adopting optical technologies for nondestructive testing. Nondestructive testing of surface coatings has a promising future [21]. One of the most important methods for differentiating engineered products in terms of performance, quality and life cycle cost is surface engineering. To determine the predominant wear mechanisms, SEM is used to characterize the scars created by the microabrasive machine. SEM images for the abrasive slurry concentrations of 0.25 g/cm^3, 0.35 g/cm^3, 0.45 g/cm^3, and 0.2 N load are shown in Figure 3.25a–c. The SEM images for the abrasive slurry concentrations of 0.25 g/cm^3, 0.35 g/cm^3, 0.45 g/cm^3, and 1 N load are shown in Figure 3.25d–f. SEM analysis of the photographs revealed the same trend for the 0.5 N load [21].

The surface engineering is the backbone of modern manufacturing and materials science. It meets the growing demands for increased productivity, efficiency and sustainability in industrial applications through continuous change and adoption of new technologies. This chapter provides a broad overview of the advances, approaches, backgrounds, characteristics, methodologies and creative

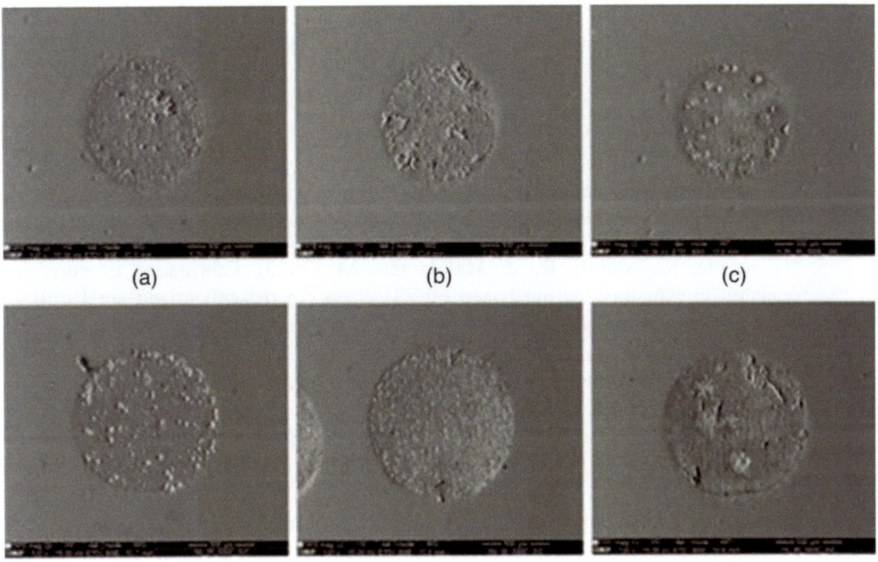

FIGURE 3.25 SEM images of craters: there are five different ways of loading a given amount of N: (a) 0.2 N, (SC) 0.25 g/cm^3; (b) 0.2 N, (SC) 0.35 g/cm^3; (c) 0.2 N, (SC) 0.45 g/cm^3; (d) 1.2 N, (SC) 0.25 g/cm^3; (e) 1.2 N, (SC) 0.35 g/cm^3; and (f) 1 N loading, (SC) 0.45 g/cm^3 [21] [open access].

applications of surface engineering, highlighting its benefits and guiding principles. Surface engineering will remain at the forefront of industrial breakthroughs and innovations in manufacturing processes and component durability as industries strive for improvement.

3.13 PROBLEMS

1. Explain the different types of surface geometries of ductile materials.
2. Name several interfaces and surfaces that are critical to coating procedures and technologies from a tribological perspective.
3. Explain how that the materials have different macro- and microstructures.
4. Explain how that the apply thermal spray in conditions where there is little or no adhesion.
5. How can surface coatings be checked if the mechanical strength of the two contacting bodies is different?

REFERENCES

1. Huang, N., Wang, Y., Zhang, Y., Liu, L., Yuan, N., & Ding, J. (2023). Multifunctional coating on magnesium alloy: Superhydrophobic, self-healing, anti-corrosion and wear-resistant. *Surf. Coat. Technol*, *463*, 129539.

2. Sultana, A., Zare, M., Luo, H., & Ramakrishna, S. (2021). Surface engineering strategies to enhance the in situ performance of medical devices including atomic scale engineering. *Int. J. Mol. Sci*, *22*(21), 11788.
3. Kozaderov, O., Swiatowska, J., Dragoe, D., Burliaev, D., & Volovitch, P. (2021). Effect of Cr(III) passivation layer on surface modifications of zinc-nickel coatings in chloride solutions. *J. Solid State Electrochem*, *25*, 1161–1173.
4. Su, Y.L., Kao, W.H., & Chang, G.Y. (2023). Tribological, anti-corrosion, and electrical conductivity properties of CrCx coatings deposited on stainless steel 316L and used as metal bipolar plates for fuel cells. *J. Mater. Eng. Perform*, *32*, 3739–3754.
5. Nazarzade, K., Naderi, R., & Mahdavian, M. (2023). Fabrication of corrosion-resistant conversion coating based on ZIF-8 on electro-galvanized steel utilizing pulse electrodeposition technique. *Surf. Coat. Technol*, *457*, 129292.
6. Mo, Y., Ma, X., Wang, S., Zhong, H., Yang, J., & Cao, Z. (2023). In-situ constructing and EDTA-enhancing of meta-aluminate passivation films on electroplated zinc surfaces. *Colloids Surf. A Physicochem. Eng. Asp*, *657*, 130573.
7. Cui, G., Song, Y., Liu, K., & Tan, M. (2021). Interaction of carbon dots from grilled Spanish mackerel with human serum albumin, γ-globulin and fibrinogen. *Foods*, *10*(10), 2336.
8. Kanazawa, T., Hayakawa, M., Vinas, D., Tahara, Y., Hata, N., & Yoshimoto, M. (2023). Sustainable technology for remanufacturing of carburized steels by laser hardening. *J. Mater. Res. Technol*, *24*, 39–48.
9. Bagatella, S., Ciapponi, R., & Turri, S. (2022). Nanomechanical characterization of bacterial polyhydroxyalkanoates using atomic force microscopy. *Appl. Sci.*, 12(10), 4994.
10. Zhang, J., & Zhang, W. (2022). AFM image analysis of the adsorption of xanthate and dialkyl dithiophosphate on chalcocite. *Minerals*, *12*(8), 1018.
11. Nikolova, M.P., & Apostolova, M.D. (2023). Advances in multifunctional bioactive coatings for metallic bone implants. *Materials*, *16*, 183.
12. Costa, R.C., Nagay, B.E., Dini, C., Borges, M.H., Miranda, L.F., Cordeiro, J.M., Souza, J.G., Sukotjo, C., Cruz, N.C., & Barão, V.A. (2023). The race for the optimal antimicrobial surface: Perspectives and challenges related to plasma electrolytic oxidation coating for titanium-based implants. *Adv. Colloid Interface Sci*, *311*, 102805.
13. Cui, H., Du, C., & Zhang, H. (2023). Applications of phase field methods in modeling fatigue fracture and performance improvement strategies: A review. *Metals*, *13*(4), 714.
14. Refait, P., Grolleau, A.M., Jeannin, M., Rémazeilles, C., & Sabot, R. (2020). Corrosion of carbon steel in marine environments: Role of the corrosion product layer. *Corr. Mater. Degrad* , *1*(1), 10.
15. Zhang, Z., Kitada, A., Fukami, K., & Murase, K. (2022). Aluminum electroplating on AZ31 magnesium alloy with acetic anhydride pretreatment. *Acta Met. Sin. Engl. Lett*, *35*, 1996–2006.
16. Tian, M., Jian, Z., Hai, R., & Chang, F. (2023). Non-isothermal crystallization kinetics of amorphous electroless nickel-phosphorus alloy plating. *J. Therm. Anal. Calorim*, *148*, 1959–1970.
17. Wang, C., Jiu, H., Zhang, L., Xu, R., Che, S., Guo, Z., Han, Y., Ma, J., Li, H., & Guo, F. (2023). Carbon coating hollow NiS nanoparticles anchored on Ti_3C_2 nanosheet as high-performance anode for lithium-ion batteries. *J. Alloys Compd*, *956*, 170332.
18. Dai, G., Wu, S., Huang, X., Wang, M., & Wu, R. (2023). Research on the coating formation of Al-induced electroless plating on metallic surfaces. *J. Mater. Sci*, *58*, 3768–3789.

19. Koli, P.B., Kapadnis, K.H., Deshpande, U.G., Tupe, U.J., Shinde, S.G., & Ingale, R.S. (2021). Fabrication of thin film sensors by spin coating using sol-gel LaCrO3 Perovskite material modified with transition metals for sensing environmental pollutants, greenhouse gases and relative humidity. *Environ. Chall*, *3*, 100043.
20. Hakala, T.J., Holmberg, K., & Laukkanen, A. (2021). Coupling molecular dynamics and micromechanics for the assessment of friction and damage accumulation in diamond-like carbon thin films under lubricated sliding contacts. *Lubricants*, *9*(3), 30.
21. Baptista, A., Pinto, G.F., Sousa, V.F., Silva, F.J., & Fernandes, F. (2023). Study on the micro-abrasion wear behavior of PVD Hard coating under different SiC abrasive–particles/ distilled water ratios. *Materials*, *16*(8), 2939.
22. Chen, H., Qin, Z., He, M., Liu, Y., & Wu, Z. (2020). Application of electrochemical atomic force microscopy (EC-AFM) in the corrosion study of metallic materials. *Materials*, *13*(3), 668.
23. Sánchez-España, J., Ilin, A., & Yusta, I. (2022). Metallic copper (Cu[0]) obtained from Cu2+-rich acidic mine waters by two different reduction methods: Crystallographic and geochemical aspects. *Minerals*, *12*(3), 322.
24. Su, W., Cui, X., Jin, G., Liu, C., Liu, E., Zhao, H., & Cao, B. (2023). Effects of different Y compounds on the microstructure and properties of laser cladding NiTi coatings in a semi-open environment. *Intermetallics*, *160*, 107936.
25. Lidsky, D., Cain, J.M., Hutchins-Delgado, T., & Lu, T.M. (2023). Inverse metal-assisted chemical etching of germanium with gold and hydrogen peroxide. *Nanotechnology*, *34*, 065302.
26. Kennedy, D., Xue, Y., & Mihaylova, E. (2005) Current and future applications of surface engineering.
27. Rengarajan, S., Ramaraj, B.K., Paramashivan, S.S., & Velusamy, S. (2023). Oxidation behavior of near nanostructured coating developed by the HVOF process. *Mater. Corros*, *74*, 482–492. [CrossRef] Li, Y.R.; Su, C.C.; Chang, S.H. Applying Aluminum-Vertically-Aligned Carbon Nanotube Forests Composites for Heat Dissipation. Nanomaterials 2019,*9*, 758.
28. Zhu, P., Zhang, Q., Xia, Y., Sun, K., Lin, X., Gou, H., … & Wu, G. (2023). Effect of nanoscale W coating on corrosion behavior of Diamond/Aluminum composites. *Nanomaterials*, *13*(2), 307.
29. Bonu, V., & Barshilia, H.C. (2022). High-temperature solid particle erosion of aerospace components: Its mitigation using advanced nanostructured coating technologies. *Coatings*, *12*, 1979.
30. Zhang, S.N., Chen, Z., & Arfaei, B. (2021). Graphene coating as a corrosion protection barrier for metallic terminals in automotive environments. *SAE Int. J. Adv. Curr. Pract. Mobil*, *3*, 3176–3183.
31. Wasekar, N.P., Bathini, L., Ramakrishna, L., Rao, D.S., & Padmanabham, G. (2020). Pulsed electrodeposition, mechanical properties and wear mechanism in Ni-W/SiC nanocomposite coatings used for automotive applications. *Appl. Surf. Sci*, *527*, 146896.
32. Wang, J., Wang, L., Su, X., Xiao, R., & Cheng, H. (2022). Polarization maintaining fiber temperature and stress gradient sensitization sensor based on semiconductor-Metal–Polymer three-layer film coating. *ACS Appl. Mater. Interfaces*, *14*, 20053–20061.
33. Li, C.-J., Luo, X.-T., Dong, X.-Y., Zhang, L., & Li, C.-X. (2022). Recent research advances in plasma spraying of bulk-like dense metal coatings with metallurgically bonded lamellae. *J. Therm. Spray Technol*, *31*, 5–27.

34. Liu, B., Xiao, F., Zhu, H., & Tang, M. (2023). Promising WC-30WB-10Co cemented carbide coating with improved density and hardness deposited by high velocity oxy-fuel spraying: Microstructure and mechanical properties. *J. Mater. Eng. Perform*, *32*, 6405–6411.
35. Lee, H.-S., & Singh, J.K. (2022). Deposition and corrosion studies of plasma arc thermal sprayed Zn and 85Zn–15Al films on steel surface. *J. Mater. Sci*, *57*, 19650–19665.
36. Zou, H., Yan, S., Shen, T., Wang, H., Li, Y., Chen, J., & Meng, Y. et al. (2021). Efficiency of surface texturing in the reducing of wear for tests starting with initial point contact. *Wear*, 482, 203957.
37. Li, H., Chen, Z.W., & Ramezani, M. (2021). Wear behaviours of PVD-TiN coating onTi-6Al-4V alloy processed by laser powder bed fusion or conventionally processed. *Int. J. Adv. Manuf. Technol*, *113*, 1389–1399.
38. Zhang, M., Zhou, T., Li, H., & Liu, Q. (2023). UV-durable superhydrophobic ZnO/SiO2 nanorod arrays on an aluminum substrate using catalyst-free chemical vapor deposition and their corrosion performance. *Appl. Surf. Sci*, *623*, 157085.
39. Venkatachalapathy, V., Katiyar, N.K., Matthews, A., Endrino, J.L., & Goel, S. (2023). A guiding framework for process parameter optimisation of thermal spraying. *Coatings*, *13*(4), 713.
40. Brover, G.I., & Shcherbakova, E.E. (2022). Aspects of structure formation in surface layers of steel after laser alloying from various coatings. *Metallurgist*, *66*, 672–680.
41. Dashtbozorg, B., Penchev, P., Romano, J.-M., Li, X., Sammons, R.L., Dimov, S., & Dong, H. (2021). Development of surfaces with antibacterial durability through combined S phase plasma hardening and a thermal femtosecond laser texturing. *Appl. Surf. Sci*, *565*, 150594.
42. Wang, L., Liu, Y., Chen, H., & Wang, M. (2022). Modification methods of diamond like carbon coating and the performance in machining applications: A review. *Coatings*, *12*(2), 224.
43. Inyushkin, A.V., Taldenkov, A.N., Ralchenko, V.G., Bolshakov, A.P., Koliadin, A.V., & Katrusha, A.N. (2018). Thermal conductivity of high purity synthetic single crystal diamonds. *Phys. Rev. B*, *97*, 144305.
44. Zhang, Y., Li, J., Zhao, L., & Wang, X. (2014). Optimisation of high thermal conductivity Al/diamond composites produced by gas pressure infiltration by controlling infiltration temperature and pressure. *J. Mater. Sci*, *50*, 688–696.

4 Surface Engineering for Wear Test

4.1 INTRODUCTION

The problem of global warming affects everything. To stop the global warming, carbon dioxide (CO_2) emissions must be reduced. Although there are several ways to reduce CO_2 emissions, the most important is by improving machine efficiency by reducing wear and friction [1]. Due to its exceptional properties, especially high hardness and oxidation resistance, SiC coatings are often used in gas turbine engines and heat exchangers [2]. This typically applies to high-temperature applications where the temperature stability of around 1400°C is required [3]. They are reported to have moderate frictional properties under lubrication conditions. As a result, they are commonly used in coatings that require high corrosion resistance [4]. However, in dry conditions, SiC coatings have less-than-ideal frictional properties, which significantly reduces the lifespan of heat exchangers and gas turbine engines as well as the tribological behavior of SiC coatings [3]. The amorphous SiC coating has showed greater corrosion resistance due to its lower coefficient of friction (CoF) compared to crystalline SiC coating [5]. Surface texturing has many other applications, including solar cells [6], biomedicine, mechanical seals, machining operations and bearings, and technological devices [5].

In addition, the ease of fabrics handling is another factor in their widespread use. There are many methods for processing surface composition, such as chemical etching, laser processing, indentation and electrodeposition [6]. Among these methods, the laser surface texturing method used in this study is very popular due to its fast processing time, low environmental impact and excellent control of fabric size and shape [7]. In 1966, the Etsion group devised an experimental method to reduce the friction coefficients between matching contact points using fabric. Texture modeling research conducted by Etsion group in 1996 led to a comprehensive analysis of surface composition. Furthermore, reinforcement affects the mechanical properties by improving stability and adaptability at both room and elevated temperatures, thus enhancing the friction properties [5]. However, in addition to the methods discussed previously, the mechanical surface treatments applied after curing are used to improve the overall properties of thermal spray coatings (TSC), which are necessary to achieve the increased efficiency and reduced emissions [7]. Previous studies [8–11] have investigated the wear mechanisms of sintered SiC and various silicon-free ceramic coatings after ultrasonic nanocrystal surface modification (UNSM) treatment [6].

DOI: 10.1201/9781003584735-4

On another hand, we believe that no research has been conducted on the fatigue and corrosion performance of SiC coatings after curing under dry and non-dry conditions. Therefore, the experimental results of this study have shed light on the surface modification of SiC coatings and confirm whether UNSM technology can outperform sputtering technology in terms of friction and fatigue performance.

This texture is discovered in reciprocating tests with pins (phosphor bronze) and blocks (cast iron), it is found that the presence of small round holes measuring 60 × 10 microns and a textured area reduces friction coefficients by 35%. The improvement in tribological performance is due to following; higher hydrodynamic pressure, debris collection and secondary oil exposure. Silicon carbide and silicon nitride (a self-consistent pair) due to chemical friction reactions [6] have relatively low friction coefficients in aqueous conditions [7]. The critical load of a ring-on-disc experiment could be increased by 2.5 times by optimizing the settings of single texture (diameter 350 μm, depth 3.2 μm, texturing area 5%). This will improve the load-bearing capacity of self-conjugated silicon carbide. They have proposed a method by which coarse textures can improve load-bearing capacity and increase hydrodynamic pressure and the tribochemical reaction can be accelerated by accumulating water in fine textures [2].

Texturing can successfully lower wear and friction in steel-ZrO couples. Using a nanosecond laser to create microporous patterns on 100Cr6 steel discs, this phenomenon is caused by the balls wearing down more quickly due to roughness, which alters the contact geometry and speeds up the release of lubricant from the mixed-mode boundary [1].

Test results from a ball on an oil-lubricated disc have showed that surface texturing exacerbates ball wear. Regardless of oil or lubricant, they have found that roughness can successfully reduce the friction coefficient while increasing the diameter of wear surface of the ball. The results have showed that the tribological performance of fluoroelastomer could not be improved by texture. When the texture size is equal to the contact area size [3]. The effect of texture on a friction performance can found two competing mechanisms. The two mechanisms are the accumulation of wear debris, which results in less wear, and the concentration of stress at the edges of texture, which increases wear. Several studies have shown that texture can reduce wear at first contact. The case of initial point contact, compounding can reduce wear [4]. They have found that changing the surface roughness of steel has changed the surface properties of material, making it more or less resistant to corrosion. The fabric can successfully reduce wear under lubrication conditions because it can effectively retain lubricating oil [7]. The above tests have some limitations. The effect of texture on metallic materials immersed in oil is studied in the first contact test, but the effect of texture on ceramic materials immersed in water is not studied. No studies have been conducted on lubrication. In this study, more detailed study of texture parameters is carried out to investigate the effect of laser surface texturing on the tribological properties of self-fitting SiN samples subjected to initial contact sliding tests in water.

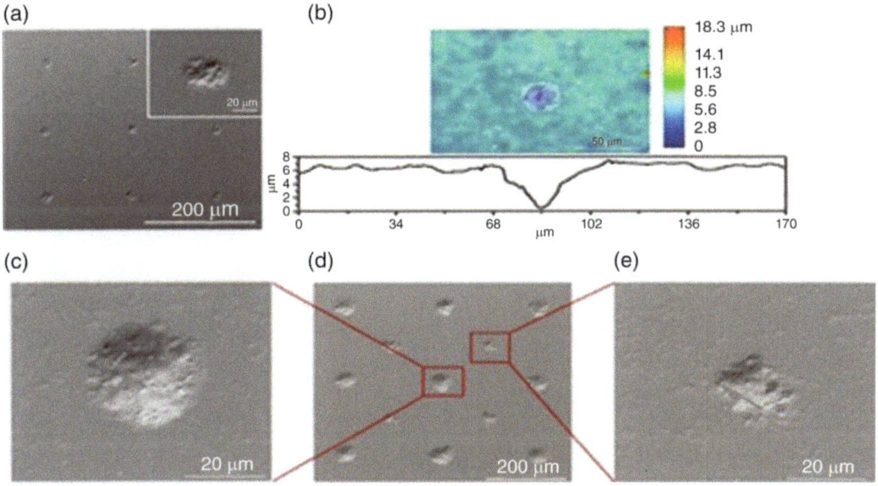

FIGURE 4.1 (a) Optical profiler measurements and SEM image of single texture sample, (b) SEM images of three different surface textures: (c) large-scale, (d) mixed texture with two different types of textures, and (e) small-scale [12].

4.2 MATERIALS

Balls with a diameter 9.525 mm are manufactured using SiN (China National, Corporation). The hardness of the material is 1586 HV. The roughness of balls is 0.02 μm. Discs with diameter 56 mm and thickness 6.6 mm are fabricated using SiN (China National Machinery Industry Corporation, China). The hardness of disc is 1678 HV. Scanning electron microscopy (SEM) in Figure 4.1a shows the images of specific textures. SEM images with mixed texture are shown in Figure 4.1d.

After laser treatment, the surfaces are polished for 2 minutes with 3000 mesh sandpaper due to debris accumulation. Diamond lapping fluid with diameter 10 μm is used to grind the disc surfaces. After this, they are polished with a cerium oxide-based grinding fluid with a particle size 0.4 μm and roughness 0.016 μm. The unique qualities are shown in Figure 4.2.

Tribological properties can be effectively studied using a unidirectional sliding running-in test [13]. Si N is tested by rolling in deionized water on a friction and wear test bench UMT-TriboLab (Bruker, Germany) [1, 3, 13]. Figure 4.2a illustrates a schematic diagram of the contact configuration. The disc rotated while the ball remained stationary during the friction experiment. The test disc has a rotation radius 16 mm. At the start of experiment, the ball-on-disc device has produced point contact; however, as time went on, the ball's wear grew, and ball and disc have made flat contact. A friction meter is used to record the friction coefficient at each step of procedure. The test loads are N = 10 and 15. At 1.33, 1.524 and 1.056 GPa, the highest Hertzian contact pressures at the initial point are recorded. The diameters of round pads are 0.137 mm, 0.12 mm and 0.095 mm.

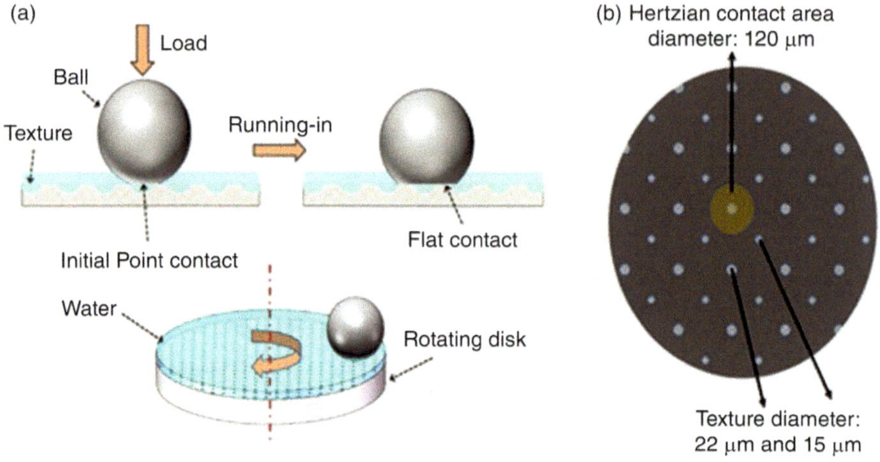

FIGURE 4.2 (a) Schematic diagram of the contact configuration. (b) Texture diameters and Hertzian diameter ratios of the circular contact area [12].

Figure 4.2b shows the relationship between the texture diameter and Hertzian diameter of circular pad. All test speeds are: 0.25, 0.5, and 0.75 m/second [4]. Although the viscosity in the experiment has a significant effect, the temperature of experiment has changed very little, indicating that the effect of lubricant viscosity is not investigated [5].

4.2.1 Laser Surface Texturing

The two variables for the laser parameter are number of marks and laser power density. The number of marks is varied from 1 to 4, and the laser energy density varies from 0.18 J/cm to 0.91 J/cm. Changing the correlation parameter made it possible to process tissues of different sizes. Information on laser properties and processing aspects can be found in [4]. Ring tissues with a diameter 15 to 85 μm are created on the surface of the discs. Tissue depth ranged from 5 to 16 μm and tissue area ranged from 0.5% to 8%. When choosing texture settings, we are guided by references [5–7]. In recent years, the study of multilevel and hierarchical tissues has received more attention [6–8]. The created three biomimetic hierarchical textures on Ti-6Al-4V surfaces to study their effect on frictional properties. The results have showed that the hierarchical structure could reduce the friction coefficient by 79% [7]. The multiscale textures can reduce the friction coefficient by 70% at higher contact pressures [8]. Therefore, in the light of this study, both single-band and multi-band textures are considered. A later paper replaced the multiscale textures with mixed textures, combining them with two separate textures of different depths and diameters.

4.2.2 Tribological Testing

This study has used the standard ASTM G99 test procedure, which has good reproducibility and repeatability [2]. An experiment is conducted using a ball-on-disc device to see if sliding silicon nitride in water could cause surface texturing, exacerbating initial point contact wear. The results of the two distinct methods for estimating contact temperature using thermocouple measurements are presented in this section. Figure 4.3 shows the maximum and average friction coefficients of the thermoplastic material moving across stainless steel at stroke lengths of 22.5 mm and 4.5 mm. Especially for the maximum friction coefficient, the thermoplastic material shows a slightly decreasing friction coefficient with increasing stroke length. A deeper sample texture can improve performance by promoting lubricant accumulation [13], making it more effective in initiating tribochemical reactions [6]. Consequently, the deeper texture and roughness of worn surface are smaller [3], which allows the feed pressure to be increased minimize deterioration [7].

Using a constant load (10 N) and constant speed (0.5 m/second), the effect of different texture depths on the tribological performance of self-conjugated SiN is studied. This is achieved by maintaining a constant tissue area, 2% and diameter, 22 μm. demonstrates the amount of wear on the ball, which is determined by

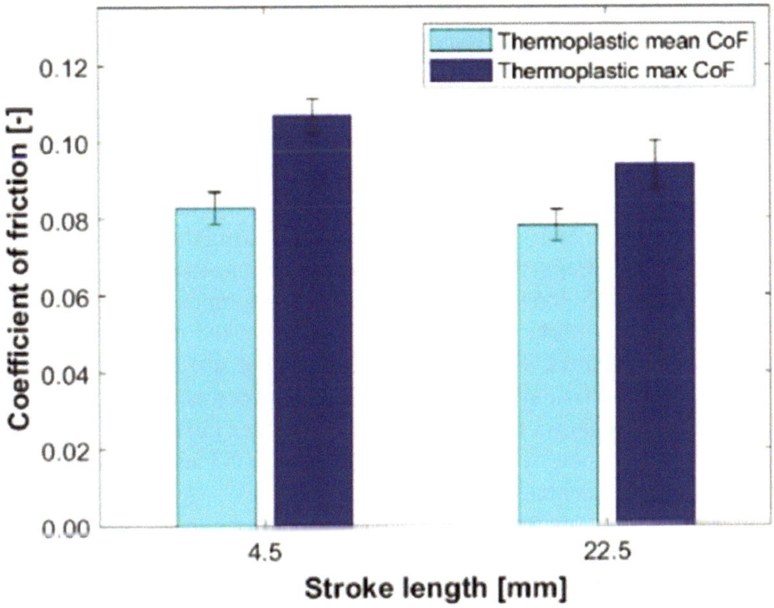

FIGURE 4.3 Average of the mean and maximum coefficients of friction as a function of stroke length for the thermoplastic bearing material [14] [open access].

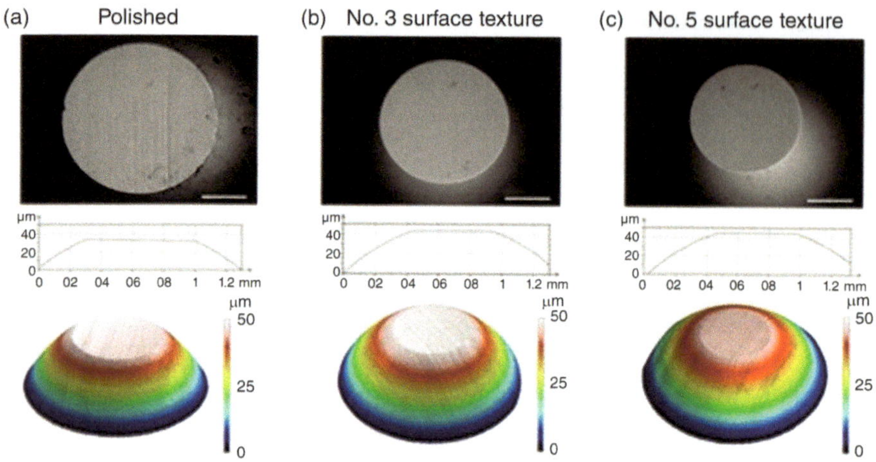

FIGURE 4.4 Optical are obtained by rubbing of the balls. (Bottom) 3D microscope images and line scans of friction ball surfaces: (a) sample (no texture) (1), (b) sample (no texture) (3), and (c) sample (textured) (5)) [12].

measuring the diameter of wear scar on ball. The sample texture depth of 11 μm ensures the least wear of ball, 7.06×10^4 mm. Compared to the non-textured sample, 26.8×10^4 mm, this indicates 74% reduction in wear. There are photographs obtained using optical microscopy of worn surfaces of balls as illustrated in Figure 4.4. To visualize the degree of ball degradation associated with wear, Figure 4.4 shows line scans and three-dimensional microscopic images obtained from the surfaces of balls. Compared to non-textured sample, the area of spherical wear marks in the textured sample is noticeably smaller. As the contact pressure increases, the height loss during the static loading test increases linearly. A large portion (35% to 59%) of the total height loss during the reciprocating sliding tests is caused by creep and compressive deformation, especially at low contact pressures. Figure 4.5 shows the worn surface topography of the stainless-steel surface and the thermoset polymer pin. The wear of the thermoset polymer pin is uniformly distributed, although the stainless-steel surface was uneven before the test. This is particularly evident at the lowest contact pressure of 26 MPa (Figure 4.5a). This shows that the thermoset polymer pin and the stainless-steel surface are well aligned. At higher contact pressures (53 and 80 MPa), the edge is more worn than the center.

Initially, all samples underwent a 600-second load test (10 N) at a constant speed, 0.5 m/second. The sample are then used in studies to see if the former could be used under constant and variable speed loading could function effectively at various loads and speeds. The exact experimental parameters to ensure statistical reliability, the experiment is carried out at least three to four times.

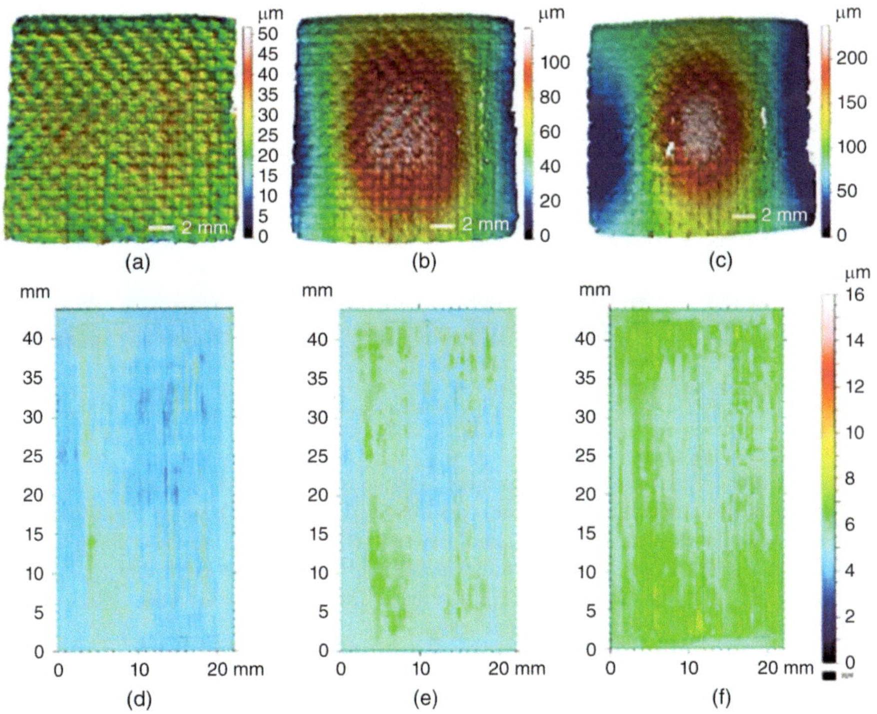

FIGURE 4.5 Surface topography of the load-carrying surface of the thermoset after sliding against stainless steel at a nominal contact pressure of (a) 26 MPa, (b) 53 MPa and (c) 80 MPa. Objective: 2.75×, field of view: 1×. Observe that the scale on the y-axis differs between the different pressures. Surface topography of the wear track formed on the stainless-steel surface after sliding against the thermoset at a nominal contact pressure of (d) 26 MPa, (e) 53 MPa and (f) 80 MPa. Objective: 2.75×, field of view: 0.5×. Black areas in figure (d–f) are missing data points and correspond to regions with thick layers of polymer wear debris, which have poor reflectivity during surface topography measurements. The sliding direction in all figures is parallel to the vertical axis [14] [open access].

4.2.3 Texture Depth and Effect

In visual media, art and design, the term "texture depth and effect" is frequently used to describe how texture produces a sense of dimension and realism. It refers to the properties of an object's surface, including roughness, smoothness, softness, gloss and opacity. Texture can be inferred (as in a painting or digital art, where the illusion of texture is created) or real (as in the roughness of sandpaper). Depth describes how three-dimensional or distant a visual composition appears. When it comes to establishing perspective and contrast, texture depth is crucial. The interplay of light and shadow on surfaces or layered textures can intensify this impression [15–17].

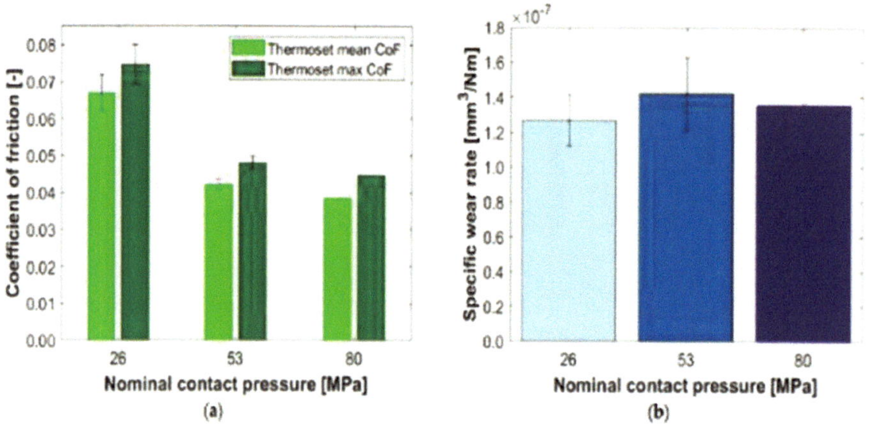

FIGURE 4.6 Variation of (a) the average of the mean and maximum coefficients of friction as a function of nominal contact pressure for the thermoset bearing material and (b) the average specific wear rate calculated between 80 h and the end of the test as a function of nominal contact pressure for the thermoset bearing material. Note that the standard deviation bars are missing for the reciprocating sliding tests at 80 MPa due to interruption during the repetitive test. [14] [open access].

4.2.4 Textured Area

This examines the influence of different textured regions on the tribological properties of self-conjugated SiN. There are no changes in diameter, 15 μm or texture depth, 5 μm. The friction coefficient as a function of test time is shown in Figure 4.6a. The burn-in operation stops after 600 seconds if the textured area reaches 6% or less. The usefulness of textured Si-DLC in oil lubrication at different temperatures is investigated, and it is discovered that the tribological performance of textured Si-DLC is better at different temperatures compared to untextured Si-DLC coating.

The tribological performance of amorphous hydrogenated diamond-like carbon (DLC) coatings under oil-lubricated sliding conditions can vary significantly depending on factors such as coating thickness, contact surface diameter and type of oil used for lubrication. When investigating the tribological performance of composite DLC coatings, most studies kept the geometric properties of formulation unchanged. However, the effects of textural parameters such as density, diameter and dimple depth of amorphous hydrogenated DLC have not received much attention in the literature [13].

4.3 TEXTURE EFFECT

For the thermoset sliding on stainless steel at different nominal contact pressures ranging from 26 to 80 MPa, Figure 4.6a shows the mean and maximum friction coefficients. As the nominal contact pressure increases, the thermoset shows a decrease in both mean and maximum friction coefficients. While the difference between the mean and highest nominal contact pressures (80 MPa) is only 9%

and 8%, respectively, the decrease between the lowest nominal contact pressures (26 MPa) and the mean (53 MPa) is significantly larger, at 59% and 55% for the mean and maximum friction coefficients, respectively. This suggests that, unlike other reports of thermoset at contact pressures, the decrease in friction does not follow a linear trend 9–28 MP. Compared to samples without texture, samples with texture show less wear on ball. According to the data obtained, the total wear on 2% woven area is 73% less than on non-woven area. Extra wide textured area leading to a decreasing in the contact area of friction zone and an increasing in contact pressure [7]. An area that is too small cannot create sufficient hydrodynamic pressure [8]. For this reason, samples with 2% woven areas show better wear characteristics than samples with 0.5% and 6% woven areas.

4.4 DIAMETER AND IMPACT OF TEXTURE

This section investigates the effect of fabric diameter on the performance of self-conjugated SiN with a fixed fabric thickness, 5 μm and fabric area, 1%. The results of experiment are shown in Figure 4.7. A non-composite sample requires

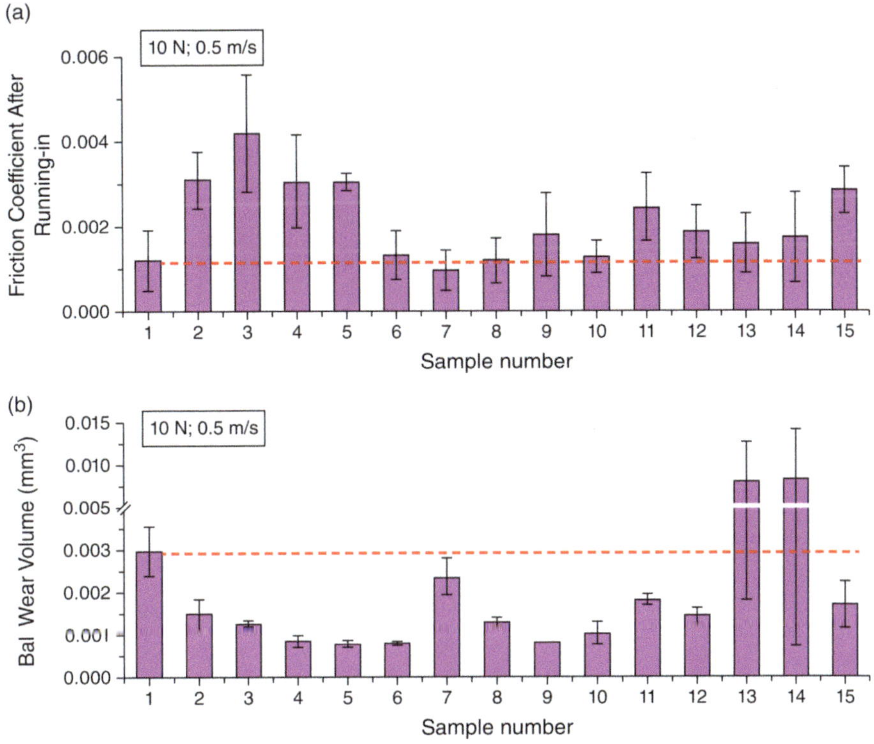

FIGURE 4.7 A comparison of wear volumes of (a) the sample balls (b) with the friction coefficients [12].

more analysis time than a composite sample, as shown in Figure 4.7a. A friction coefficient of at least 0.0012 is found for both samples with and without composition. With a diameter, 15 μm as shown in Figure 4.7b. All machined composite specimens have friction coefficients ranging 0.001 to 0.0035, but there is no obvious correlation. To take advantage of these benefits, scientists optimize the geometric properties of fixture, such as density, diameter, depth and placement angle for specific operating conditions [8]. Several mechanisms have been proposed to improve tribological performance [5], including increasing load-bearing capacity by providing lubrication at high loads, acting as small wear traps to reduce third-body wear and providing additional fluid pressure in hydrodynamic lubrication [9].

The effect of fabric diameter on load-carrying capacity is shown in Figure 4.7c and d. Typical bearing pressure increases then decreases and then increases again as wall diameter increases. Samples with diameter, 15 μm outperform undecorated samples by 51% (Figure 4.7c). The 15 μm diameter sample has the best bearing performance value, 11.81×10^4 ηN/P as shown in Figure 4.7d. This result indicates that this sample has the maximum load-bearing capacity among the test sets. They have found that 20 μm diameter texture exhibits an optimal friction property, which is consistent with the findings of this study [4]. In numbers, a relatively new method for improving the tribological performance of mechanical parts is surface mounting. This method involves creating micro- and nanotextures on the surface of mechanical components to reduce contact temperature, wear and friction. Also, they have found that small diameter texture can increase lubricant viscosity and reduce wear in a point contact experiment [12]. Samples with fabric diameters, 15, 25 and 85 μm show smaller ball wear size compared to undecorated sample, while samples with fabric diameters, 45 and 65 μm show larger ball wear size (Figure 4.7e). When using the sample with the best texture (texture diameter 15 μm), the wear volume is reduced by 56%.

4.4.1 Complex Effects of Various Textural Characteristics

The average and maximum friction coefficients of the thermosetting material sliding against stainless steel at different nominal contact pressures are shown as a function of time over the test duration in Figure 4.8. The curves for each contact pressure have comparable patterns and are close to each other, while the surrounding curves show the dispersion between repeated tests. Note that the test continued for the remainder of its duration after the sudden peak at 113 hours at a contact pressure of 80 MPa due to a power outage. The surrounding curves show that the friction readings of the test device have good long-term reproducibility across all contact pressures.

Compared to undecorated specimen, the wear rates of balls in twelve decorated specimens are lower. Depending on how much the fabric area changes, the amount of wear will vary greatly. However, samples 4 and 6 show that a large change in fabric depth will result in only a small change in wear size, indicating

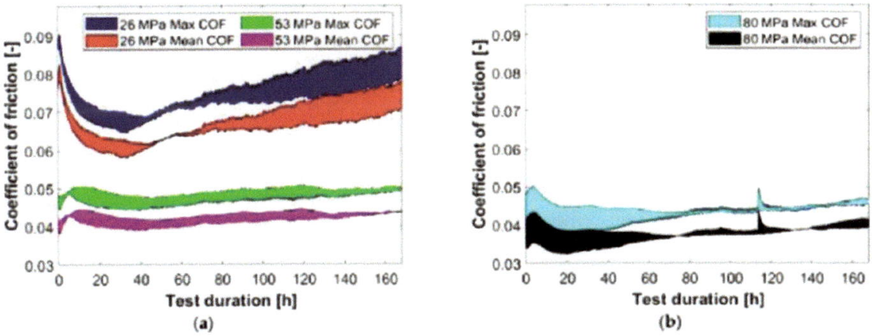

FIGURE 4.8 Average and maximum coefficients of friction versus time for the two repeat tests with the thermoset at a contact pressure of (a) 26 and 53 MPa and (b) 80 MPa. The enveloping curves indicate the scatter between repeat tests. Note that the second test at 80 MPa was interrupted after 113 h due to power failure and thereafter restarted for the remaining test duration [14] [open access].

that fabric area influences wear size more than fabric depth [3, 7]. It can be seen that all samples (samples 4, 5, 6, 9), which showed the least amount of corrosion, have a coefficient, 2%. This result indicates that the tissue area must reach an appropriate value before maximum wear reduction occurs [3, 18]. The balls of sample 5 have the least wear and dimensions, 7.062 × 10 mm. 26 percent of sample has no texture, d = 11 μm, a = 2% and w = 22 μm. The optimal tissue diameter-to-depth ratio is 0.5, which is consistent with the study findings [9–11, 19, 20].

4.5 COMBINED TEXTURE EFFECT

4.5.1 Comparing Textures: Single and Mixed

Samples 5 (2% tissue area, 11 μm tissue depth, 22 μm tissue diameter) and 8 (1% tissue area, 5 μm tissue depth, 15 μm tissue diameter) performed best, following good results in previous tests. The textural composition is shown in Figure 4.1c. A constant load, constant speed experiment is conducted on mixed material specimens. The experimental results contrast with results for samples 1, 5 and 8 as shown in Figure 4.9.

Friction coefficients after operation are shown on several samples; Samples 1 and 8 have the lowest friction coefficients, 0.0012. Friction coefficients ranged from 0.001 to 0.0031 for all samples, but no significant correlation were observed. The mixed fabric sample has ideal abrasion resistance as seen in Figure 4.9b. Its textured sample has 84% less ball wear than the non-textured sample. The mixed-texture sample in Figure 4.9c has the highest average support pressure tolerance, with an average support pressure 145% higher than the textured sample. They have found through numerical analysis that a multiscale composite specimen

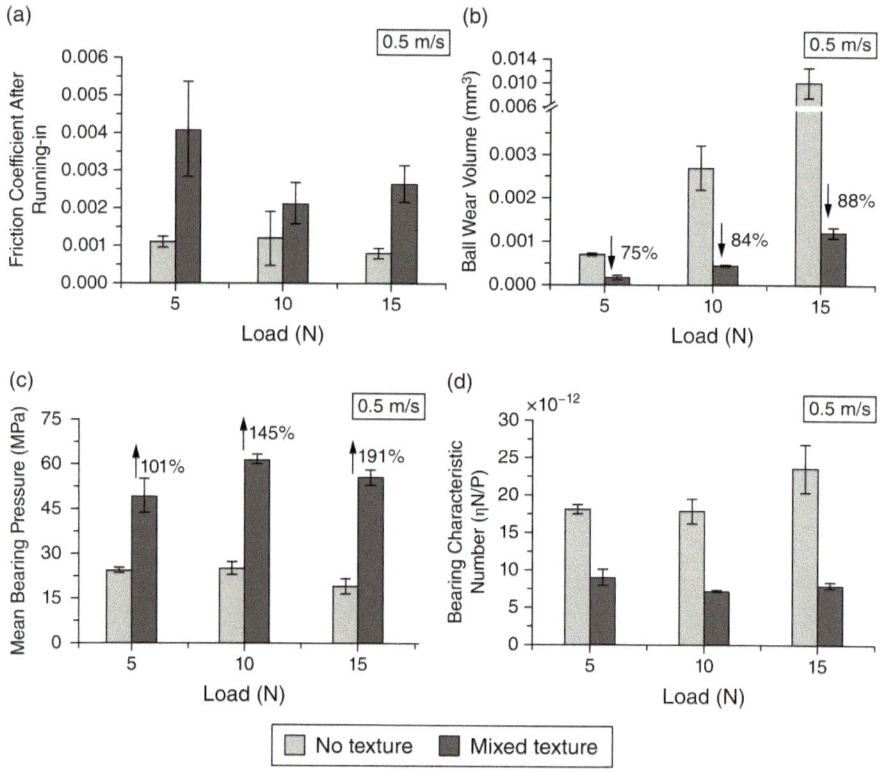

FIGURE 4.9 Depending on test load displays the following values: (a) Friction coefficient after running-in; (b) wear volume of Si_3N_4 balls; (c) Mean scar contact pressure; and (d) bearing characteristic number ηN/P [12].

could have 80% less corrosion than non-composite specimen [21–24]. Also, they have found that a smoother bearing surface reduced contact stress compared to bearings with different textures. For example, our results show that multiscale texture can successfully reduce wear.

Because textures of different sizes are consistent, mixed textures have better wear resistance than single textures. While a small texture stores lubricant, a huge single texture creates dynamic pressure [25–27]. In addition, compared to a single texture, a sample with a mixed texture can generate greater dynamic pressure [28]. The mixed texture significantly reduces wear due to the interaction of the two textures.

4.5.2 Test Load's Impact on Samples with Variable Texture

The impact of the experimental loading on the mixed texture sample's tribological performance is covered in this section. These trials maintained a

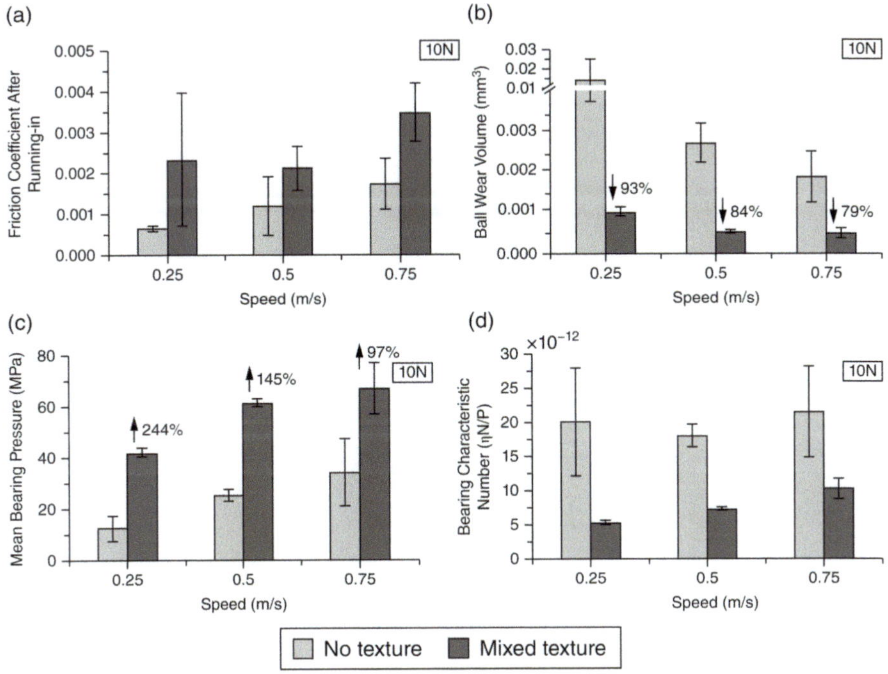

FIGURE 4.10 (a) Friction coefficient variation in textured and non-textured samples. (b) Wear reduction rate peaking at 88% under 15 N load. (c) Average contact pressure of scar, with a 191% lift at 15 N. (d) Relationship between load and bearing characteristic number, showing improved load-bearing capacity in mixed textures.

constant speed of 0.5 m/s while varying the weight, 5, 10, 15 N. The experiment's results are presented in Figure 4.10. In Figure 4.10a, the friction coefficient is displayed. The sample with texture exhibits a drop and growth of the mixed texture, whereas the sample without texture displays a mild variation in friction coefficient at different weights. The blended texture can greatly decrease wear under different loads. Figure 4.10b illustrates that under a high load, 15 N, the wear reduction rate reaches its maximum of 88%. The average contact pressure of scar as a function of test load is shown in Figure 4.10c. Figure 4.10d illustrates the relationship between load and bearing characteristic number. Two figures show how mixed textures depending on load can significantly improve load-bearing capacity. The highest degree of lift under 15-N weight is 191% (Figure 4.10c). In addition, the characteristic number of mixed textures bearing, 8.15 ± 10 is significantly less than non-textured sample, 19.90 ± 10.

The more wear particles generated during the running-in process, the higher the load. Wear can be reduced by using mixed textures to effectively retain wear debris [27, 28]. The greater the ability of texture to reduce wear at high loads, the greater its effect on the accumulation of wear residues as load increases.

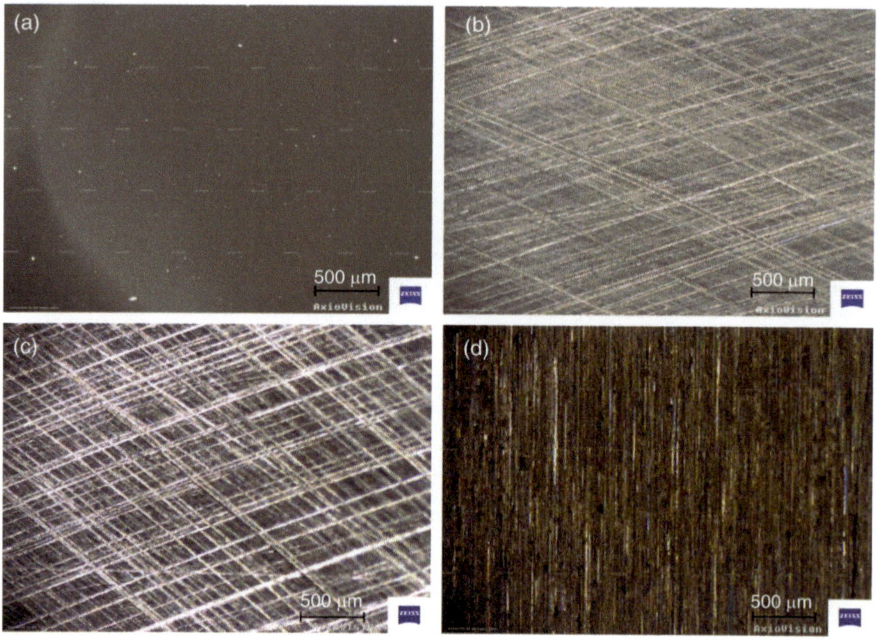

FIGURE 4.11 Original morphologies of four kinds of surfaces: (a) polished surface; (b) textured surface 1; (c) textured surface 2; (d) ground surface [29] [open access].

4.5.3 Testing Speed Effect on Various Textures

Tests are carried out at different speeds, 0.25, 0.5 and 0.75 m/second and with a constant load, 10 N. Figure 4.11 shows the surface topography of the original polished, finished, and milled workpiece.

4.6 MECHANISMS OF WEAR

Figure 4.12 depicts the friction coefficient's typical progression under various pv values, ranging from 0.185 to 1.85 MPa·m/s. The friction coefficient's mean value and standard deviation. Figure 7 shows that, at about 0.40, the polished surface has the maximum friction coefficient. A significant sticky component of the tribological coupling is caused by the polished surface's low surface roughness (Ra = 0.02 μm) [28]. The distribution of peaks and valleys on the surface for various structured surfaces is also responsible for this outcome. Most of textures can reduce wear, while the textures in other literature increase wear due to sliding wear process of self-conjugated silicon nitride in water, which will be discussed in the next section. The wear and friction characteristics of a material's surface can be modified by texturing the surface. With conformal plane contact, it reduces wear and friction rates exceptionally well. On other hand, most studies have shown that it increases wear and reduces friction coefficient during

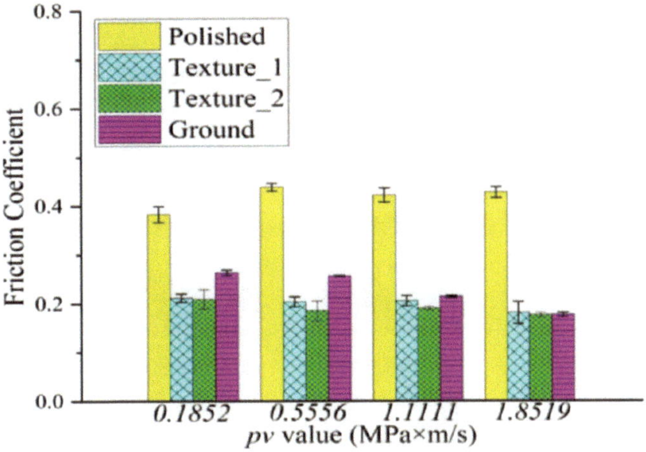

FIGURE 4.12 Mean and standard deviation of the friction coefficient of four kinds of surface patterns in three repeated experiments under different *pv* values [29] [open access].

initial point contact. To determine whether surface texturing could reduce the wear of silicon nitride when sliding in test water, starting from first point contact, ball-on-disc sliding tests are conducted in this study. The study has examined the influence of three texture parameters: depth, texture area and diameter. Most textures are found to be able to reduce the wear of Si_3N_4 after the running-in procedure [30]. The results in Figure 4.13 and 4.14 comparison show that the wear scar diameter in oil is smaller than in water, since oil has a higher viscosity than water, higher dynamic pressure can be generated in the oil experiment to reduce wear [31]. The following formula is used to calculate wear efficiency [12], which helps to more clearly illustrate how texture affects wear. If the texture efficiency is greater than zero, it can effectively reduce wear; if the texture has an efficiency less than zero, wear will be increased.

4.6.1 Brief Description of How Each Texture Affects Wear

In addition, the oil experiment experiences less wear than the water experiment due to the extractive ingredient that can form a protective film layer under high load [32–36]. As shown in Figure 4.15, some of the CF-PEEK wear debris, especially the CF debris, is captured by the grooves, which reduces the wear of the three-body scraper. As a result, the PEEK particles are transferred to a friction layer on the textured stainless-steel surface and act as a solid lubricant, reducing the friction coefficient on the textured surface. Moreover, since the textured surface created with higher texture densities and wider valley depth has lower Mr2, the friction coefficient of tuexured_2 is slightly smaller than that of textured_1 under the conditions of large pv value ranging from 1.11 to 1.85 MPa m/s. It is believed that the groove texture on surfaces with low Mr2 value during movement is more capable of capturing the wear debris.

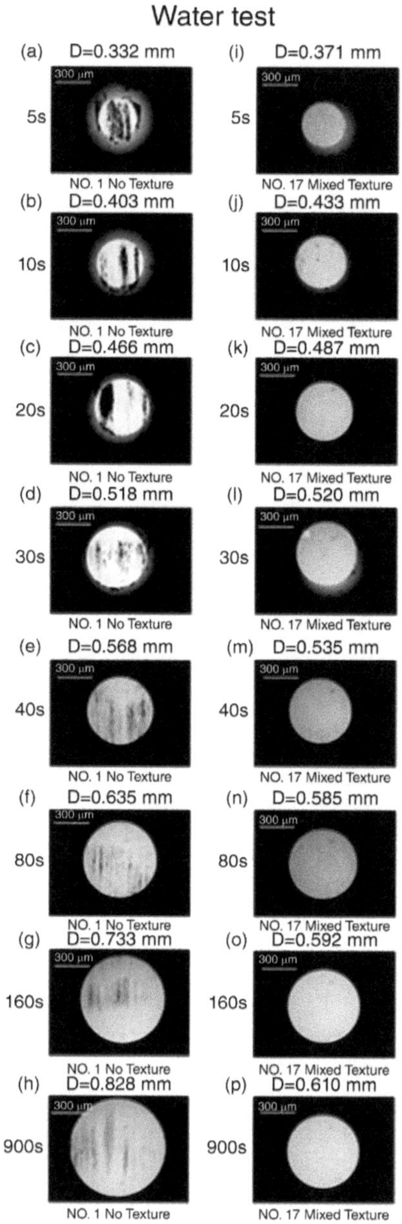

FIGURE 4.13 Optical micrograph showing wear marks on ball sample used to test oil at different times without texture and with mixed texture: no texture in (a)–(h), (i)–(p) uneven texture (c) Wear scar diameter in oil vs. water. (d) Effect of texture parameters on Si_3N_4 wear. (e) Dynamic pressure influence. (f) Wear efficiency calculation. (g) Texture efficiency and wear reduction. (h) Negative efficiency and increased wear. (i) Wear scar evolution. (j) Textured vs. non-textured wear. (k) Texture area ratio impact. (l) Texture depth influence. (m) Wear trends in oil vs. water. (n) Overall effect of texturing on wear [12].

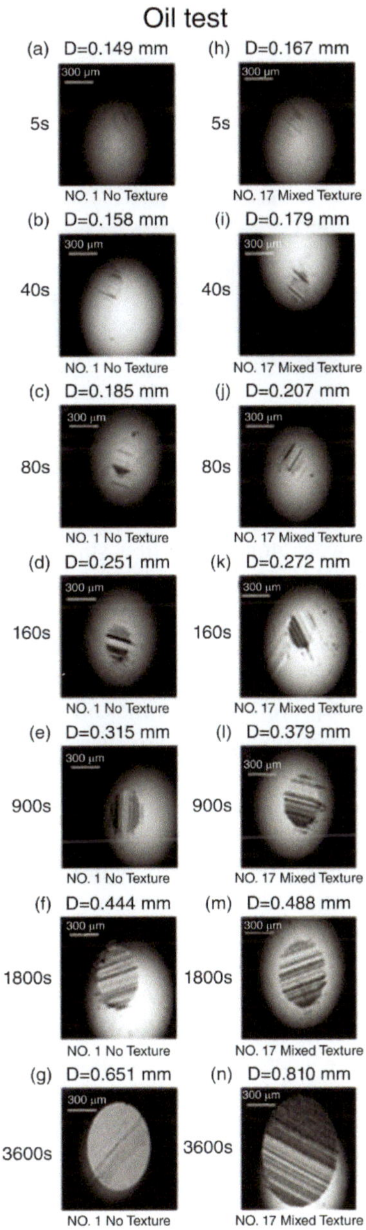

FIGURE 4.14 Wear volumes of mixed texture balls and smooth Si_3N_4 balls depending on test time: Water (a) and oil (b) can be used as lubricants (c) Wear scar diameter in oil vs. water. (d) Effect of texture parameters on Si_3N_4 wear. (e) Dynamic pressure influence. (f) Wear efficiency calculation. (g) Texture efficiency and wear reduction. (h) Negative efficiency and increased wear. (i) Wear scar evolution. (j) Textured vs. non-textured wear. (k) Texture area ratio impact. (l) Texture depth influence. (m) Wear trends in oil vs. water. (n) Overall effect of texturing on wear [12].

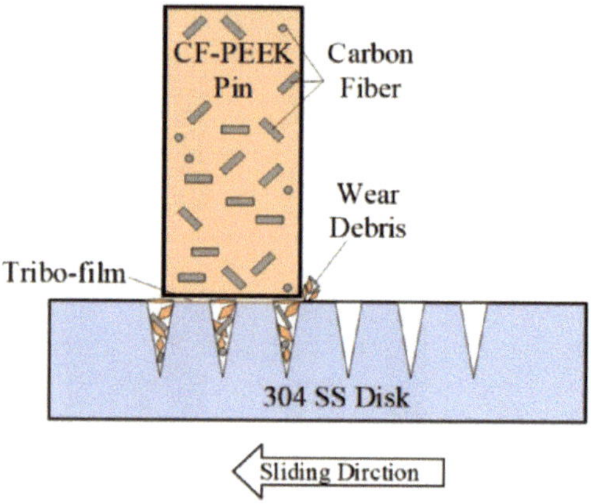

FIGURE 4.15 Schematic diagram of the friction reduction mechanism by surface texture [29] [open access].

Texturing increases wear during the first stage of running-in when exposed to water and soil, as can be seen in Figure 4.16. However, the influence of texture in an oil environment continues to increase, and in a water environment, it changes from increasing wear to decreasing wear, as wear progresses [37]. Under oil and water conditions, the two slightly different terminals suggest different wear mechanisms [38, 39]. For the first time, oil is used in the experiment. The textured sample wears more severely due to stress concentration around the edge of the texture [15, 40, 41]. Abrasive wear appears as a result of fact that there are several scratches on the surface of the ball from the oil

For a sample of different materials, there is first slight decreasing then increasing. Although the maximum friction coefficient of mixed texture is only 0.0035, it is not able to significantly reduce the friction coefficient. Using mixed textures can significantly reduce wear. The highest percentage of wear reduction, 93% occurs at low speed, i.e., 0.25 m/second. According to this result, texturing has the greatest effect on reducing wear at low speeds (Figure 4.11b). More pressure can be applied to a sample with a non-uniform texture.

At low speed, 0.25 m/s, where this effect is most pronounced, the mixed-texture sample can withstand stresses that are 244% greater than the maximum value for the untextured sample (Figure 4.11c). This result is consistent with the findings of Wang et al. [42]. Figure 4.11d shows the bearing performance evaluated at different speeds. The characteristic indicator for the sample with different textures (7.59 × 10) is significantly lower than that of the non-composite sample (19.74 × 10). In this way, the fabric can effectively increase the load-bearing capacity.

The textured sample has more wear debris, which aggravated the wear as its initial wear is higher than that of the non-textured sample [16, 17, 43]. The surface roughness is too high to generate the additional dynamic pressure required to minimize wear [44].

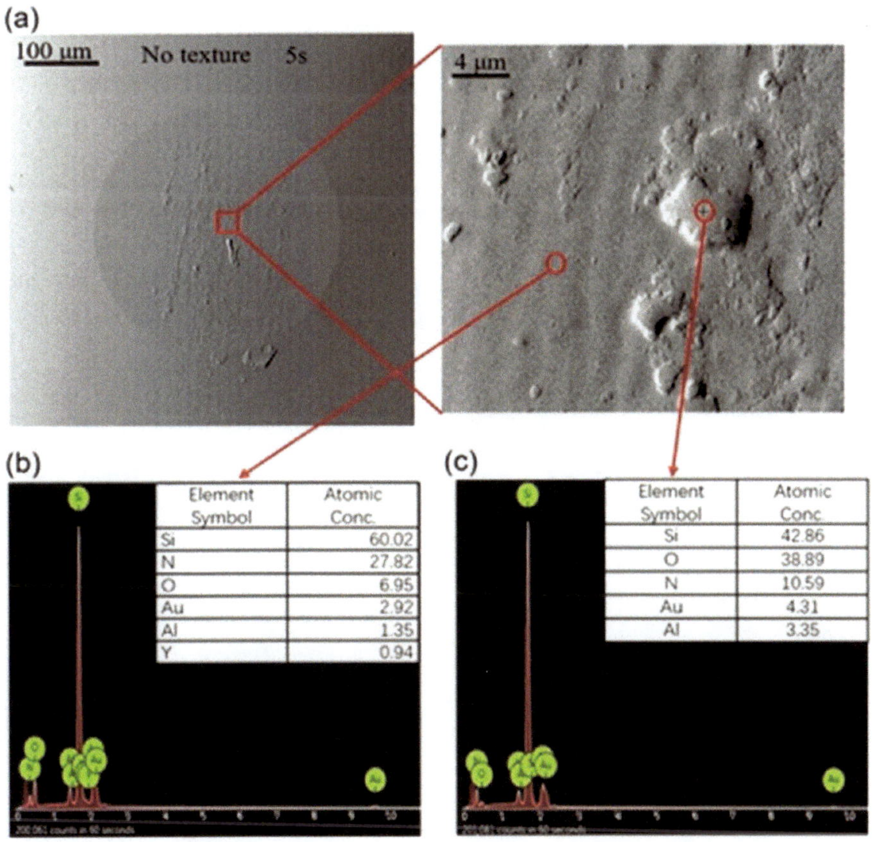

FIGURE 4.16 (a) SEM image of untextured Si_3N_4 ball wear in water (5 seconds, 10 N, 0.25 m/s). (b) EDS analysis showing surface composition changes. (c) Comparison of wear progression in textured vs. untextured samples, with wear initially increasing and then decreasing in water [12].

Signs of wear from textured and non-textured balls immersed in water. As can be seen in Figure 4.17 and 4.18, the samples are examined using an energy dispersive spectrometer (EDS) and SEM. SEM (Figure 4.19) and EDS (Figure 4.20) are used to examine the wear marks on the discs of textured and non-textured samples immersed in water [12]. According to SEM results in Figure 4.17a, numerous inserts can be seen on the worn surface of ball [45–48]. There is a lot of oxygen in the attachments (Figure 4.17b).

$$Si_3N_4 + 6H_2O = 3SiO_2 + 4NH_3 \tag{4.1}$$

$$SiO_2 + H_2O - Si(OH)_4 \tag{4.2}$$

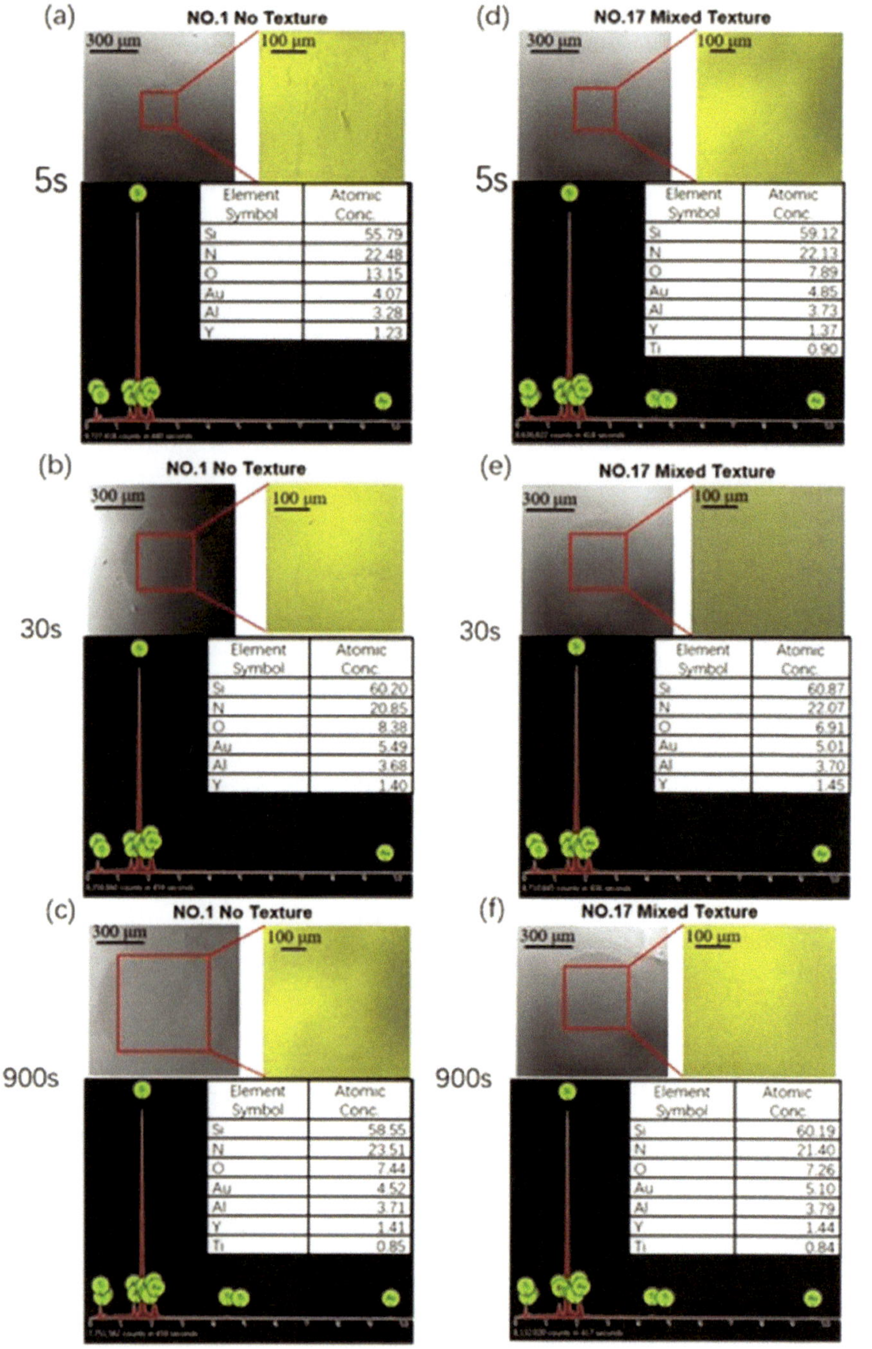

FIGURE 4.17 SEM and EDS data displaying a worn track's surface scan of a ball sample taken during testing in water at various dates, both with and without texture: (a)–(c) are texturally empty; (d)–(f) heterogeneous texture [12].

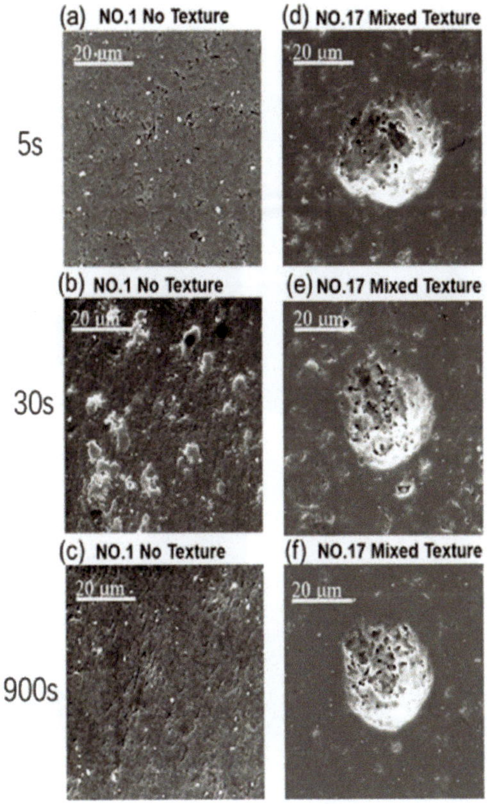

FIGURE 4.18 SEM photographs showing surface scans of a worn track with mixed texture and without texture on a disc sample obtained by testing in water at different times. No texture in (a)-(c); (d)-(e) uneven texture. (f) a different stage in the analysis, possibly related to further changes in the worn surfaces or a comparison between different types of wear textures [12].

Figure 4.18, which presents the results of SEM and EDS of non-textured and differently textured samples at different stages of running-in process, illustrates the change in components of worn surface of ball during the procedure. A textured sample always contains some silica associated with wear pattern, whereas a mixed-texture sample has a smoother wear pattern. This result suggests that because texture can concentrate lubricant, silica formed on mixed-texture surfaces can react rapidly with water to form water-dissolved silicic acid [12, 43–48]. On another hand, on non-textured surfaces, silica does not have time to react with water and remains on worn surface. During the running-in phase, when the two worn surfaces are increasingly in contact, silica and water react faster on the non-textured surface due to the greater water content between the two surfaces [44, 49]. On non-textured surface, this results in a decrease in oxygen content as shown in Figures 4.18a–c. For mixed texture

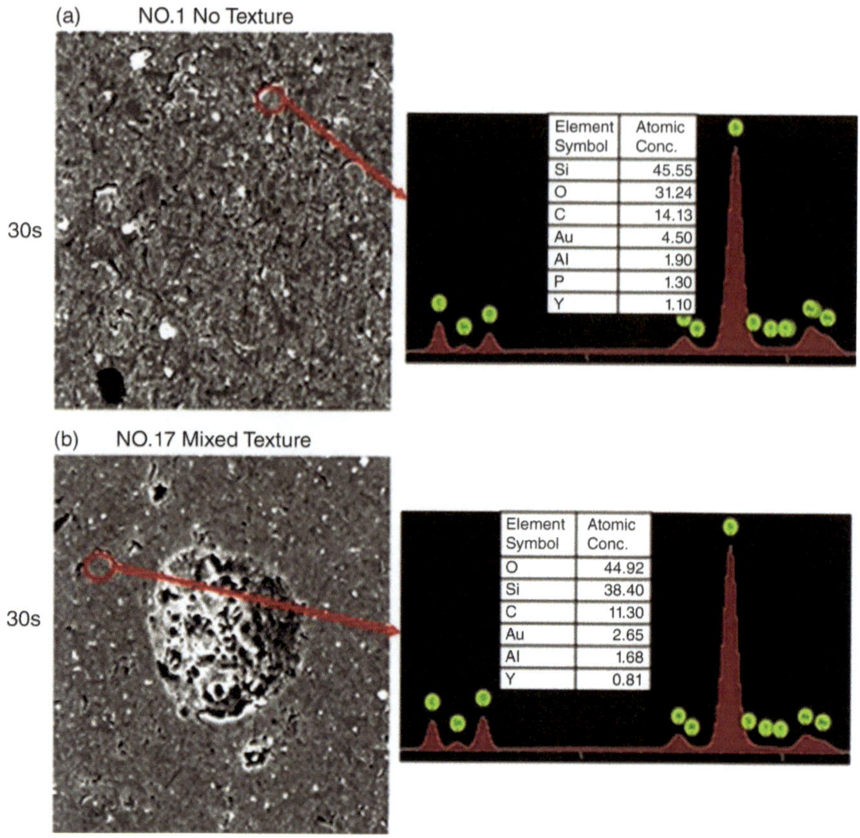

FIGURE 4.19 EDS images of disc specimen from a water a running-in time of 30 seconds, showing surface scanning of worn track with mixed texture (b) and no texture (a) [12].

samples, the oxygen content on the worn surface remains virtually unchanged during the running-in process due to the rapid reaction of silica and water (Figure 4.18d–e).

The wear of disc samples with mixed and non-textured texture at different running-in periods is shown in Figure 4.19. On a non-textured and mixed-textured sample, it is impossible to distinguish noticeable signs of wear during running-in. The surprisingly short break-in period, 5 to 30 seconds, suggests a rapid evaluation method, potentially involving thermal or mechanical testing. Figure 4.19 shows the lamellar attachment of silica, indicating that the running-in process occurs when the chemical equation (4.1) is satisfied [42, 49–52]. Both this conclusion and result are shown in Figure 4.17 are confirmed by each other. Visible scratches appear on the surface of samples without texture at a running-in time, 900 s, while on the surface of samples with a mixed texture, there are no scratches at all. The idea that texture can successfully reduce break-in wear is

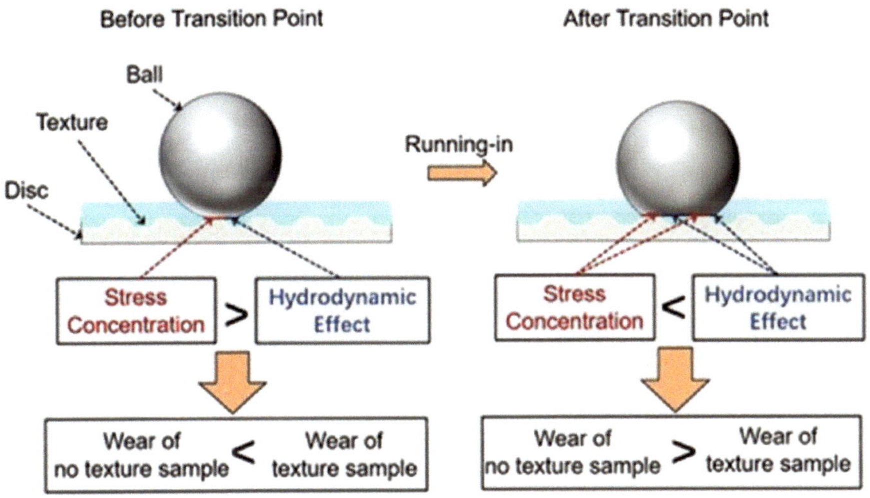

FIGURE 4.20 Schematic representation of wear during water testing during the running-in process [12].

supported by this result. Figures 4.17–4.19 show that although the initial wear of the textured sample in water is also higher than that of the non-textured sample (Figure 4.13), tribochemical reactions lead to a smoothing of the worn surface during the running-in process [53–57]. If the surface is smooth enough, it can withstand more dynamic pressure than a non-textured surface. The texture helps reduce wear in situations where the impact of dynamic pressure exceeds that of stress concentration. The underwater mechanism involved in the running-in process is shown in Figure 4.20. There are two types of surface texturing effects; the hydrodynamic effect [58, 59] and stress concentration effect [18, 60–65] created by the surface texturing edge. The influence of stress concentration exceeds the influence of dynamic pressure before the transition point of the running-in process [66–68].

The purpose of this study is to investigate the possibility of reducing silicon nitride wear during ball-on-disc sliding tests in test water by applying surface texturing starting from the initial point of contact. It has examined three aspects of texture; diameter, texture area and depth. During the process, it is found that most fabrics are less corroded by Si_3N_4. To further determine the wear mechanism, the wear of balls at different times and lubricants during the running-in process is investigated. An overview of specific results is presented below:

- Most single-texture samples can improve the load-bearing capacity and successfully reduce the wear of the balls. The best tribological performance is shown by a sample with a texture thickness of 11 microns. Compared to the non-textured sample, it shows 74% reduction in wear and 94% increasing in load-bearing capacity.

- The mixed-texture sample showed superior wear resistance and load-bearing capacity compared to the single-texture samples. Compared to untextured sample, the mixed-texture sample exhibited 244% greater load-bearing capacity and 93% less wear.
- Oil-water lubricant comparison tests were carried out to investigate the process of lowering textural wear during running-in.

In this study, the titanium–graphite composites treated with low-temperature metal powder are processed by changing the output power and scanning speed of a nanosecond laser and then analyzed for surface morphology, roughness and wettability. Experimental results have showed that surfaces processed with the same ET values exhibit different surface morphology and surface roughness parameters when using different combinations of laser power and scanning speed. Figure 4.21 provides a documentation by displaying 3D maps of surfaces P2 and V2 processed with ET = 1 mJ and surfaces P4 and V1 processed with ET = 2 MJ. Surface V1, is created with a higher output power and scan speed to show Rsm parameter, 93.62 μm, while ET = 2 mJ on surface P4 gave a low value, 66.08 μm, likely due to low scan speed and 50% pulse overlap. The ratio of trench void volume (Vvv) to peak material volume (Vmp) shows significant variations, reaching values, 2.1 for surface P4 and 4.6 for surface V1 [69].

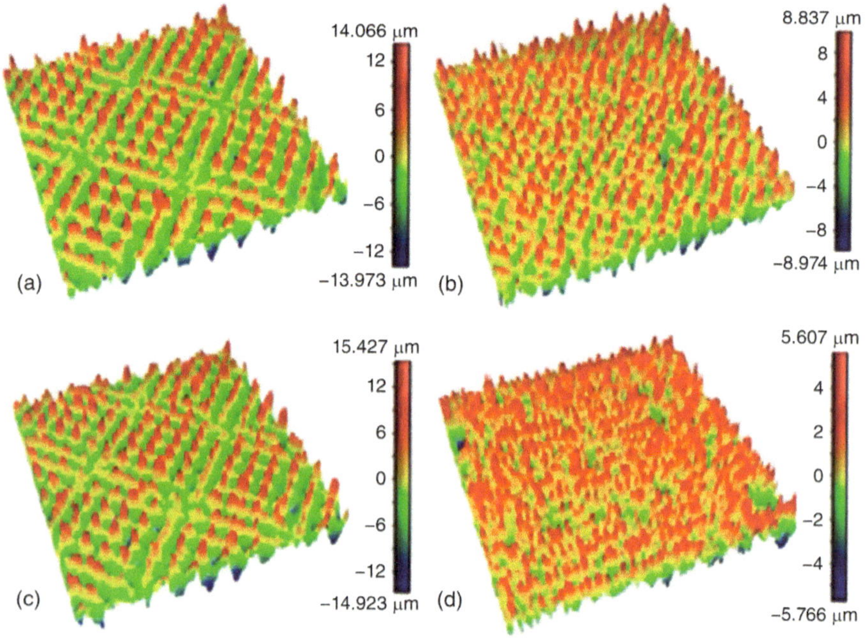

FIGURE 4.21 3D Surface Maps of P2 (a), V2 (b), P4 (c), and V1 (d) Showing the Effects of Varying ET, Laser Power, and Scan Speed on Morphology and Roughness [69] [open access].

Figure 4.21 shows the surface morphology and 3D reconstruction of specimens. The specimens with P180 and P400 showed grinding textures in a regular direction, according to representative SEM image (Figure 4.1a). In addition, several deep cavities were visible between the grinding textures and P4000 specimen have showed a small scratch. In addition, the P4000 surface has a narrower interval between scratches than its equivalent P180 and P400 surfaces. In addition, 3D reconstructed surfaces of specimens with different roughness scales are shown in Figure 4.1b. The P180 and P400 surfaces have showed various valleys and protrusions relative to the reference plane. The P180 surface is ranged from −11.7% to 13.5%, and P400 surface is ranged from -15.3% to 15.3% between μm and 5.4 μm [69].

Later, some abrasive particles form a transfer layer and act as solid lubricants, while other particles settle in the groove texture with the machining stage, reducing the effects of micro-cutting. As shown in Figure 4.22, the maximum amplitude of the AE signal of textured_1 is smaller than that of textured_2 because the Ra of textured_1 is smaller than that of textured_2. As a result, the micro-cutting effect of the textured surface removes relatively little wear debris, reducing the AE RMS value [70].

The Sa and Sq values of sample surfaces are shown in Figure 4.22c. In a long-term immersion test conducted under quasi-static conditions, the long-term degradation of pure magnesium samples with different roughness levels in simulated biological fluids is evaluated. At the same time, the degradation kinetics of samples in fake saliva is observed. Figure 4.23 shows the surface morphology and chemical composition after immersion in different fluids. As shown in

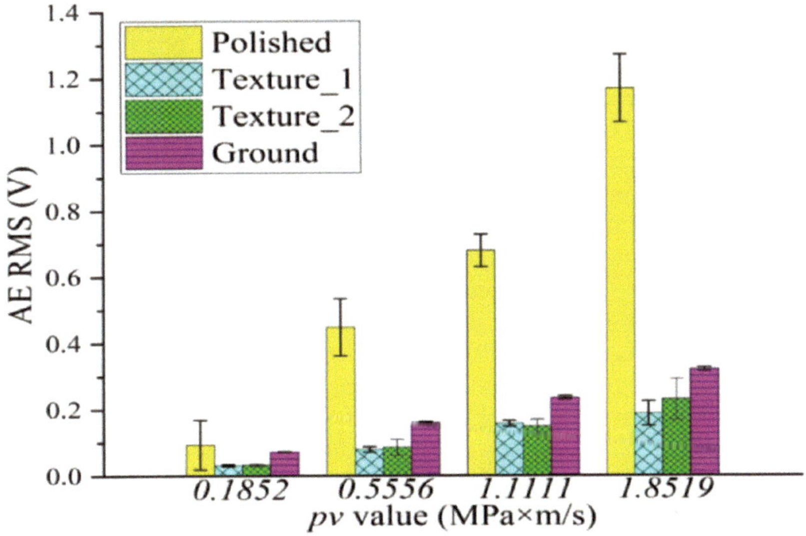

FIGURE 4.22 Mean and standard deviation of AE RMS values of four kinds of surface patterns in three repeated experiments under different *pv* values [29] [open access].

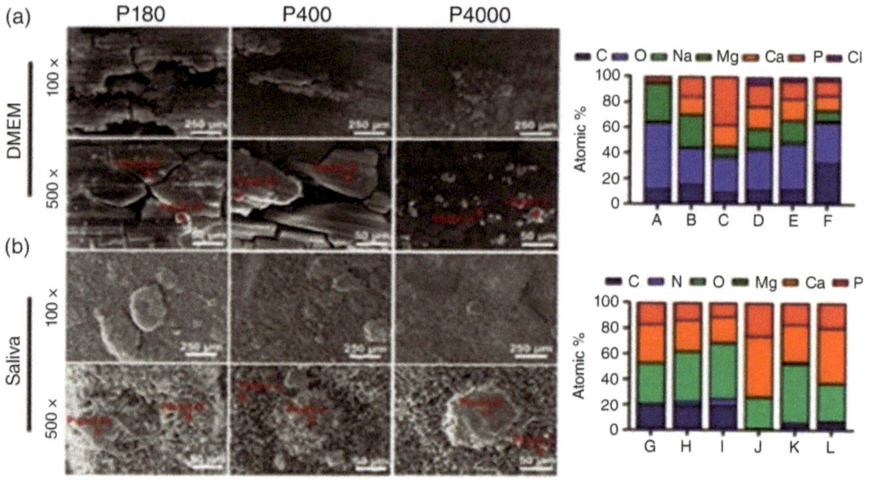

FIGURE 4.23 Chemical composition of samples and surface morphological properties after immersion test. (a) Representative SEM images at 100× and 500× magnification and scale bars at 250 μm and 50 μm, respectively. The surface chemical composition concerning immersion in simulated body fluid is marked with red dots A-F. (b) Representative SEM images at 100× and 500× magnification and scale bars at 250 μm and 50 μm, respectively. The surface chemical composition is measured by EDS, corresponding to red dot G-L for immersion in simulated saliva [70] [open access].

Figure 4.23a, after 30 days of immersion in simulated biological fluids (DMEM + 10% FBS), different surface roughnesses have showed multi-layered morphology, while the surfaces of P180 and P400 groups have showed grinding texture. Interestingly, compared with their other roughness counterparts, P400 and P4000, the platelet structure area on surface of P180 is larger. At the same time, cavities and gaps appeared on substrate of P180 and P400 groups [70].

The microstructure of surface with and without UV treatment is remarkably similar, as shown in Figure 4.24c, d.

The surface of the titanium dioxide wear scar is possible source of friction force change due to microscopic examination is conducted in a sealed environment, which means that the friction process involves direct contact between the silicone beads and the titanium dioxide wear scar plate. During the macroscopic friction experiment of UMT5, the only possible lubricant source is PAO4 oil. Therefore, it is speculated that the phenomenon of decreasing friction coefficient after UV treatment in the macroscopic and microscopic oil lubrication experiments is due to increased adsorption of PAO base oil molecules on surface of photosensitive inorganic materials, which forms a lubricating layer with good lubricating effect after UV treatment [71].

This chapter shows how surface texturing can significantly improve the performance of silicon nitride under water lubrication. It can also reduce wear. Wear reduction and load-bearing capacity can be significantly increased by carefully

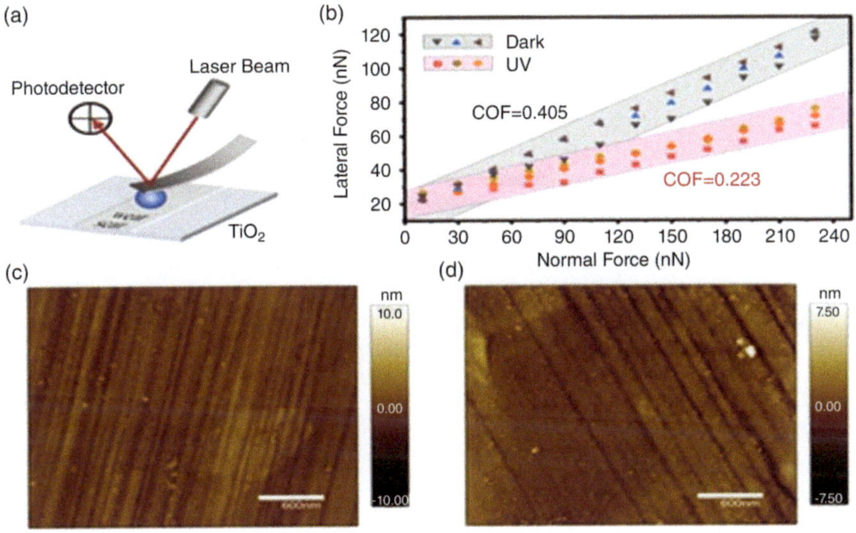

FIGURE 4.24 Schematic diagram of AFM test and friction coefficient values of TiO_2 wear microcrack after AFM studies. The microstructure of wear zone (c) with and without UV treatment (d) [71] [open access].

selecting and optimizing textural properties. The knowledge gained from this study advances surface engineering and materials science by providing a useful basis for further research and real-world applications.

4.7 PROBLEMS

1. Explain why the wear rate of flexible materials is significantly higher than wear rate of laser surface.
2. Which interfaces and surfaces are critical to the depth and impact of fabric from tribal perspective?
3. How would the relative wear rates of these materials differ if the textures are compared; single versus mixed versus spiral and if the abrasive particles are garnet versus silicon carbide?
4. Why does choosing a different fabric affect the test speed?
5. When two objects in contact have different mechanical strengths, what causes the transfer of wear mechanisms?

REFERENCES

1. Monti, J.M., Stewart, J.A., Custer, J.O., Adams, D.P., Depla, D., & Dingreville, R. (2023). Linking simulated polycrystalline thin film microstructures to physical vapor deposition conditions. *Acta Mater, 245*, 118581.
2. Chen, J., An, Y., Liu, G., Chen, G., Zhao, X., & Jia, L. (2022). Tribological performance and thermal stability of a novel cold sprayed nanostructured Ni-based lubrication coating. *J. Therm. Spray Technol, 31*, 1702–1711.

3. Uno, K., & Ohta, M. (2023). Control of Al composition of mist chemical vapor deposition grown -(AlGa)2O3 alloy thin films by acetylacetonation of Al ion. *Jpn. J. Appl. Phys*, *62*, SF1026.
4. Zhao, Q., Hu, G., Huang, R., Qiang, L., & Zhang, X. (2022). Iodide-induced galvanic replacement of nickel film on copper as activator for electroless nickel-phosphorus plating. *Mater. Lett*, *314*, 131833.
5. El Hadad, A., García-Galván, F., Mezour, M., Hickman, G., Soliman, I., Jiménez-Morales, A., Barranco, V., Galván, J., & Perry, C. (2020). Organic-inorganic hybrid coatings containing phosphorus precursors prepared by sol–gel on Ti6Al4V alloy: Electrochemical and in-vitro biocompatibility evaluation. *Prog. Org. Coat*, *148*, 105834.
6. Perrin, F., Ziarelli, F., & Dupuis, A. (2020). Relation between the corrosion resistance and the chemical structure of hybrid sol-gel coatings with interlinked inorganic-organic network. *Prog. Org. Coat*, *141*, 105532.
7. Samanta, S., Jin, S., Lee, C.-H., Lee, S.-S., Struyf, H., Kim, T.-G., & Park, J.-G. (2023). Effect of ammonium halide salts on wet chemical nanoscale etching and polishing of InGaAs surfaces for advanced CMOS devices. *Mater. Sci. Semicond. Process*, *161*, 107469.
8. Li, H., Chen, Z.W., Fiedler, H., & Ramezani, M. (2021). Wear behaviour of n ion implanted Ti-6Al-4V alloy processed by selective laser melting. *Metals*, *11*, 1639.
9. Watanabe, Y., Capaz, R.B., & Simao, R.A. (2023). Surface characterization using friction force microscopy and the Jarzynski equality. *Appl. Surf. Sci*, *607*, 155070.
10. Kuznetsova, T., Lapitskaya, V., Warcholinski, B., Gilewicz, A., & Chizhik, S. (2020). Friction and wear of ZrN coatings under conditions of microcontact using atomic-force microscopy. *J. Frict. Wear*, *41*, 287–294.
11. Mehta, A., Vasudev, H., & Singh, S. (2024). Sustainable manufacturing approach with novel thermal barrier coatings in lowering CO2 emissions: performance analysis with probable solutions. *International Journal on Interactive Design and Manufacturing (IJIDeM)*, *18*(8), 5427–5439.
12. Zou, H., Yan, S., Shen, T., Wang, H., Li, Y., Chen, J., & Meng, Y. et al. Efficiency of surface texturing in the reducing of wear for tests starting with initial point contact. Wear, *482*, 203957.
13. Mahana, D., Mauraya, A.K., Singh, P., & Muthusamy, S.K. (2023). Evolution of CuO thin films through thermal oxidation of Cu films prepared by physical vapour deposition techniques. *Solid State Commun*, *366–367*, 115152.
14. Berglund, K., Rodiouchkina, M., Hardell, J., Kalliorinne, K., & Johansson, J. (2021). A novel reciprocating tribometer for friction and wear measurements with high contact pressure and large area contact configurations. Lubricants, 9(12), 123.
15. Rashid, M.-U., Tomkowski, R., & Archenti, A. (2022). Effect of surface pre-treatment on the adhesion between HiPIMS thick Cu:CuCNx coating and WC-Co Shim. *Coatings*, *12*, 1484.
16. Liu, D., Li, J., Jin, X., Mu, R., & Yang, W. (2022). Effect of coating pre-treatment on surface recrystallization of DD6 single crystal. *Materials*, *15*, 7004.
17. Kim, H.-K., Kim, S.-M., & Lee, S.-Y. (2022). Influence of interlayer materials on the mechanical properties and thermal stability of a CrAlN coating on a tungsten carbide substrate. *Coatings*, *12*, 1134.
18. Binding, S.C., Pernik, I., Gonçales, V., Wong, C.M., Webster, R., Cheong, S., Tilley, R.D., Garcia-Bennett, A., Gooding, J.J., & Messerle, B.A. (2019). Simultaneous functionalization of carbon surfaces with rhodium and iridium organometallic complexes: Hybrid bimetallic catalysts for hydroamination. *Organometallics*, *38*, 780–787.

19. Mehta, A., Vasudev, H., Singh, S., Prakash, C., Saxena, K.K., Linul, E., Buddhi, D., & Xu, J. (2022). Processing and advancements in the development of thermal barrier coatings: A review. *Coatings*, *12*, 1318.
20. Park, S.M., Nahm, S., & Oh, Y.S. (2021). Thermal durability of ytterbium silicate environmental barrier coating prepared by suspension plasma spray. *J. Korean Ceram. Soc*, *58*, 192–200.
21. Molak, R., Topolski, K., Spychalski, M., Dulińska-Molak, I., Morończyk, B., Pakieła, Z., Nieuzyła, L., Mazurkiewicz, M., Wojucki, M., & Gebeshuber, A.; et al. (2019). Functional properties of the novel hybrid coatings combined of the oxide and DLC layer as a protective coating for AZ91E magnesium alloy. *Surf. Coat. Technol.*, *380*, 125040.
22. Marin, E., Lanzutti, A., Nakamura, M., Zanocco, M., Zhu, W., Pezzotti, G., & Andreatta, F. (2019). Corrosion and scratch resistance of DLC coatings applied on chromium molybdenum steel. *Surf. Coat. Technol*, *378*, 124944.
23. Kai-Hui, D., Ying-Wei, S., & En-Hou, H. (2021). Research progress on the preparation of wear-resistant micro-arc oxidation coatings on titanium alloys. *Surf. Technol*, *50*, 57–65.
24. Yang, C., Zhu, J., Cui, S., Chen, P., Wu, Z., Ma, Z., Fu, R.K., Tian, X., Chu, P.K., & Wu, Z. (2021). Wear and corrosion resistant coatings prepared on LY12 aluminum alloy by plasma electrolytic oxidation. *Surf. Coat. Technol*, *409*, 126885.
25. Peñas-Caballero, M., Martín-Cordón, J., Barranco, V., Galván, J.C., Santana, M.H., Lopez-Manchado, M.A., & Verdejo, R. (2023). Corrosion control by autonomous self-healing epoxy coatings based on superabsorbent healing agents. *Prog. Org. Coat*, *182*, 107600.
26. Huangfu, H., Guo, X., Li, N., Xiong, Y., Huang, Y., Zhang, J., & Wang, L. (2023). A smart composite coating with self-reporting and self-healing functions to enhance corrosion protection for magnesium alloys. *Prog. Org. Coat*, *181*, 107598.
27. Prochazka, J., Studeny, Z., & Nguyen, C.H. Wear resistance enhancement of case-hardening steels by utilizing plasma nitriding in military applications. In Proceedings of the (2021) METAL 2021—30th Anniversary International Conference on Metallurgy and Materials, Brno, Czech Republic, 26–28 May 2021; pp. 612–620.
28. Garois, S., Daoud, M., & Chinesta, F. (2023). Data-driven inverse problem for optimizing the induction hardening process of C45 spur-gear. *Metals*, *13*, 997.
29. Jin, Z., Zhang, X., Hou, Z., Yao, Z., & Shen, H. (2022). Tribological properties of carbon fiber-reinforced PEEK against 304 stainless-steel with reticulate surface texture. Materials, 15(24), 8789.
30. Aprilia, A., Maharjan, N., & Zhou, W. (2023). Decarburization in laser surface hardening of AISI 420 martensitic stainless steel. *Materials*, *16*, 939.
31. Zhang, H., Wu, S., Ao, N., Zhou, L., Wang, L., & Liu, D. (2023). The effect of gradient order on impact toughness of carbon S38C axle steels with induction hardening. *Eng. Fail. Anal*, *149*, 107254.
32. Pei, L., Lin, D., Yuan, S., Lu, R., Bai, Z., Sun, Y., Zhu, Y., Jiang, Y., Zhu, J., & Wang, H. (2023). A multifunctional and long-term waterborne anti-corrosion coating with excellent 'hexagonal warrior' properties. *Chem. Eng. J*, *457*, 141158.
33. Zhang, Y., Wu, H., & Guo, S. (2022). Sandwich-Structured surface coating of a silver-decorated electrospun thermoplastic polyurethane fibrous film for excellent electromagnetic interference shielding with low reflectivity and favorable durability. *ACS Appl. Mater. Interfaces*, *14*, 40351–40360.
34. Hernán, D.M.V., Aida, M.E., & Gilberto, B.G. (2020). Detailed study of the electrochemical behavior of low-reflectivity TiAlN coatings. *Surf. Innov*, *9*, 293–307.

35. Picas, J., Menargues, S., Martin, E., & Baile, M. (2023). Cobalt free metallic binders for HVOF thermal sprayed wear resistant coatings. *Surf. Coat. Technol, 456*, 129243.
36. Ganvir, A., Jahagirdar, A.R., Mulone, A., Örnfeldt, L., Björklund, S., Klement, U., & Joshi, S. (2021). Novel utilization of liquid feedstock in high velocity air fuel (HVAF) spraying to deposit solid lubricant reinforced wear resistant coatings. *J. Mater. Process. Technol, 295*, 117203.
37. Bouabibsa, I., Lamri, S., & Sanchette, F. (2018). Structure, mechanical and tribological properties of Me-doped diamond-like carbon (DLC) (Me = Al, Ti, or Nb) hydrogenated amorphous carbon coatings. *Coatings, 8*, 370.
38. Anderson, N., Zhang, B., Meng, A., Zhang, X., Lijesh, K., Khonsari, M., & Meng, W. (2023). Cr-containing diamond-like carbon coatings deposited on 316 stainless steel substrates: Characterization and interfacial fracture toughness measurements. *Surf. Coat. Technol, 462*, 129462.
39. Peng, J., Liao, J., Zhang, G., Huang, J., & Qiu, X. (2023). Microstructure evolution and mechanical performance of tetrahedral amorphous carbon coatings with dense droplets during annealing. *Mater. Chem. Phys, 301*, 127697.
40. Kiryukhantsev-Korneev, P.V., Sheveiko, A.N., & Petrzhik, M.I. (2019). Approaches to increasing the adhesion strength of hard wear-resistant nanostructured coatings based on the Ti–B–(Cr, Si, C)–N system. *Prot. Met. Phys. Chem. Surfaces, 55*, 502–510.
41. Vereschaka, A., Tabakov, V., Grigoriev, S., Aksenenko, A., Sitnikov, N., Oganyan, G., Seleznev, A., & Shevchenko, S. (2019). Effect of adhesion and the wear-resistant layer thickness ratio on mechanical and performance properties of ZrN—(Zr,Al,Si)N coatings. *Surf. Coat. Technol, 357*, 218–234.
42. Wang, X., Kato, K., Adachi, K., & Aizawa, K. (2003). Loads carrying capacity map for the surface texture design of SiC thrust bearing sliding in water. *Tribol. Int, 36*(3), 189–197.
43. Sizova, O.V., Teryukalova, N.V., Leonov, A.A., Denisova, Y.A., Novitskaya, O.S., & Kolubaev, A.V. (2022). Influence of the intermediate layer on the adhesion and friction of titanium-based nitrided coatings on a copper substrate. *Russ. Phys. J, 65*, 1123–1129.
44. Meng, R., Deng, J., Duan, R., Liu, Y., & Zhang, G. (2019). "Modifying tribological performances of AISI 316 stainless steel surfaces by laser surface texturing and various solid lubricants. *Optics & Laser Technology, 109*, 401. [Avcı, B.; Kazmanlı, K.; Evren, B.; Ürgen, M. Contribution of galvanic coupling with TiN, TiAlN, and CrN to the corrosion of steel in neutral and acidic chloride solutions. Mater. Corros. 2023.]
45. Burliaev, D.V., Kozaderov, O.A., & Volovitch, P. (2021). Zinc-nickel alloy coatings: Electrodeposition kinetics, corrosion, and selective dissolution. A review. *Kondens. Sredy Mezhfaznye Granitsy, 23*, 3–15.
46. Shekhanov, R.F., Gridchin, S.N., & Balmasov, A.V. (2020). Electroplating and physicochemical properties of Zinc–Nickel alloy coatings from ammonium oxalate electrolytes. *Surf. Eng. Appl. Electrochem, 56*, 427–431.
47. Pinto, J., Quiroz, D., Delvasto, P., & Blanco, S. (2018). Characterization of a zinc-nickel alloy coating obtained from an electrolytic bath produced with spent batteries as raw materials. *J. Phys. Conf. Ser, 1119*, 012005.
48. Holmberg, K., & Erdemir, A. (2017). Influence of tribology on global energy consumption, costs and emissions. *Friction, 5*, 263–284.
49. Zhang, Z., Yan, D., Liu, X., Li, W., Wang, Z., Wang, Y., Song, D., Zhang, T., Liu, J., & Wang, J. (2023). Self-healing nanocomposite coatings containing organic-inorganic inhibitors functionalized dendritic silica nanocontainers for synergistic corrosion protection of carbon steel. *Colloids Surf. A Physicochem. Eng. Asp, 656*, 130430.

50. Xu, L., Pi, L., Dou, Y., Cui, Y., Mao, X., Lin, A., Fernandez, C., & Peng, C. (2020). Electroplating of thick Hard chromium coating from a trivalent chromium Bath containing a ternary complexing agent: A methodological and mechanistic study. *ACS Sustain. Chem. Eng*, *8*, 15540–15549.
51. Ramelan, A., Fauzi, A., Rachmantyo, R., Sasmita, F., Lidya, L., Eko, A.Z., & Nugroho, A.S. (2020). Effect of PEG addition and temperature on hard chromium coating using eco-friendly trivalent chromium. *AIP Conf. Proc*, *2262*, 040019.
52. Lin, X.-Y., Yu, Y.-H., Tse, M.-M., Cheng, S.-H., Liu, Y.-H., & Lu, K.-L. (2023). Semiconductive (Cu–S)n metal–organic framework incorporated polyimide nanocomposite coatings forming an oxide barrier for durable anticorrosion effects. *Prog. Org. Coat*, *180*, 107562.
53. Ramesh, C., Tyagi, P., Kaswan, J., Yadav, B.S., Shukla, A.K., Senthil Kumar, M., & Kushvaha, S.S. (2020). Effect of surface modification and laser repetition rate on growth, structural, electronic and optical properties of GaN nanorods on flexible Ti metal foil. *RSC Adv*, *10*, 2113–2122.
54. Deng, K., Wang, X., Huang, S., Li, P., Jiang, Q., Yin, H., Fan, J., Wei, K., Zheng, Y., & Shi, J.; et al. (2023). Effective suppression of amorphous Ga2O and related deep levels on the GaN surface by high-temperature remote plasma pretreatments in GaN-based Metal–Insulator–Semiconductor electronic devices. *ACS Appl. Mater. Interfaces*, *15*, 25058–25065.
55. Zhang, S., Huang, M., Wu, Y., Yang, M., Lin, T., He, P., & Paik, K.-W. A study on the oxygen plasma treatment on the peel adhesion strength and solder wettability of SnBi58 based anisotropic conductive films. In Proceedings of the 2019 Electronic Components and Technology Conference, Las Vegas, NV, USA, 28–31 May 2019; pp. 2022–2028.
56. Dong, M., Li, M., & Hang, T. Application of Ultrathin Organic Insulating Films for Electronic Package. In Proceedings of the 2019 IEEE International Conference on Integrated Circuits, Technologies and Applications (ICTA), Chengdu, China, 13–15 November 2019.
57. Kencana, S.D., Kuo, Y.-L., Yen, Y.-W., Schellkes, E., & Chuang, W. Improving the solder wettability via atmospheric plasma technology. In Proceedings of the 2019 IEEE 69th Electronic Components and Technology Conference (ECTC), Las Vegas, NV, USA, 28–31 May 2019; pp. 2067–2071.
58. Yan, C., Byrne, D., Ondry, J.C., Kahnt, A., Moreno-Hernandez, I.A., Kamat, G.A., Liu, Z.-J., Laube, C., Crook, M.F., & Zhang, Y.; et al. (2022). Facet-selective etching trajectories of individual semiconductor nanocrystals. *Sci. Adv.*, *8*, eabq1700.
59. McGillivray, R. Tribology Part 1 The Basics of Tribosystemsand Part 3: Standards, testing, and classification in Tribology, December 12, 2021.
60. Cheng, L., Adinolfi, V., Weeks, S.L., Barabash, S.V., & Littau, K.A. (2019). Conformal deposition of GeTe films with tunable Te composition by atomic layer deposition. *J. Vac. Sci. Technol. A*, *37*, 020907.
61. Ma, Y., Song, J., Wang, X., Liu, Y., & Zhou, J. (2021). Synthesis, microstructure and properties of magnetron sputtered lead zirconate titanate (Pzt) thin film coatings. *Coatings*, *11*, 944.
62. Shamma, K., Albrithen, H., & Alodhayb, A. (2022). Thickness dependence of the response of metal-dioxide-based micromechanical sensors for sensitive gamma-ray detection. *Appl. Radiat. Isot*, *186*, 110225.
63. Beaucejour, R., D'agati, M., Kalyan, K., & Olsson, R.H. (2022). Compensation of the stress gradient in physical vapor deposited All xScxN films for microelectromechanical systems with low out-of-plane bending. *Micromachines*, *13*, 1169.

64. Bai, J., Zhang, J., Eiler, K., Yang, Z., Fan, L., Yang, D., Zhang, M., Hou, Y., Guan, R., & Sort, J.; et al. (2022). Electrochemically fabricated surface-mesostructured cuni bimetallic catalysts for hydrogen production in alkaline media. *Nanomaterials, 12*, 118.
65. Sun, X., Shao, X., Yi, J., Zhang, J., & Liu, Y. (2022). High-efficient carbon dioxide-to-formic acid conversion on bimetallic PbIn alloy catalysts with tuned composition and morphology. *Chemosphere, 293*, 133595.
66. Yang, C., Zhang, B., Xie, X., Li, C., Xu, Y., Wang, H., & Wang, L. (2021). Three-dimensional independent CoZnAl-LDH nanosheets via asymmetric etching of Zn/Al dual ions for high-performance supercapacitors. *J. Alloys Compd, 861*, 157933.
67. Arora, H., Jung, Y., Venanzi, T., Watanabe, K., Taniguchi, T., Hübner, R., Schneider, H., Helm, M., Hone, J.C., & Erbe, A. (2019). Effective hexagonal boron nitride passivation of few-layered InSe and GaSe to enhance their electronic and optical properties. *ACS Appl. Mater. Interfaces, 11*, 43480–43487.
68. Kavinkumar, T., Sivagurunathan, A.T., & Kim, D.-H. (2023). Construction of highly transparent, flexible, and robust solid-state symmetric supercapacitors using NiO electrodes roughened by conformal atomic layer deposition. *Appl. Surf. Sci, 616*, 156453.
69. Chen, H., Qin, Z., He, M., Liu, Y., & Wu, Z. (2020). Application of electrochemical atomic force microscopy (EC-AFM) in the corrosion study of metallic materials. *Materials, 13*(3), 668.
70. Chen, J., Dai, J., Qian, J., Li, W., Li, R., Pang, D., & Xu, S. (2022). Influence of surface roughness on biodegradability and cytocompatibility of high-purity magnesium. *Materials, 15*(11), 3991.
71. Sang, X., Han, K., Zhu, M., & Ma, L. (2024). Friction reduction achieved by ultraviolet illumination on TiO2 surface. *Materials, 17*(7), 1680.

5 Tribological Studies of AISI 316 Stainless Steel Surfaces

5.1 INTRODUCTION

Marine environment is more demanding and complex than the terrestrial environment. Materials used in marine engineering equipment may be subjected to various stresses during installation and maintenance. In addition, corrosion from oil and gas environments, seawater and potential exposure to low or high temperatures play important role. Due to its excellent corrosion resistance, good resistance to chloride erosion and resistance to high temperatures and pressures, 316L stainless steel has extensive application in marine engineering equipment [1]. However, the application of friction parts is limited by poor wear resistance and other problems are due to low surface hardness and load-bearing capacity of 316L stainless steel [2]. Various surface problems have been observed in recent years. Engineering techniques such as laser melting, ion implantation, physical vapor deposition, high-current pulsed ion beam surface modification and laser surface texturing have been used to improve the wear resistance of metal surfaces.

The proposed surface texturing technology challenges the conventional wisdom that "a smoother surface is beneficial for improving the tribological performance of engineering parts", among other surface treatment technologies. Due to its fast processing time, high throughput and excellent controllability, laser surface structuring, also known as laser surface texturing (LST), has attracted much interest among various surface texturing techniques [2–4]. To create regular microtextures, such as micropits, on the surface of a material, a focused high-energy laser beam irradiates the material along a specified path creating microbumps and microgrooves. These texture forms influence the load-bearing capacity of contact surface and hydrodynamic lubrication during friction. They serve as a lubricant reservoir to improve the dynamic pressure lubrication effect, which reduces wear due to accumulation of wear particles when surface is not lubricated [5], high reproducibility, exceptional flexibility [6] and precise control of pattern shape [7]. Periodic surface structures have the ability to generate hydrodynamic pressure [2], accumulate lubricants and trap wear particles [8]. A number of preliminary studies have shown the influence of surface texture orientation and distance and geometry to reduce wear and friction [6]. And, Wang et al. [5] have created microgroove patterns on steel surfaces in a picosecond.

DOI: 10.1201/9781003584735-5

In metal manufacturing industries, machining processes are among the most important manufacturing processes. In technical terms, this procedure involves using a cutting tool to remove material from a workpiece to obtain the desired shape and size. The most important factor in performing effective machining procedures is the cutting tool. Common workpiece materials for cutting tools include ceramics, carbide, high-speed steels and coated cutting tools. Each cutting tool has a unique set of properties, and the way that is used depends on the processing parameters and flow stress of workpiece. Two other deciding factors are price and description of intended use of these cutting tools [9]. High-speed steel (HSS) and carbide-cutting tools are the primary materials used in most industrial components in machining process. The reason for this is that HSS and carbide are hard enough to cut metal. Moreover, carbide and high-speed steel are inexpensive and widely available. To facilitate lubrication and cooling, coolant is used in combination with the use of high-speed steel and carbide.

Although using cutting fluids to reduce temperatures during machining has benefits, it is also harmful to the environment, puts the skin of machine tools at risk and increases operating costs [6]. When the groove spacing is uniformly distributed, the textured surfaces have exhibited good wear and friction properties [7]. The newly created samples with a Penrose-type pattern have shown a slight decreasing in friction coefficient. The tribological performance of textured surfaces can be significantly improved by combining surface texturing with solid lubrication. For example, MoS_2 is deposited on microtextured titanium alloy surfaces [10]. They have concluded that the friction coefficient of titanium alloys can be significantly reduced by creating a MoS_2 coating. In MoS_2 films are also deposited on a rough surface [7]. They have found that the capacitance of MoS_2 increases on textured surfaces occupying half of convex shape. Graphene is obtained in many layers on a rough surface [8]. They have showed that textured surfaces are superior to uncoated steel sections in terms of wear resistance [11]. They have noted that friction decreases steadily as temperature increases. A review of the literature shows that solid lubricants can be used on textured surfaces to significantly reduce wear and friction. WS_2, MoS_2 and other solid lubricants are the most widely used are silver, graphite and CaF_2. They are used as alloying additives in the matrix, foam plastics and plasma-deposited coatings. In this study, we have generated microtextures on the surface of austenitic stainless steel [12]. The solid lubricating particles introduced into the textured samples included MoS_2, graphite, CaF_2 and hBN. This study has investigated how different solid lubricants affect the tribological behavior of textured austenitic stainless-steel surfaces.

5.2 MATERIALS

Stainless steel 316L with dimensions 17 × 17 mm is chosen. Before laser surface weaving, the sample is ultrasonically cleaned in anhydrous ethanol for 10 minutes, then dried and set aside. In this case, the surface of sample is polished to a roughness of approximately Ra 0.05 μm. Mobil PAO8 base oil is subjected to the oil lubrication test and reef crystals synthetic salt water is used for the seawater lubrication test. ASTM 1141–98 is used for the preparation of synthetic seawater. The ratio of water and sea salt is 1:30 by weight.

5.2.1 Recent Progresses and Applications in Laser-Based Surface Texturing Systems

One of the well-known methods for improving the surface topography of engineering materials is laser surface texturing. Recently, several surface texturing methods such as picosecond, femtosecond and nanosecond laser texturing have been used to modify surfaces and investigate the deterministic behavior of textured patterns with different sizes and shapes. Laser surface texturing has been used to improve the lubricity and wear properties, mechanical adhesion at coating-substrate interface and to create superhydrophobic surfaces due to greater accuracy and flexibility of texturing.

The purpose is to discuss the improved tribological performance of laser surface texturing for engineering surfaces. Surface texturing is applied to the samples using a pulsed Nd:YAG laser with a wavelength of $\lambda = 1064$ nm. The surface of sample is processed with a laser operating at a power, 12 W, frequency, 30 kHz and scanning speed, 130 mm/second. The sample's whole surface is covered in grooves spaced equally apart from one another. The chosen groove densities are (in that sequence) 0%, 1%, 2%, 4%, 7%, 14%, 29% and 57%. On rough surfaces, rubbing is utilized to apply solid lubricating coats. The microtextures are filled with solid lubricants. This study used a variety of lubricants, including graphite, calcium fluoride (CaF_2), tungsten disulfide (WS_2) and hexagonal boron nitride (h-BN) (see Figure 5.1).

FIGURE 5.1 SEM images of graphite (a), h-BN (b), CaF_2 (c), and WS_2 (d) [2]. The purpose is to discuss the improved tribological performance of laser surface texturing for engineering surfaces [2].

5.2.2 Wear and Friction Test

The alignment of the microscopic patterns is perpendicular to the sliding direction. The corresponding ball is an AISI 316 steel ball with a diameter of 9.525 mm. The tests have used a standard load of 20 N, a sliding speed of 10 mm/second and a stroke length of 10 mm. All experiments are carried out at room temperature for sixty minutes. Several tests are performed multiple times to ensure reproducibility. The surface morphology is studied using scanning electron microscope (SEM). Energy dispersive X-ray spectroscopy (EDS) is used to study the chemical composition of damaged surfaces. To assess the surface profile using 3D wear tracks, Zeta-20 laser microscope (USA) is used. Then, using the ratio of wear volume to normal load and distance to determine the wear rate.

5.3 MICROTEXTURE CHARACTERISTICS

A typical micropatterned surface topography is shown in Figure 5.2. High-energy laser pulses are used to vaporize materials during laser beam processing. The microscale groove is created by melting and evaporating the target material. The surface has a uniform distribution of microscale grooves (see Figure 5.2a). Along the edges of grooves, laser micromachining creates tiny projections of re-solidified material (see Figure 5.2b). The microchannel in Figure 5.2 is approximately

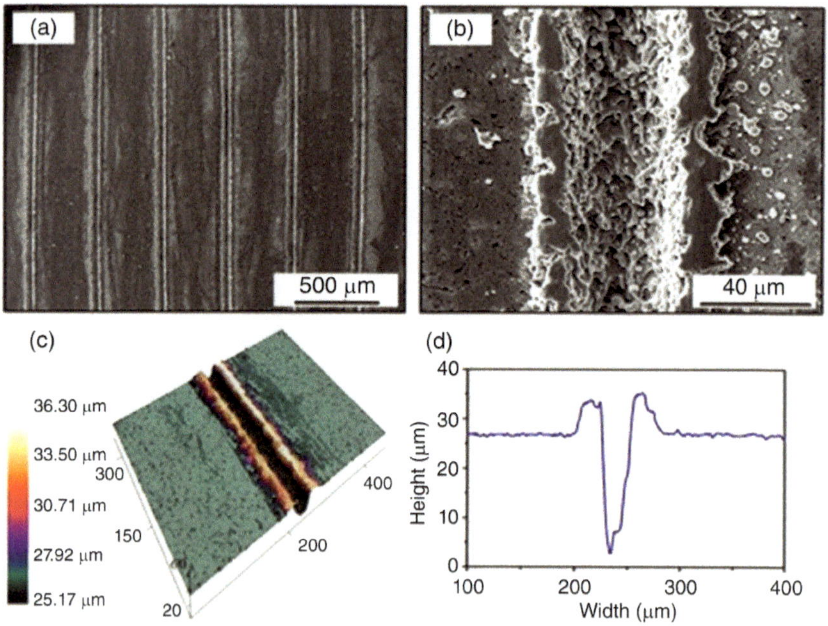

FIGURE 5.2 Micrographs of composite sample: groove surface in three dimensions, (a) microscopic grooves, (b) individual grooves, (c) and (d) groove intersection [2].

57 μm wide and 22 μm deep. Various textured surfaces are coated with several types of solid lubricants. Solid lubricants such as graphite, h-BN, CaF_2 and MoS_2 are used in this study. The rough surface is polished by hand using solid lubricating granules. The characteristic shape of MoS_2 particles polished on an ablation surface is visible in Figure 5.3. MoS_2 powders are poured into microgrooves. Bumps along the edges of the groove may cause differences in measured heights.

Characterization of the morphology and dynamics of biomass materials at nanoscale simultaneously remains a challenging task despite the state-of-the-art characterization techniques. Atomic force microscopy (AFM) offers a unique advantage to overcome this limitation. AFM can produce high-resolution images of biomass materials. Moreover, it can detect mechanical and topographical features on the material surface in real-time due to the nanoscale contact force with the object. This allows for detection without damaging the sample. To study the

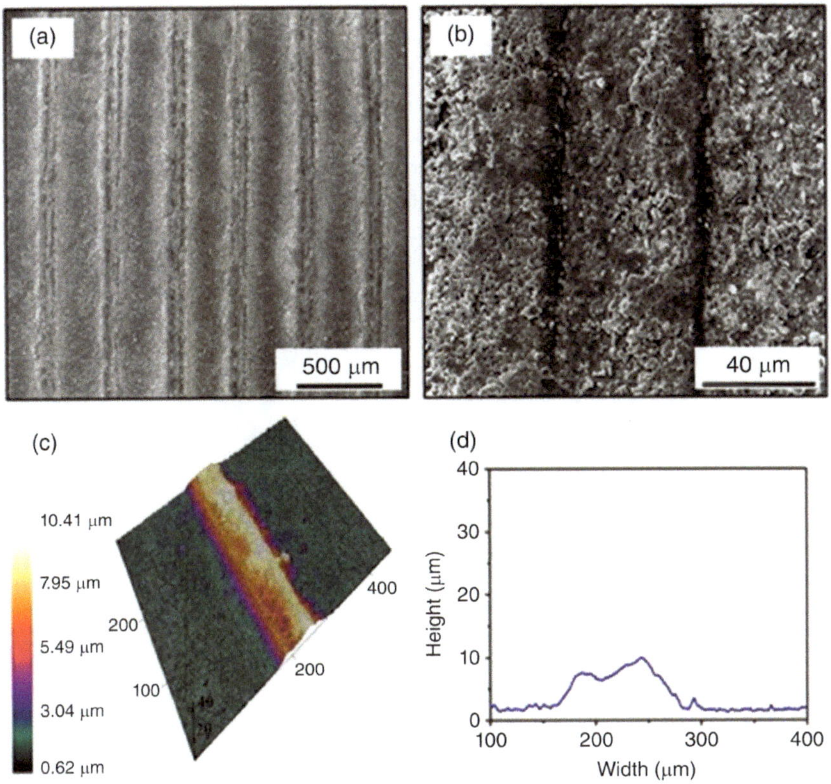

FIGURE 5.3 (a) Polished surface with MoS_2 particles: Refers to a smooth surface where MoS_2 particles are visible, polished with solid ubricating granules. (b) Microgrooves filled with MoS_2 powders: Shows grooves filled with MoS_2 powders, illustrating how the granules settle. (c) Bumps along groove edges: Refers to bumps or irregularities at the groove edges, affecting height measurements. (d) AFM image of rough surface with biomass: AFM image showing detailed surface and mechanical properties of biomass materials at the nanoscale [2].

microscopic characterization of biomass materials, the ability of AFM probes to interact with biomass materials under mild conditions (such as normal temperature, normal pressure, or biological fluids) is of paramount importance. This study reviews the fundamentals of AFM and advances in its application to study the microscopic characterization of biomass materials (including surface adhesion, intermolecular interactions, morphology and micromechanical properties), as well as the use of AFM in combination with Fourier transform infrared spectroscopy (FTIR) and Raman spectrometry [13, 14].

5.4 TRIBOLOGICAL PROPERTIES OF EMBOSSED SURFACES

The graphs in Figures 5.4a and b illustrate the time evolution of friction coefficients for surfaces that are patterned simultaneously. This study is based on the sliding of seven differently patterned surface samples at speeds of up to 3.02 m/second on an abrasive material over a distance of up to 3948.9 m. The seven examples consist of 20-mm wide convex domes attached to flat steel substrates using Bakelite and calcium carbonate. Three particles of quartz sand of different sizes are used as an abrasive in the experiments, reaching a maximum of approximately 0.96 in less than 10 seconds. The friction coefficient gradually falls over time. At final stage, the friction coefficient is equalized [1–4, 15].

Figure 5.4c shows the steady-state friction coefficients for surfaces with and without patterns. Friction is greater on a stable surface without patterns. The error

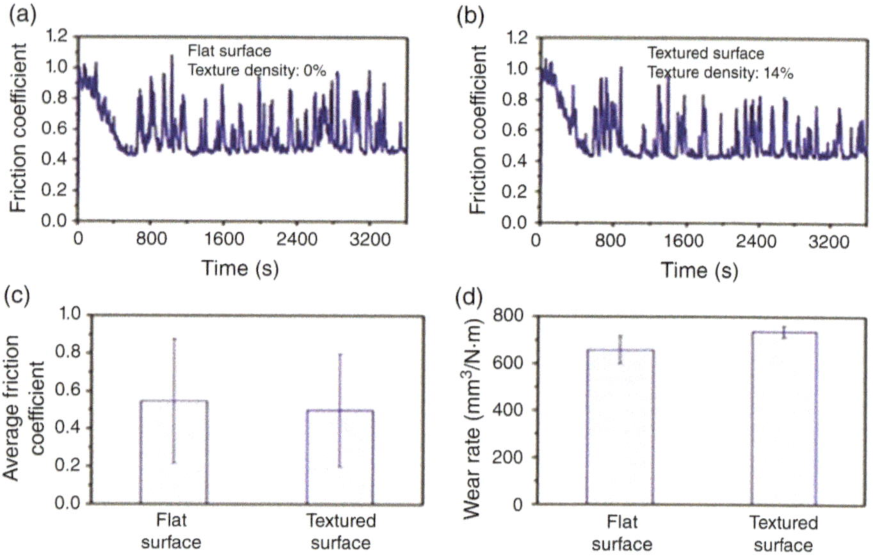

FIGURE 5.4 Comparison of average friction coefficient and wear rate for smooth and textured surfaces, as well as the friction coefficient of textured (a) and (b) flat surfaces, (c) shows the steady-state friction coefficients for surfaces with and without patterns. (d) Microgrooves, the wear rate is slightly higher [2].

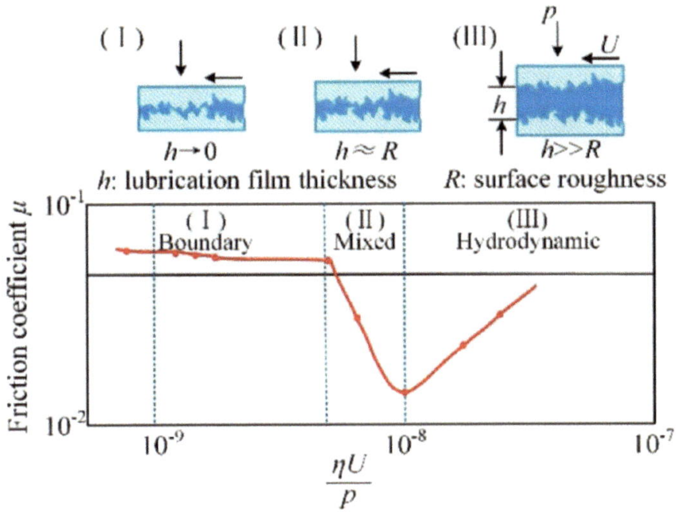

FIGURE 5.5 Stribeck curve [16].

bar represents the random variability of friction coefficient. In addition, vibrations may be less noticeable due to surface texture. However, in case of micro-grooves, the wear rate is slightly higher (Figure 5.4d).

The average friction coefficient and standard error for the different surface friction pairs at each speed must then be determined. Figure 5.5 shows the curved relationship between speed and average friction coefficient with an error bar. The average friction coefficient for the hemispherical hole surface increases monotonically with increasing rotation speed, while that for the smooth surface fluctuates.

5.5 TRIBOLOGICAL CHARACTERISTICS OF PATTERNED SURFACES POLISHED WITH VARIOUS TYPES OF SOLID LUBRICANTS

If we combine Figure 5.6 and four-wear rate conditions of samples, it can be seen that although the wear rates of h-BN, WS_2, CaF_2, and WS_2 interfaces are higher, the conditions of seawater lubrication of surfaces of non-textured and textured wear samples. This may be because, in places of wear, an oxidative friction process of mating GCr15 ball occurs, leading to the formation of brittle oxides on the surface and affecting the mechanical properties of mating surface. 316L stainless steel has good corrosion resistance, so the mechanical properties of sample are not changed significantly, which reduces the corrosion rate of the 316L stainless steel sample (Figure 5.6a). Similar to the results in Figure 5.4b, the friction coefficient of patterned surfaces polished with h-BN solid lubricants shows a similar trend. Approximately 0.6 is the initial friction value. The friction coefficient increases noticeably in Figure 5.6b.

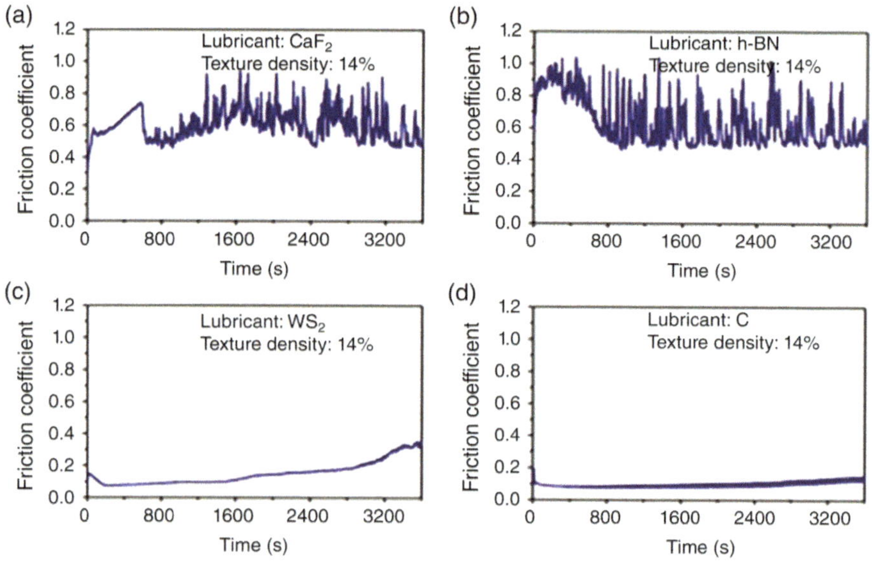

FIGURE 5.6 Relationship between the friction coefficient and time for a relief surface treated with lubricants such as CaF_2 (a), h-BN (b), WS_2 (c), and graphite (d) [2].

Conversely, samples filled with WS_2 lubricants or hard graphite may have low friction coefficients throughout the experiment. The specimen polished with WS_2 solid lubricant has a friction curve starting at about 0.10, as shown in Figure 5.6c. The friction coefficient then stabilizes at 0.16. This is achieved over time by gradually reducing friction. The sample exhibits stable sliding behavior between 240 and 1600 seconds. Then the friction begins to increase steadily. Conversely, the sample is polished with hard graphite [1, 3, 15]. Initially, the friction coefficient of lubricant decreases sharply (see Figure 5.6d). During the remaining service life of sample, the friction coefficient, 0.10 is low. The wear rate and average friction coefficient for patterned surfaces polished with various solid lubricants are shown in Figure 5.7.

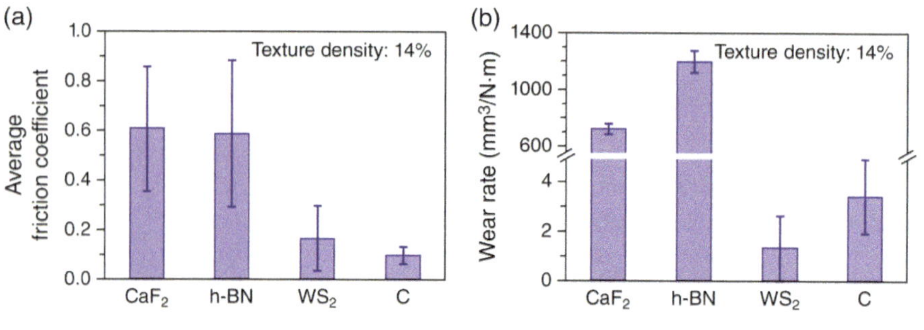

FIGURE 5.7 Analysis of wear rate and friction coefficient of polished patterned surface using lubricants such graphite, (a) Hexagonal Boron Nitride (h-BN), (b) Calcium Fluoride CaF_2, and WS_2 [2].

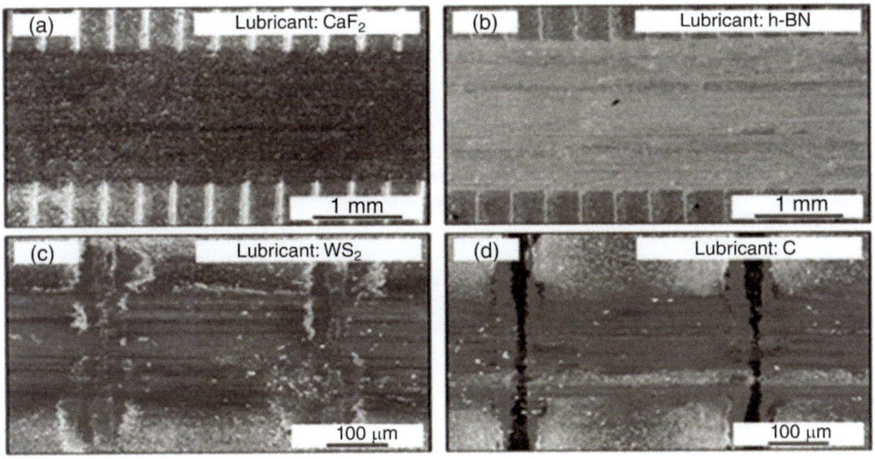

FIGURE 5.8 SEM images of wear patterns on textured surfaces polished with solid lubricants of (a) graphite, (b) CaF_2, (c) h-BN, and (d) WS_2 [2].

The average friction coefficient is significantly lower when polishing patterned surfaces with WS_2 or graphite solid lubricants than when polishing patterned surfaces with CaF_2 or h-BN solid lubricants. The surface ground and polished with graphite have the lowest average friction coefficient in this study. Moreover, the sample treated with CaF_2 or h-BN solid lubricant shows a much higher wear rate compared to the samples treated with WS_2 or graphite solid lubricant [2–4, 15].

Wear traces of patterned samples polished with solid graphite lubricants h-BN, CaF_2 and WS_2 are shown in Figure 5.8. Significant wear damage occurs on a patterned surface polished with hard graphite lubricants CaF_2 or h-BN (see Figure 5.8a, b). Although there are signs of wear, the fine surface texture has completely disappeared. In addition, large amounts of corrosion residues are still present on wear marks. On another hand, the patterned surface polished with WS_2 luminescent materials or hard graphite still shows some signs of wear (Figure 5.8c, d).

The hemispherical hole surface is in a state of hydrodynamic lubrication within the test speed range, while the smooth surface is in a state of transition from mixed lubrication to hydrodynamic lubrication, as shown by the classic Strebek curve (Figure 5.9). The evolution trend and shape of the curve are identical to those for the friction pairs; the lower sample is made of 316 L, while the upper sample is made of CF/PEEK – under identical working conditions as those in reference [17], but because the sample material was switched, the values decreased by 0.03 to 0.04.

In addition, EDS analysis shows a powerful C signal at position C. According to Figures 5.9b and c of the EDS study, a continuous film of solid lubricant covers the entire wear path. The weathered surface has several brittle cracks. Figure 5.9c shows more localized peeling of the solid lubricant layer than Figure 5.9b.

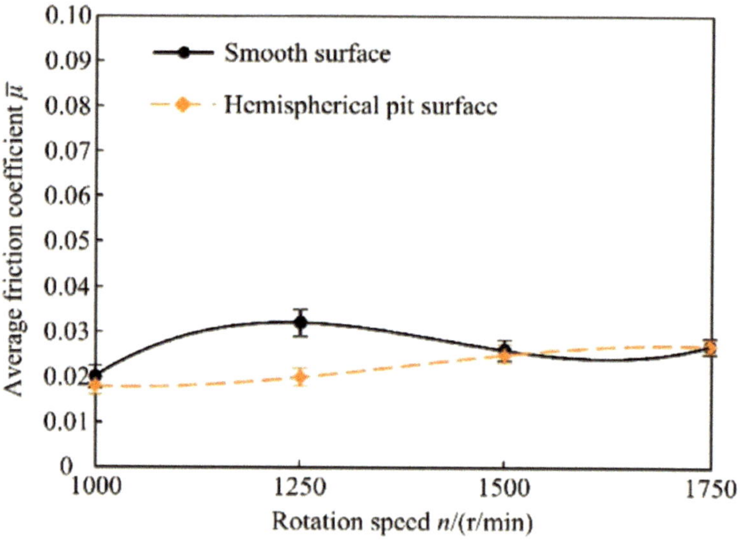

FIGURE 5.9 Curve dependence of the average friction coefficient on the rotation speed for different friction pairs [16].

The higher the degree of solids separation, the higher the wear rates associated with a thicker lubricant layer (see Figure 5.7b). By integrating the pressure distribution on the upper surface of the seawater film, the bearing capacity can be determined. The bearing capacity of the hemispherical pit surface is 3.5% larger than that of the smooth surface, as shown in Figure 5.10. This indicates that the pits increase the bearing capacity and reduce the resistance through the hydrodynamic lubrication effect [18–21].

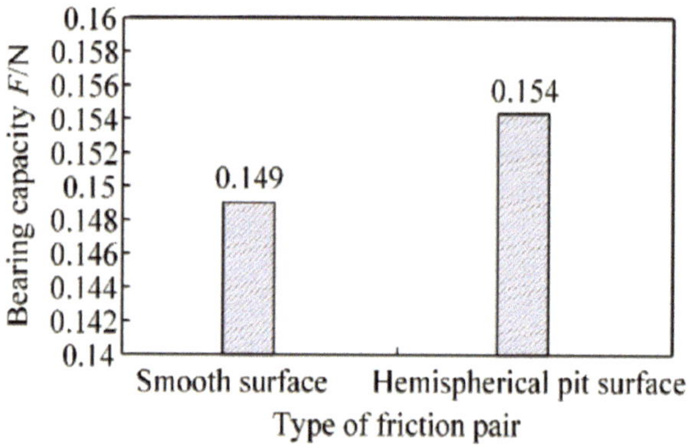

FIGURE 5.10 Histogram of the bearing capacity for different types of friction pairs [16] [open access].

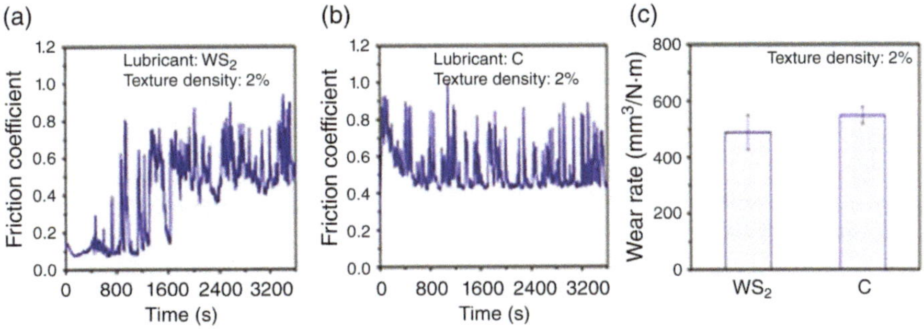

FIGURE 5.11 Comparison wear rate and friction coefficient. A comparison of wear rate and change in friction coefficient over time is shown for relief surface polished using (a) and (b) graphite lubricants (c) wear rate and tungsten disulfide (WS_2) [5].

5.6 EFFECT OF FINE GROOVE DENSITY ON TRIBOLOGICAL PERFORMANCE

Figure 5.11 illustrates the change in friction coefficient. In this case, the density of microgrooves is set to 2%. The friction coefficient of a polished specimen using WS_2 solid lubricant starts at about 0.10 as shown in Figure 5.11a. Initially, the friction coefficient is low and stable for approximately 850 seconds. In this case, the friction coefficient increases sharply. When conflict arises, the process becomes unstable. Final stage friction coefficients typically range from 0.48 to 0.49. Friction curve for a sample polished with graphite and vice versa. When using solid lubricants, a clear trend can be seen (see Figure 5.11b). The friction coefficient at the beginning of test is about 0.1. Then the friction coefficient increases rapidly; In the first 80 seconds, it reaches a value, 0.85. However, the value of friction coefficient steadily decreases during the operation stage. Finally, at this stage, the average sliding friction value is about 0.51. Figure 5.11 compares the wear of two samples mentioned above (c). The wear rate of WS_2 solid lubricant sample is 11% lower than sample polished with graphite solid lubricant [2, 5, 6, 9]. The cross-section shown in Figure 5.12 confirms this. The worn surface of the drilled sample, polished with solid graphite lubricant, shows an increasingly deeper wear [17, 22–24].

The effect of groove density on the average friction coefficient of patterned specimens is shown in Figure 5.13. The error bar displays the friction coefficient between the maximum and minimum values during the test. The patterned sample gives a lower value than full-width (FW) at sample when the material density is low. The average friction coefficient of the patterned sample is 9% lower than machined sample at a texture density, 14%. The Basic method for abrasion or corrosion may be caused by when a given force is applied repeatedly, the contact component the material and isotopes produce abrasive particles, fragmentation and crushing [7, 11, 25–28]. As the groove density increases, the average friction coefficient increases [8, 29, 30]. The dependence of texture density on the

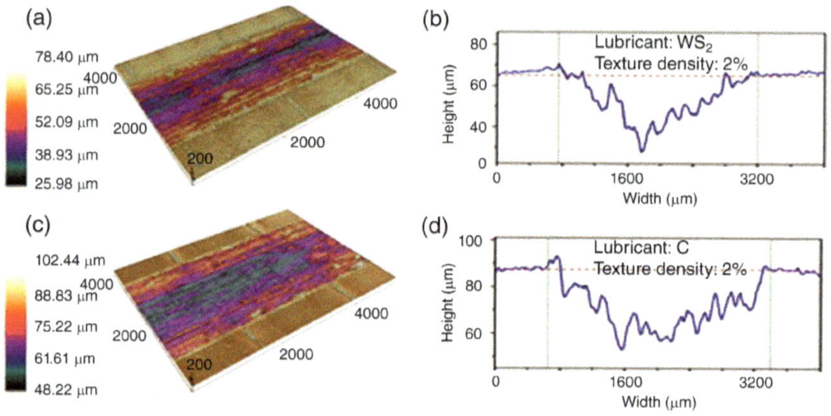

FIGURE 5.12 (a) The friction coefficient starts at around 0.1, then increases rapidly, reaching 0.85 in the first 80 seconds. It then steadily decreases during the operation stage, with an average sliding friction value of about 0.51 by the end. (b) The wear comparison between the two samples mentioned above shows that the wear rate of the WS_2 solid lubricant sample is 11% lower than that of the graphite solid lubricant sample. (c) The cross-section image confirms this, showing deeper wear on the drilled sample polished with graphite solid lubricant. (d) This figure also illustrates the progressively deeper wear observed in the sample polished with graphite solid lubricant [5].

average friction coefficient for patterned samples polished with various lubricants is shown in Figure 5.14. The average friction coefficient of a flat specimen is lower than raised specimen with the corresponding texture density (see Figure 5.14a). In Figure 5.14b, it can be seen that the average friction coefficient of samples operates with h-BN solid lubricant that is higher than flat samples, despite the change in texture density. Both samples are polished with WS_2 graphite or lubricant [2].

Compared to flat sample, it shows a significant reduction in the average friction coefficients (Figure 5.14c and d). Meanwhile, the average friction coefficients of

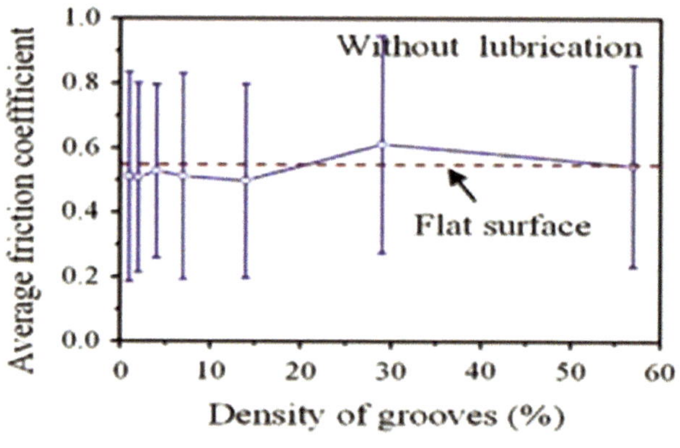

FIGURE 5.13 How the average friction coefficient of surface varies with texture density. The average friction coefficient of a flat surface is shown as a dotted horizontal line [2].

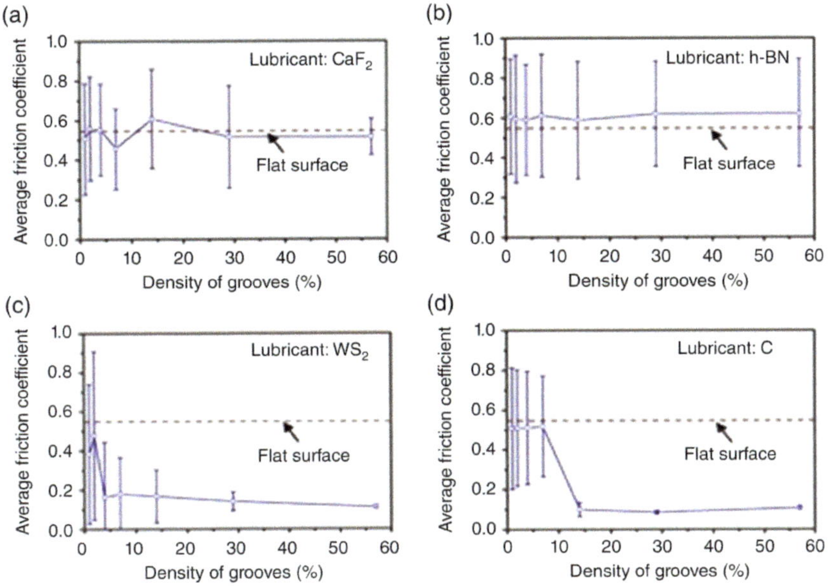

FIGURE 5.14 How the average friction coefficient of surface polished with CaF_2 (a), h-BN (b), WS_2 (c) and solid lubricant (d) depends on texture density. The average coefficient of friction for a flat surface is shown as a dotted horizontal line [2].

two samples paint a similar picture. In addition, relief specimens polished with WS_2 solid lubricant provide excellent results in terms of friction reduction with low texture density [12, 31–35].

5.7 SURFACE TEXTURE

Cracks in the material's surface texture at edges inevitably occur as a result of laser surface texturing process. Laser-processed irregularities are now usually polished directly, but this approach does not take into account the effect on friction pairs [36–38].

This study used three different texture patterns grooves, square pits and round pits prepared on the surface of 316L stainless steel to study the tribological performance of polished and unpolished surfaces under dry friction, oil and seawater conditions. The results have showed that the polished textured surface of 316L stainless steel suffered more abrasion and primarily caused adhesive wear, while the unpolished textured surface abraded less and primarily caused abrasive wear. The friction behavior of smooth and soft samples is similar (see Figure 5.4). AISI 316 Because stainless steel is not very hard, it is easy to wear. When sliding, a lot of wear residues are formed. As a result, noticeable changes in friction are observed in both samples. However, the average friction coefficient of the composite sample is still lower than flat sample [2].

This helps explain the thin fabric that can collect wear and tear. The microtexture suffered significant damage during corrosion tests due to their low corrosion resistance (see Figure 5.5). This process produces a large number of wear particles.

The friction coefficient becomes unstable due to particles of third body. Wear particles may be trapped by the thin fabric. Some wear debris has been removed from the sliding surface. The average friction coefficient of the patterned surface is slightly higher than flat surface when the pattern density is high as seen in Figure 5.13. This can be explained by a linear decreasing in the contact area with increasing [2].

Contact pressure increases simultaneously with a decreasing in contact area. The friction coefficient and surface pattern deteriorate faster due to increased contact pressure. It is interesting to note that the sample polished with different lubricants (see Figure 5.6). The average friction coefficient did not decrease for the samples polished with CaF_2 solid lubricant, as shown by curve in Figure 5.14a. Since CaF_2 is brittle at room temperature, the average friction coefficient is high [39–47]. As temperature increases, atomic force decreases. Then the sliding plane transformation is simply performed in the sliding direction. But displacement of CaF_2 slip plane at room temperature is relatively difficult. The result is a higher friction coefficient [48–54]. It should be noted that the average friction coefficient of the samples polished with h-BN lubricant is significantly higher than flat samples (see Figures 5.6 and 5.14).

The ability to obtain a uniform coating on a part is one of the challenges in electroplating process. Lack of uniformity results from uneven current density across the surface, which can be affected by much smaller features (such as the threads on a threaded fastener) or the macroscopic shape of the component. The throwing ability of bathtub means its ability to create a uniform coating on parts of complex geometry. It can be controlled to some extent by both chemical additives and coating conditions [55–61]. Despite this, uneven coating thickness is still a problem in many systems; for example, maintaining uniform coating thickness on a fastener's base, flank and thread flange continues to be a major hurdle in mass machining of parts. For friction applications, electroplated chrome plating is durable, 850–1250 V when applied and generally resistant to corrosion in a variety of typical environments. They have a much thicker coating than decorative chrome plating, typically 10 to 500 microns, and applied from a chromic acid bath. Electroplated hard chromium has a fine-grained microstructure, 10 nm with oxide inclusions and often visible microcracks [51–54, 62].

The sliding mechanism wears the surface, preventing the formation of low-friction layer. In addition, the microtextured area is filled with h-BN solid lubricant. Since there is not enough space, it is not possible to store the waste to prevent erosion. Consequently, the contact surfaces experienced significant plastic deformation (Figure 5.8b). It is noteworthy that both samples polished with solid graphite lubricant or WS_2 have showed a relatively low friction coefficient (see Figure 5.6c and d). The patterned surface polished with solid graphite lubricant has showed a greater number of local chips and cracks (see Figure 5.9b and c). Graphite had a lower adhesive strength value than WS_2. In addition, Figure 5.7b shows an example of blue color and wear rate of WS_2. Compared to graphite solid lubricants, the level of solid lubricants is lower. A thick layer forms on the contact surface of steel balls (see Figure 5.10). This triple layer has been removed to improve the lubrication and protect the ball surface from wear during sliding.

On another hand, the sample polished with WS_2 solid lubricant has a lower average friction coefficient than sample polished with graphite solid lubricant when the texture density is low (see Figure 5.14c and d). This helps explain the relatively low bond strength of graphite. This is supported by the early decreasing in friction coefficient of a sample of hard graphite polished with a lubricant (see Figure 5.11). When WS_2 solid lubricant is removed, the friction curve increases significantly from the contact point (Figure 5.11a).

Microgrooves are created using laser ablation on an AISI 316 stainless steel substrate. There are four types of solid lubricants; Rubs into graphite, CaF_2, h-BN, WS_2 and other microscopic patterns [54, 62–64]. These samples are prepared and tribological studies are carried out on them. As a result, the following deductions are made:

- At low texture density, the average friction coefficient of embossed sample is lower than "full-field" (FF) sample. Both samples experienced severe plastic deformation during sliding. The friction coefficients deteriorated as a result of gradual wear of microgrooves in the contact zone.
- Regardless of change in density, textures polished with CaF_2 solid lubricant do not appear to show a decreasing in the average friction coefficient. This makes sense since CaF_2 is brittle at ambient temperatures. At room temperature, displacement of CaF_2 slip plane turns out to be a relatively difficult task.
- The average coefficient of friction is significantly higher for patterned specimens polished with h-BN solid lubricant than for the flat specimens.

The friction coefficients of two patterned specimens polished with hard luminescent graphite or WS_2 are relatively low. The reason for this may be the formation of triple layer of lubricant on the worn surface. In addition, due to relatively weak adhesive strength of graphite, wear and friction increased.

Figure 5.15 shows the fluctuations of the sample contact temperature for different surface friction pairs over the test period at different speeds. During the first

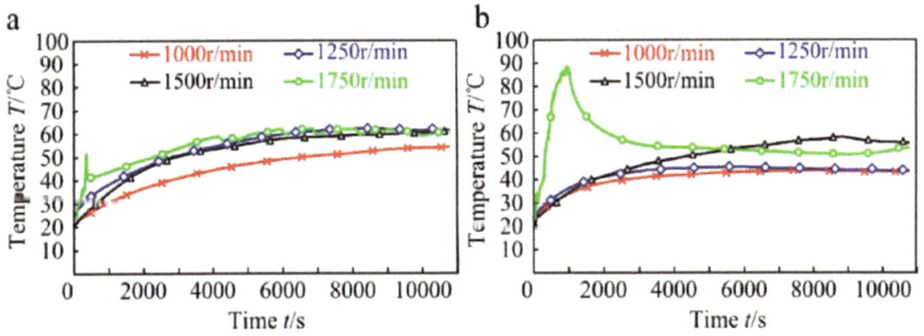

FIGURE 5.15 Variation in the specimen contact temperature with the test time for different friction pairs: (a) smooth surface; (b) hemispherical-pit surface [16] [open access].

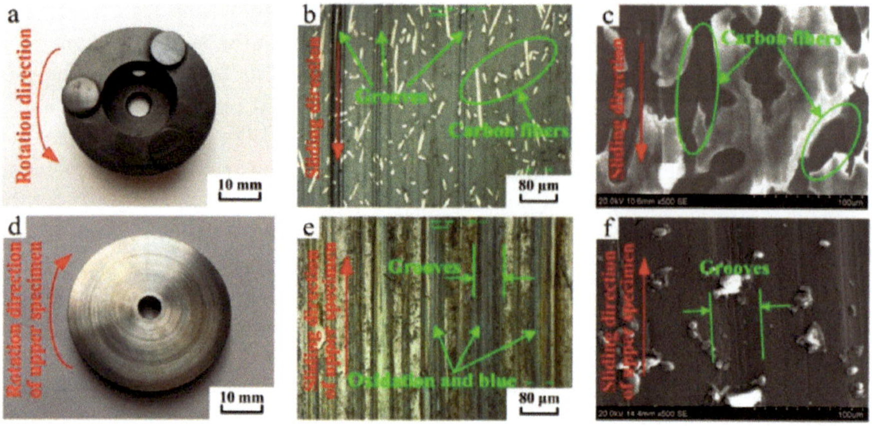

FIGURE 5.16 Wear surface morphology of the smooth-surface friction pair (P = 300 N, n = 1500 r/minute): (a) photograph of CF/PEEK, (b) LSCM image of CF/PEEK, (c) SEM image of CF/PEEK, (d) photograph of 316 L, (e) LSCM image of 316 L, and (f) SEM image of 316 L [16] [open access].

stage, the contact temperature of each bottom sample was lowered from the ambient temperature to a constant value, which was maintained until the end of the tests. The adjustment time for the smooth surface and the steady-state period were 5000 s and 54–62°C, respectively, while those for the hemispherical pit surfaces were 3000 s and 44–58°C. The sample surface temperature rises due to frictional heating, and this rise is reflected in the initial stage of the temperature curve. If the sample temperature is higher than the ambient temperature, it will release its heat as thermal radiation to the saltwater in the area. Figure 5.16 provides a quantitative representation of surface changes analyzed by summing the macrotexture parameters obtained after grinding and ball polishing [64].

There are some important considerations when discussing the corrosion pattern and rate of 316L stainless steel, especially when comparing etched and full field (FF) samples. Under the operating conditions of P = 300 N and n = 1500 r/minute, Figure 5.16 shows the wear surface morphology of the friction pairs with a hemispherical pit and smooth surface. The micro-roughness collided violently when the upper and lower samples slid relative to each other, which corresponds to the high friction coefficient and sample contact temperature during the operating period. Since CF/PEEK has a lower hardness than 316 L, the micro-roughness on its surface was gradually ground to become equal to that on the 316 L surface. As a result, the material on the CF/PEEK surface was favored to push, create plastic flow, and plow new grooves, which changed the original marks left by the machining tool. The plowing effect is the name given to this microscopic cutting action [64].

There are a number of variables that can affect wear behavior when comparing the wear scar morphology on high-luster surface (HLS) 316L stainless steel

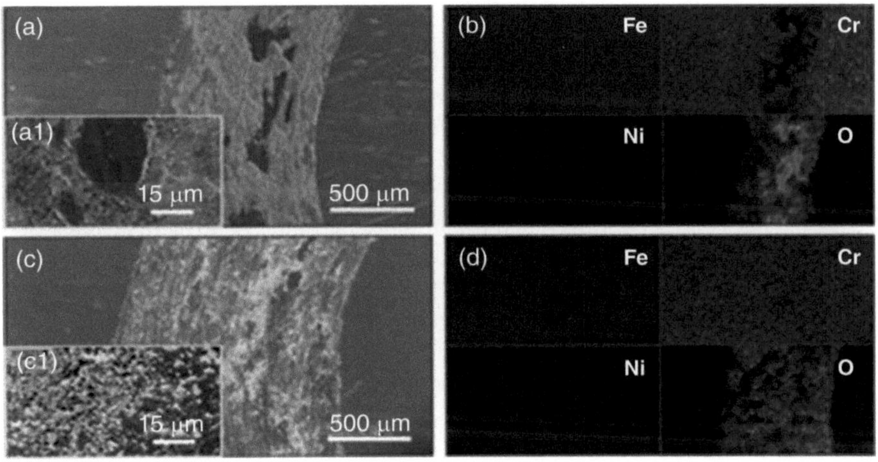

FIGURE 5.17 SEM morphologies and corresponding EDS mapping results for HLS 316L stainless steel and obtained wear scars. The obtained wear scars of 316L stainless steel are shown in (a), (a1): The enlarged wear scar morphology of the 316L stainless steel specimen (b), and (c) together with the corresponding EDS mapping images (a), (c), and (d) of the wear scar of HLS 316L stainless steel and the enlarged wear scar morphology shown in (c1) and (d), respectively [64] [open access].

specimens to as-received stainless steel. After tribotesting, SEM and EDS map measurements are performed to obtain further information about the wear scar properties under 5 N normal load. The adhesive wear phenomena shown in Figure 5.17a are characterized by the significant deformation of counterpart materials and their adhesion to the worn surface of wear scar. Meanwhile, the inset of Figure 5.17a shows that some small particles also covered the worn surface near adhesion phenomenon. Figure 5.17b may indicate that this material is oxidative. Therefore, it may indicate the predominantly adhesive wear mode combined with moderate tribooxidative wear mode for the 316L stainless steel specimen, which is obtained under dry sliding conditions. Figure 5.17c illustrates the unique wear scar morphology of specimens made of HLS 316L stainless steel. This indicates a serious tribe-oxidative wear mode since many oxidized particles are formed on the wear track surface and no material adhesion occurred [64].

The tribological performance of AISI 316 stainless steel can be improved using solid lubricants and laser-ablated microtextures, as demonstrated in this study. The results show that although some lubricants and textures show potential, further optimization is needed to fully realize their full potential. Improved performance in tribological applications is possible due to proposed approach to increase wear resistance by using protrusions during machining. Stronger and more reliable materials will be developed as a result of ongoing research and development in this area, ultimately reducing failure rates and maintenance costs in critical industrial applications.

5.8 PROBLEMS

1. Explain why hard angular particles will grind brittle materials so quickly. Justify why ductile materials will wear out much more slowly than brittle materials.
2. Can you name several surfaces and tribological properties of polished patterned surfaces using different types of solid lubricants?
3. What is possible difference might there be in the relative wear rates of materials if the effect of fine groove density on tribological performance is true?
4. How are wear and friction tests carried out on two metal bodies?
5. Why do different types of solid lubricants and laser surface texturing change the tribological characteristics of AISI 316 stainless steel surfaces?

REFERENCES

1. Kottarathil, S., Narayanaru, S., Unni, S.M., Tamaki, T., Kuroki, H., Hareesh, U.N.S., Anilkumar, G.M., & Yamaguchi, T. (2022). Selective borohydride oxidation reactions of zeolitic imidazolate framework-derived bimetallic carbon alloy electrocatalysts for alkaline fuel cell applications. *ACS Appl. Energy Mater, 5*, 12571–12582.
2. Meng, R., Deng, J., Duan, R., Liu, Y., & Zhang, G. (2019). Modifying tribological performances of AISI 316 stainless steel surfaces by laser surface texturing and various solid lubricants. *Optics & Laser Technology, 109*, 401–411.
3. Gao, X., Wen, Y., Tan, R., Huang, H., & Kawi, S. (2023). A review of catalyst modifications for a highly active and stable hydrogen production from methane. *Int. J. Hydrogen Energy, 48*, 6204–6232.
4. Zhang, Y., Zhao, Y., Hu, W., Wang, X., & Yang, R. (2023). Microstructure and surface engineering through indium modification on Ni-rich layered cathode materials for enhanced electrochemical performance of lithium-ion batteries. *J. Alloys Compd, 934*, 167862.
5. Sun, X., Jing, M., Dong, H., Xie, W., & Luo, F. (2023). CuO-ZnO submicroflakes with nanolayered Al2O3 coatings as high performance anode materials in lithium-ion batteries. *J. Alloys Compd, 953*, 170137.
6. Huang, J.Y., Fei, G.T., Xu, S.H., & Wang, B. (2023). ZnO–SiO2 composite coating with anti-reflection and photoluminescence properties for improving the solar cell efficiency. *Compos. Part B Eng, 251*, 110486.
7. Singh, N., Malik, A., Nohwar, S., Jana, R., & Mondal, P.C. (2023). Covalent surface modification of nickel ferrite nanoparticles for electrochemical supercapacitor performance. *N. J. Chem, 47*, 5308–5315.
8. Chiu, L.H., Yang, C.F., Hsieh, W.C., & Cheng, A.S. (2002). Effect of contact pressure on wear resistance of AISI H13 tool steels with chromium nitride and hard chromium coatings. *Surf. Coat. Technol, 154*, 282–288.
9. Wang, C., Gann, E., Chesman, A.S.R., & McNeill, C.R. (2019). Residual solvent additive enables the nanostructuring of PTB7-Th:PC71BM solar cells via soft lithography. *AIP Adv, 9*, 065024.
10. Guo, C., Jia, R., Li, X., Tian, X., Chen, J., Ge, H., Huang, H., Yuan, L., & Xu, C. (2023). Influence of backside surface morphology on passivation and contact characteristics of TOPCON solar cells. *Sol. Energy, 258*, 278–288.

11. Aghazadeh, M., Ganjali, M.R., Morad, M.M., & Gharailou, D. (2020). Saccharide-capped superparamagnetic copper cations-doped magnetite nanoparticles for biomedical applications: A novel and simple synthesis procedure, in-situ surface engineering and characterization. *Curr. Nanosci*, *16*, 770–778.
12. Sun, J., Shi, Z., Li, M., Chen, S., Zhong, M., Liu, X., Sha, J., & Jia, S. (2021). Theoretical investigation of the microfluidic and magnetic field-assisted self-assembly of colloidal magnetic-plasmonic nanoparticles. *J. Phys. D Appl. Phys*, *54*, 325004.
13. Hivart, P., Hauw, B., Crampon, J., & Bricout, J.P. (1998). Annealing improvement of tribological properties of manganese phosphate coatings. *Wear*, *219*, 195–204.
14. Gabe, D.R. (2002). Hard Anodizing—what do we mean by Hard? *Met. Finish*, *100*, 52–57) 21.
15. Mncwabe, Z., Farahani, M.D., & Friedrich, H.B. (2020). Switching between oxidation types using molybdenum phosphate catalysts for paraffin activation using doped Fe as surface acidity modifier and MoOx as an oxygen insertion tool. *Catal. Lett*, *150*, 728–737.
16. Liang, Y., Gao, J., Gao, D., Sun, Y., Zhang, Z., & Zhao, J. (2021). Effect of micro-textured surfaces and sliding speed on the lubrication mechanism and friction-wear characteristics of CF/PEEK rubbing against 316L stainless steel under seawater lubrication. *Applied Sciences*, 11(21), 9915.
17. Geng, Z., Li, X., Ji, L., Li, Z., Zhu, S., Cui, Z., Wang, J., Cui, J., Yang, X., & Liu, C. (2021). A novel snail-inspired bionic design of titanium with strontium-substituted hydroxyapatite coating for promoting osseointegration. *J. Mater. Sci. Technol*, *79*, 35–45.
18. De Mello, J.D.B., Goncalves, J.L., & Costa, H.L. (2013). Influence of surface texturing and hard chromium coating on the wear of steels used in cold rolling mill rolls. *Wear*, *302*, 1295–1309.
19. Juneghani, M.A., Farzam, M., & Zohdirad, H. (2013). Wear and corrosion resistance and electroplating characteristics of electrodeposited Cr-SiC nano-composite coatings. *Trans. Nonferrous Met. Soc. China*, *23*, 1993–2001.
20. Chou, C.-H., Nagarjuna, Y., Yang, Z.-C., Hsiao, Y.-J., & Wang, S.-C. (2022). Catalytic effect of Ag embedded with ZnO prepared by Co-sputtering on H2S gas sensing MEMS device. *Vacuum*, *202*, 111210.
21. Hsueh, T.-J., Li, P.-S., Fang, S.-Y., & Hsu, C.-L. (2022). A vertical CuO-NWS/MEMS NO2 gas sensor that is produced by sputtering. *Sens. Actuators B Chem*, *355*, 131260.
22. Ramezani, M., & Ripin, Z.M. (2023). An overview of enhancing the performance of medical implants with nanocomposites. *J. Compos. Sci*, *7*, 199.
23. Ding, Y., Tao, B., Ma, R., Zhao, X., Liu, P., & Cai, K. (2023). Surface modification of titanium implant for repairing/improving microenvironment of bone injury and promoting osseointegration. *J. Mater. Sci. Technol*, *143*, 1–11.
24. Lu, M., Chen, H., Yuan, B., Zhou, Y., Min, L., Xiao, Z., Zhu, X., Tu, C., & Zhang, X. (2020). Electrochemical deposition of nanostructured hydroxyapatite coating on titanium with enhanced early stage osteogenic activity and osseointegration. *Int. J. Nanomed*, *15*, 6605–6618.
25. Zhang, Y., Gao, S., Zhao, T., Chen, D., Wang, R., & Yu, Y. (2023). Ge modified by metal–organic segments as anodes for Ge-air batteries with an alkaline gel polymer electrolyte. *Arab. J. Chem*, *16*, 105021.
26. Ramezani, M., & Ripin, Z.M. (2023). 4D printing in biomedical engineering: Advancements, challenges, and future directions. *J. Funct. Biomater*, *14*, 347.

27. Mutharani, B., Ranganathan, P., Chen, T.-W., Chen, S.-M., & Vinoth Kumar, J. (2020). Thermoreversible switchlike electrocatalytic reduction of tizanidine based on a graphene oxide tethered stimuli-responsive smart surface supported Pd catalyst. *Anal. Chem*, *92*, 8965–8973.
28. Li, H., Ai, Z., Yang, L., Zhang, W., Yang, Z., Peng, H., & Leng, L. (2023). Machine learning assisted predicting and engineering specific surface area and total pore volume of biochar. *Bioresour. Technol*, *369*, 128417.
29. Rama Krishna, L., Sudha Purnima, A., & Sundararajan, G. (2006). A comparative study of tribological behaviour of microarc oxidation and hard-anodized coatings. *Wear*, *261*, 1095–1101.
30. Budinski, K.G. (1993). The wear resistance of diffusion treated surfaces. *Wear*, *162–164*, 757–762.
31. Salehi, M., Heidari, P., Ruhani, B., Kheradmand, A., Purcar, V., & Căprărescu, S. (2021). Theoretical and experimental analysis of surface roughness and adhesion forces of MEMS surfaces using a novel method for making a compound sputtering target. *Coatings*, *11*, 1551.
32. Auciello, O. (2023). Review of materials science and technological applications of transformational integrated multifunctional ox-ide/ultrananocrystalline diamond (UNCD) films/crystalline diamond for new generation of high-tech and biomedical devices. *J. Mater. Res*, *38*, 571–585.
33. You, Z.-W., Wei, L., Zhang, M.-L., Yang, F.-H., & Wang, X.-D. Design of a novel MEMS implantable blood pressure sensor and stress distribution of parylene-based coatings. In Proceedings of the 2022 IEEE 16th International Conference on Solid-State & Integrated Circuit Technology (ICSICT), Nanjing, China, 25–28 October 2022; pp. 1–3. [CrossRef]
34. Sun, D., & Böhringer, K.F. (2019). Self-cleaning: From bio-inspired surface modification to MEMS/microfluidics system integration. *Micromachines*, *10*, 101.
35. Jiang, S., Feng, Y., Chen, Z., Zhang, L., Jiang, X., Jiao, Q., Li, J., Chen, Y., Li, D., & Liu, L.; et al. (2016). Study on light extraction from GaN-based green light-emitting diodes using anodic aluminum oxide pattern and nanoimprint lithography. *Sci. Rep*, *6*, 21573.
36. Ziemba, A.M., Woodson, M.C.C., Funnell, J.L., Wich, D., Balouch, B., Rende, D., Amato, D.N., Bao, J., Oprea, I., & Cao, D.; et al. (2023). Development of a slow-degrading polymerized curcumin coating for intracortical microelectrodes. *ACS Appl. Bio Mater*, *6*, 806–818.
37. Cherian, A.M., Joseph, J., Nair, M.B., Nair, S.V., Vijayakumar, M., & Menon, D. (2022). Coupled benefits of nanotopography and titania surface chemistry in fostering endothelialization and reducing in-stent restenosis in coronary stents. *Biomater. Adv*, *142*, 213149.
38. Alcudia, A., Begines, B., Rodriguez-Lejarraga, P., Greyer, V., Godinho, V.C.F., Pajuelo, E., & Torres, Y. (2022). Development of porous silver nanoparticle/polycaprolactone/polyvinyl alcohol coatings for prophylaxis in titanium interconnected samples for dental implants. *Colloid Interface Sci. Commun*, *48*, 100621.
39. Simon, N., Schmid, M., Blendinger, F., & Bucher, V. (2022). Long term evaluation of the barrier properties of polymer/metal oxide hybrid layers for use in medical implants. *Curr. Dir. Biomed. Eng*, *8*, 435–438.
40. Visser, N.L., Verschoor, J.C., Smulders, L.C., Mattarozzi, F., Morgan, D.J., Meeldijk, J.D., van der Hoeven, J.E., Stewart, J.A., Vandegehuchte, B.D., & de Jongh, P.E. (2023). Influence of carbon support surface modification on the performance of nickel catalysts in carbon dioxide hydrogenation. *Catal. Today*, *418*, 114071.

41. Kübler, M., Wagner, S., Jurzinsky, T., Paul, S., Weidler, N., Villa, E.D.G., Cremers, C., & Kramm, U.I. (2020). Impact of surface functionalization on the intrinsic properties of the resulting Fe–N–C catalysts for fuel cell applications. *Energy Technol*, *8*, 2000433.
42. Suresh, S. (2013). Semiconductor nanomaterials, methods and applications: A review. *Nanosci. Nanotechnol*, *3*, 62–74.
43. Liu, M., Ji, Z., & Shang, L., Top-down fabrication of nanostructures nanotechnology, in: Chi Lifeng (Ed.), Nanostructured Surfaces, 8, Wiley-VCH Verlag GmbH & Co. KGaA, Weinheim, Germany, 2010, pp. 427.
44. Acikgoz, C., Hempenius, M.A., Huskens, J., & Vancso, J.G. (2011). Polymers in conventional and alternative lithography for the fabrication of nanostructures. *Eur. Polym. J*, *47*, 2033–2052.
45. Polte, J. (2015). Fundamental growth principles of colloidal metal nanoparticles: A new perspective. *Cryst. Eng. Comm*, *17*, 6809–6830.
46. Husein, M.M., & Nassar, N. (2008). Nanoparticle preparation using the single microemulsions scheme. *Curr. Nanosci*, *4*, 370–380.
47. Li, J., Zhou, P., Attarilar, S., & Shi, H. (2021). Innovative surface modification procedures to achieve micro/nano-graded Ti-based biomedical alloys and implants. *Coatings*, *11*(6), 647.
48. Habiba,K., Makarov,V.I., Weiner, B.R., & More, G. (2014). Fabrication of nanomaterials by pulsed laser synthesis, in: A. Waqar, N. Ali (Eds.), Manufacturing Nanostructures, One Central Press, Altrincham, United Kingdom.
49. Toshima, N., & Yonezawa, T. (1998). Bimetallic nanoparticles-novel materials for chemical and physical applications. *New J. Chem*, *22*, 1179.
50. Johnston, C.W., Wyatt, M.A., Li, X., Ibrahim, A., & Shuster, J. (2013). Gold biomineralization by a metallophore from a gold-associated microbe. *Nat. Chem. Biol*, *9*, 241–243.
51. Gates, B.D., Xu, Q.B., & Stewart, M., et al. (2005). New approaches to nanofabrication: molding, printing, and other techniques, *Chem. Rev.*, *105*, 1171–1196.
52. Betancourt, T., & Brannon-Peppas, L. (2006). Micro- and nanofabrication methods in nanotechnological medical and pharmaceutical devices. *Int. J. Nanomed*, *1*, 483–495.
53. Wiley, B.J., Qin, D., & Xia, Y.N. (2010). Nanofabrication at high throughput and low cost. *ACS Nano*, *4*, 3554–3559.
54. LaFratta, C.N., Fourkas, J.T., Baldacchini, T., & Farrer, R.A. (2007). Multiphoton fabrication. *Angew. Chem. Int. Ed*, *46*, 6238–6258.
55. Tran, H.A., & Tran, P.A. (2021). In situ coatings of silver nanoparticles for biofilm treatment in implant-retention surgeries: Antimicrobial activities in monoculture and coculture. *ACS Appl. Mater. Interfaces*, *13*, 41435–41444.
56. Cordeiro, J.M., Nagay, B.E., Dini, C., Souza, J.G., Rangel, E.C., da Cruz, N.C., Yang, F., van den Beucken, J.J., & Barão, V.A. (2021). Copper source determines chemistry and topography of implant coatings to optimally couple cellular responses and antibacterial activity. *Biomater. Adv*, *134*, 112550.
57. Wu, H., Xie, L., He, M., Zhang, R., Tian, Y., Liu, S., Gong, T., Huo, F., Yang, T., & Zhang, Q.; et al. (2019). A wear-resistant TiO_2 nanoceramic coating on titanium implants for visible-light photocatalytic removal of organic residues. *Acta Biomater.*, *97*, 597–607.
58. Ali, A., Chowdhury, S., Janorkar, A.V., Marquart, M., Griggs, J.A., Bumgardner, J.D., & Roach, M.D. (2023). A novel single-step anodization approach for pani-doping oxide surfaces to improve the photocatalytic activity of titanium implants. *Biomed. Mater*, *18*, 015010.

59. Qi, L., Guo, B., Lu, Q., Gong, H., Wang, M., He, J., Jia, B., Ren, J., Zheng, S., & Lu, Y. (2022). Preparation and photocatalytic and antibacterial activities of Micro/Nanostructured TiO2-based photocatalysts for application in orthopedic implants. *Front. Mater*, *9*, 914905.
60. Parfenova, L.V., Galimshina, Z.R., Gil'Fanova, G.U., Alibaeva, E.I., Danilko, K.V., Pashkova, T.M., Kartashova, O.L., Farrakhov, R.G., Mukaeva, V.R., & Parfenov, E.V.; et al. (2022). Hyaluronic acid bisphosphonates as antifouling antimicrobial coatings for PEO-modified titanium implants. *Surf. Interfaces*, *28*, 101678.
61. Akshaya, S., Rowlo, P.K., Dukle, A., & Nathanael, A.J. (2022). Antibacterial coatings for titanium implants: Recent trends and future perspectives. *Antibiotics*, *11*, 1719.
62. Von Freymann, G., Ledermann, A., Thiel, M., Staude, I., Essig, S., Busch, K., & Wegener, M. (2010). Three dimensional nanostructures for photonics. *Adv. Funct. Mater*, *20*, 1038–1052.
63. Torres, A., Cuadrado, N., Llumà, J., Vilaseca, M., & Travieso-Rodriguez, J.A. (2022). Influence of the stainless-steel microstructure on tribological behavior and surface integrity after ball burnishing. *Materials*, *15*(24), 8829.
64. Qin, W., Kang, J., Li, J., Yue, W., Liu, Y., She, D., … & Li, Y. (2018). Tribological behavior of the 316L stainless steel with heterogeneous lamella structure. *Materials*, *11*(10), 1839.

6 Tribosystems in Industrial Tribology

6.1 INTRODUCTION

The literal translation of tribology is "the science of trituration", as it comes from the Greek word "tribos" which means "to rub". The idea was first introduced by British mechanical engineer Peter H. Jost in the Jost report dated March 9, 1966. A textbook on tribodesign and technology is to be published and institutes of tribology are created. The Tribology Committee was formally created on September 26, 1966, shortly after the publication of the Jost report. The responsibilities of the committee include advising government departments and agencies on the promotion of technological progress and cost-effectiveness in the field of tribology, as well as providing advice on these issues. Annual reporting to the minister of technology on its activities, as well as trends and achievements in the field of tribology and reviewing the latest tribology techniques and suggesting them to industry is also a part of its responsibilities. The study of friction, wear, and lubrication is known as tribology [1–6]. Over the past five years, this rapidly expanding field has produced some new encouraging results and is attracting great interest from academic institutions, companies, and other research centers. Its interdisciplinary nature requires effective collaboration between chemists, chemical engineers, mechanical engineers, materials scientists, and even environmental and toxicology experts.

This chapter provides an overview of breeding systems and recent developments in breeding techniques to reduce corrosion and energy consumption. This useful reference guide presents research on the most important tertiary systems using sophisticated test equipment in the test and laboratory areas, as well as the latest discoveries in material selection and corrosion protection through surface treatments and specialty coatings, as well as lubrication and lubricants. This serves as a brief introduction for researchers in mechanical engineering, chemistry, materials science; plant maintenance personnel, engineers and chemists of automotive and transport industries; and mechanical and laboratory engineers responsible for monitoring and evaluating lubricants [7–9].

The end product is an investigation of the most important three way systems using the latest test equipment in test and laboratory areas, the latest advances in material selection and corrosion protection through surface treatments and specialized coatings, and using of lubrication and lubricants, for example, the motion patterns of the inner ball-bearing components during

DOI: 10.1201/9781003584735-6

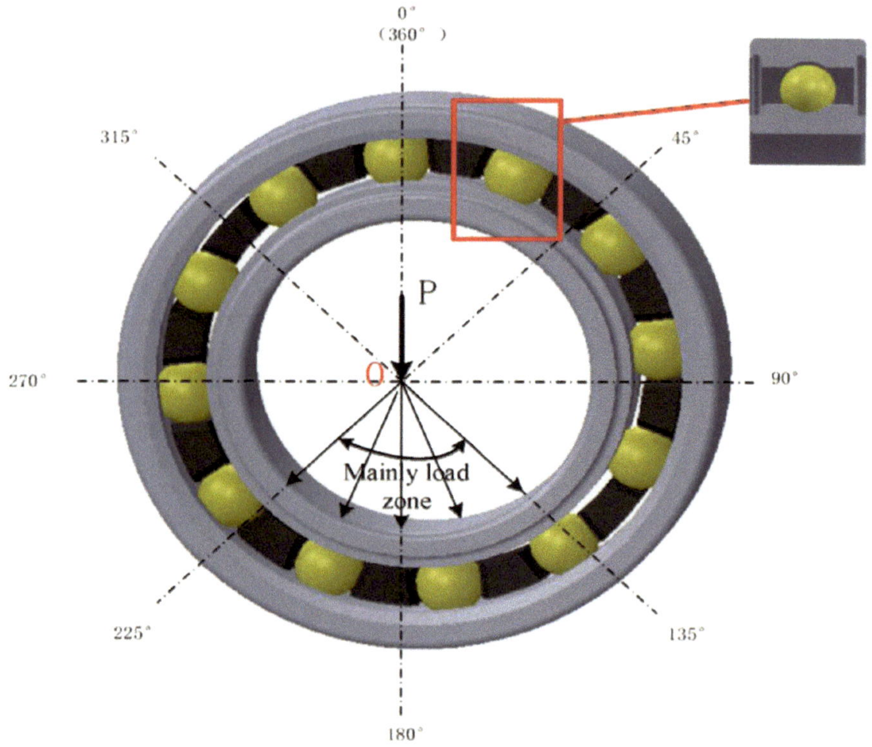

FIGURE 6.1 The structure of rolling bearing [14] [open access].

movement are depicted in Figure 6.1. The frictional heat is affected to some extent by the rotating part of the bearing motion state [5]. The rotating element in the bearing rotates around its axis and rotates around the center of the inner ring [10–13].

6.2 IMPORTANCE OF TRIBOLOGY

Tribology research initially focused on the efficient design and lubrication of machine parts such as bearings. The field of tribology has evolved to include many aspects of modern technology. Traditional applications in the field of classical tribology highlight the importance of tribology of the sliding surfaces of most mechanical components, which is necessary to increase component life and energy efficiency. In the transportation sector, tribology research has traditionally focused on reliability. But recently, improvements in energy consumption and efficiency have come to the fore, leading to development of increasingly sophisticated lubricants. For example, in this study, the 6208 silicon nitride deep groove ball bearing with P4 accuracy is used as a test object.

The test is conducted under oil lubrication conditions and it provided results on the lubricating oil film distribution law of full ceramic-coated ball bearing and an explanation of how it affects friction characteristics. The cage guide mode is the outer ring guide, and the clearance of test bearing is CN standard. The main axis oil with a viscosity value of 30 mm^2/s is the lubricating oil used in the test [9].

6.3 WORK OF TRIBOSYSTEMS

The study of tribosystems or physical systems with contacting surfaces is known as tribology. Many different aspects affect a tribosystem, so it is important to consider them when selecting materials and lubricants. Motion type, speed, temperature, load and working environment are examples of these parameters. However, in Figure 6.2, there are obvious friction areas and scratches on the side beam surface of PVX cage (the outer circular surface of the cage). Based on the scratches, it can be concluded that the outer and inner circular surfaces contact and collide during the rotation of bearing. The reason for this phenomenon is the composite structure of cage [5, 6, 10].

A tribosystem can be broadly classified into functional outputs, loss outputs, system structure, and operational inputs. Instances of each consist of operational components::

- *Motion type*: In what motion does the object move? This may include rolling, hitting, sliding and rolling, etc.
- *Magnitude of load*: What is the load and how is it applied in terms of duration, frequency and amplitude?
- *Speed*: What component of speed is at play here?
- *Temperature*: A range of variables including ambient temperature gradients, temperature caused by friction, external heating and/or cooling and flash point.

FIGURE 6.2 Removed entire silicon nitride ceramic ball bearing and all its parts [5] [open access].

- Organizational structure
- Profiles of significant elements, which include the definition the structure of a system.
- *Opposite and ground bodies*: The result can be two surfaces sliding over each other. As an example, consider two sliding metal surfaces.
- *Environment*: Atmospheric environment, such as air.
- *Medium between surfaces*: For example, lubricant in bearings located between two surfaces.

Full ceramic ball bearings are examined and tested under various lubrication and operating conditions. Among them, the bearing shown in the following figure is tested with an oil flow rate of 0.2 mL/minute, load of 5000 rpm and load of 900 N. Figure 6.2 shows that when the bearing assembly under test is disassembled, no obvious defects are found by observation. Figure 6.3 shows that the ceramic ball has high precision and low friction temperature. The friction temperature of the ceramic ball G3 fluctuates slightly. Under the same stress condition, the level of the ceramic ball decreases, its roughness decreases, and its friction temperature increases. This is caused by the increase in roughness, which raises the temperature and friction force. Figure 6.3 shows that during the friction experiment, the initial friction temperature of the

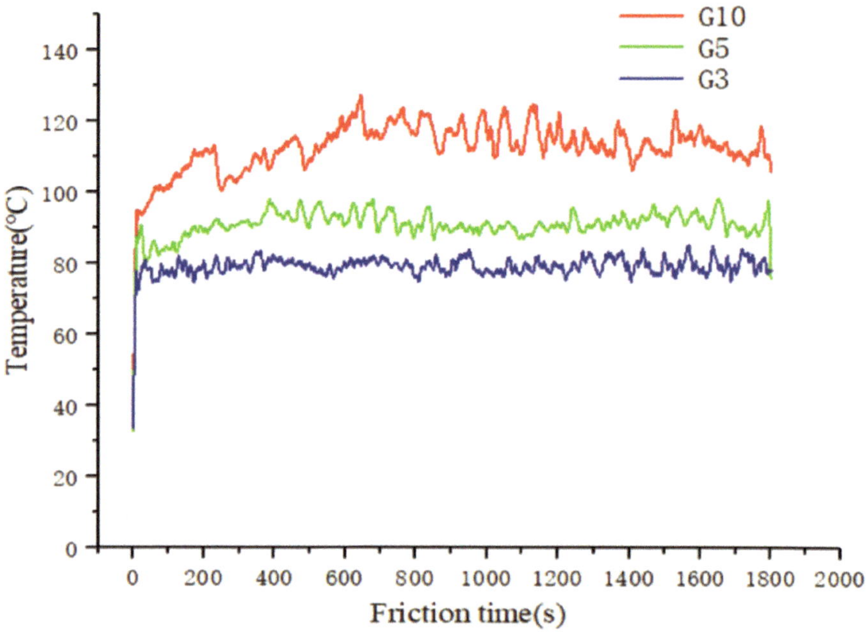

FIGURE 6.3 The friction temperature of the different level ceramic ball under the same conditions [14] [open access].

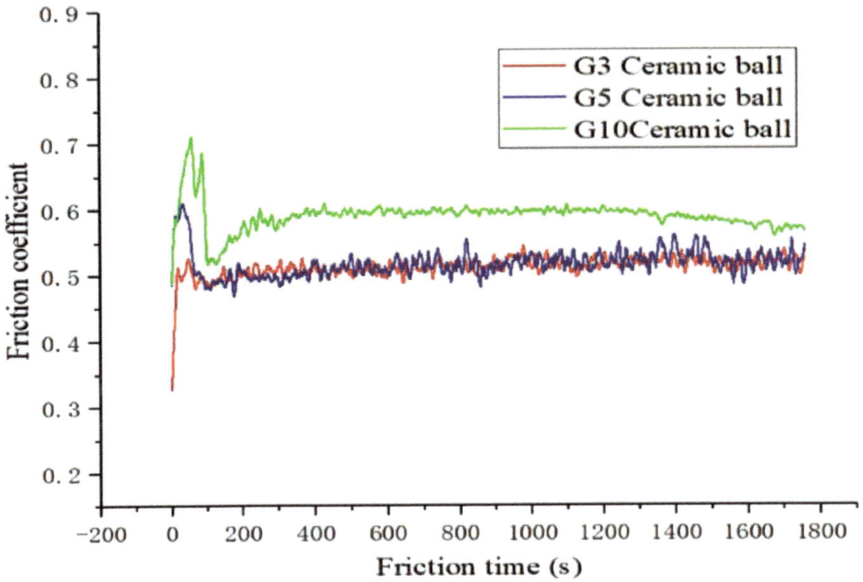

FIGURE 6.4 The friction coefficient of the different level ceramic ball under the same conditions [14] [open access].

ceramic balls was relatively high. The surface temperature ranged from 5 to 10 °C at G1 and G2 with the increase of friction, while at G10, the fluctuation range reached nearly 40 °C [2].

These photographs show spike-like structures, which are nodular growth defects or flaws common in coatings produced by industrial magnetron sputtering systems. Most likely in the context of materials science or engineering, the performance or acceptability of SS specimen for particular application may be affected if it exhibits more defects than a carbon or other composite specimen. On another hand, the CC specimen appears to have more flaws. Analysis of 2D profilometry images has showed that the nodular flaws on SS specimen are arranged in a grid pattern. The other specimens generally did not form comparable features. For each sample studied, grain analysis is performed to detect variations in grain size. 2D atomic force microscopy (AFM) images of 5 × 5 μm are used for this purpose. Figure 6.4 shows that the three layers of ceramic balls have a relatively high friction coefficient at the beginning of friction. The G3 ceramic balls move smoothly and have a relatively low friction coefficient. The G10 ceramic balls have a relatively large total change and a high friction coefficient [2].

Figure 6.5 shows how increasing the operating load of the bearing can cause its temperature to rise. Ceramic ball bearings have a relatively low-temperature rise, and the effect of load on this rise is usually linear. However, the temperature of the G10 ceramic ball initially fluctuates greatly with increasing load

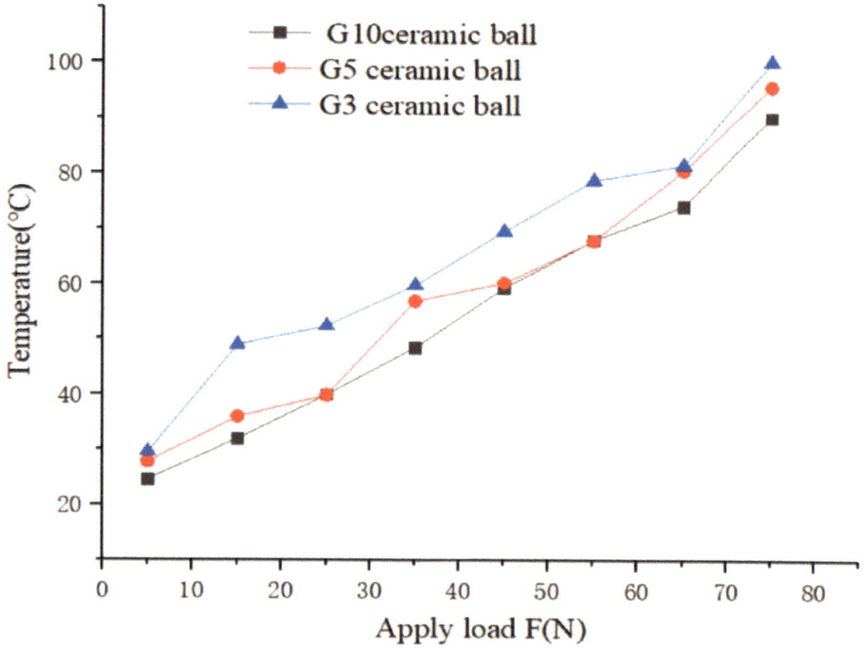

FIGURE 6.5 The change in the ceramic ball temperature under different pressures [14] [open access].

before gradually rising. This is due to the smooth surface of the G3 ceramic ball and the relatively low surface roughness. Under load conditions, the friction temperature is low and the friction force is small. However, the G10 ceramic ball has a large level of surface roughness. During the initial stage of bearing stress, it is affected by certain peaks of surface waves, which increases the friction force.

Second, it appears that sample CW-1f has the smallest grain size, while sample CW-2f has the largest grain size. Third, sample CW-1f has a very rough surface and its grains are clustered together. The study in reference [15] used in situ electrochemical (EC)-AFM to study the passive film formation in Al-Ni-Ce metallic glass. For the in situ EC-AFM measurements, the open circuit potential (OCP) settings are used and potential is converted to pitting potential (Epit). Using TM-AFM, they have found the α-Al nanophase in amorphous matrix. The phase image demonstrated that small α-Al islands are randomly distributed on the surface and their phase shift is negative concerning the matrix. Figure 6.6 shows that the slightly enlarged phase image shows that the α-Al island is surrounded by bright annular spots [1].

Figure 6.7 shows the corrosion behavior of point of slope (PS) compared to angle of slope (AS), showing how the increase in chloride ion concentration

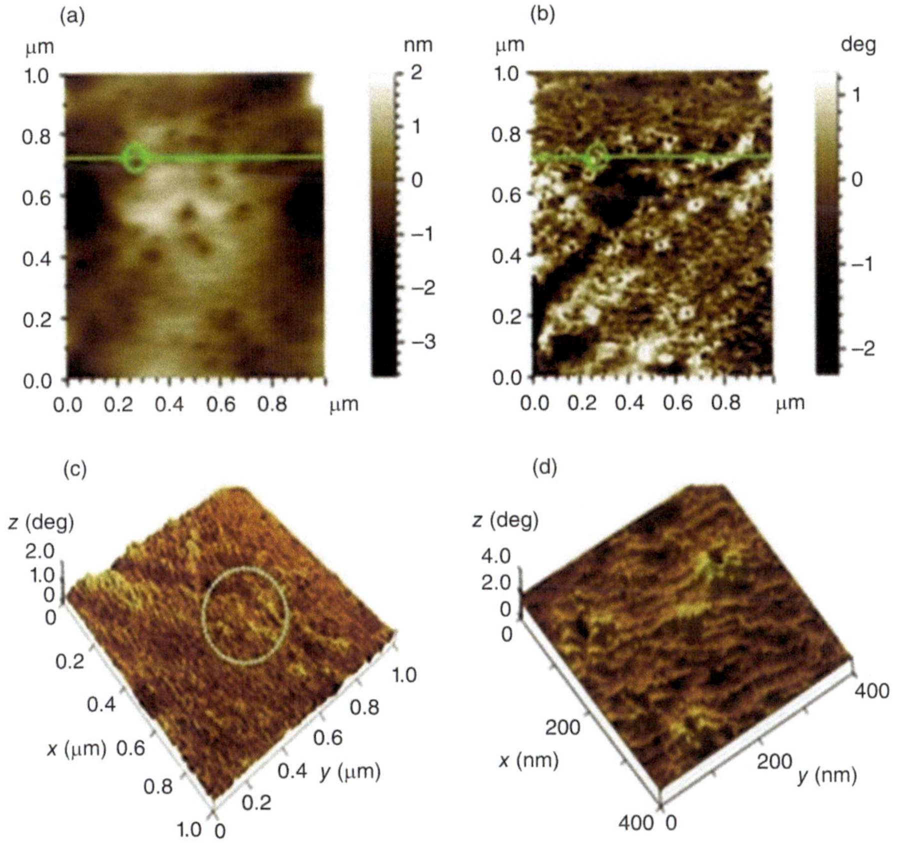

FIGURE 6.6 AFM images in touch mode (TM) of amorphous nanocrystalline annealed Al88Ni8Ce4 sample exfoliated and aged for 3 minutes in 0.01 mol/L NaCl solution at open circuit potential (OCP). Phase image (b) and topography image (a). 3D images of phase (c) and local extended phase image (d) of chain indicated in phase (c) [1] [open access].

caused a decrease in the surface change time scale of DSS 2205. In both solutions, DSS 2205 steel specimens showed high corrosion resistance despite being exposed to anodic potentials in the trans-negative oxidation zone. In contrast, AISI 316 L steel has showed a stable corrosion resistance against AS. As shown in Figure 6.8, AISI 316 L steel is easily corroded and showed pronounced corrosion with increasing chloride ion concentration when exposed to PS solution. Oval deposits are present on the surface of AISI 316 L. This is due to the chemical change in the matrix composition, which caused the formation of chromium, iron and nickel oxides, resulting in distinct growth patterns on the surface of the specimen [1].

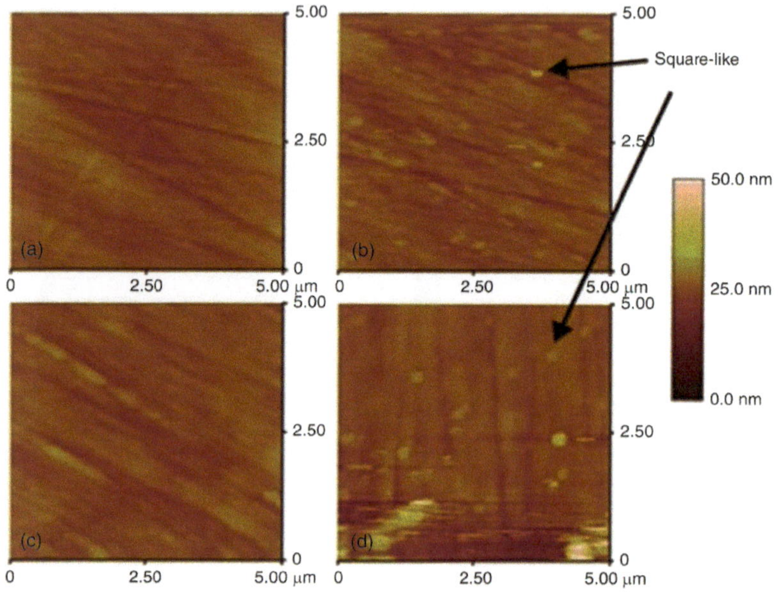

FIGURE 6.7 AFM images of DSS 2205 sample under (a) artificial saliva (AS) after 45 minutes of exposure to 0.8 V; (b) artificial saliva after an additional 11.6 minutes of exposure to 1 V; (c) saline solution (SS) after 17.2 minutes of exposure to 0.8 V; and (d) artificial saliva after an additional 7.5 minutes of exposure to 1 V [1] [open access].

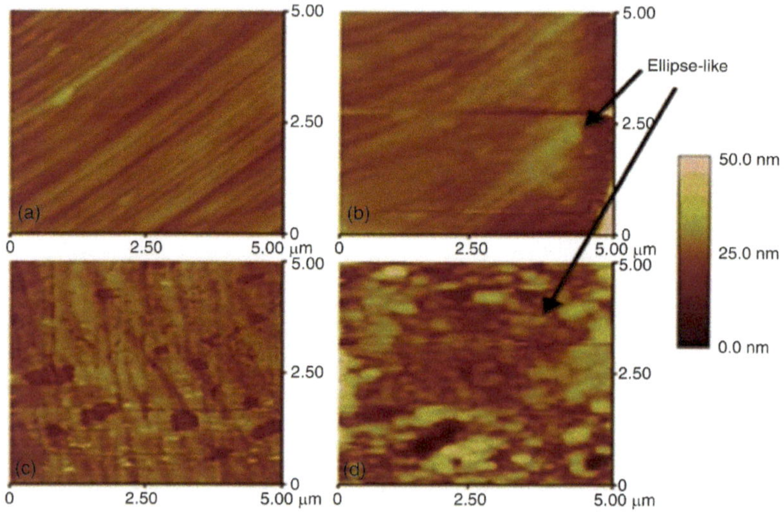

FIGURE 6.8 Surface condition evolution of AISI 316L specimen under anodic exposure: (a) initial smooth surface, (b) early corrosion Signs After 0.5 V for 10 minutes in AS solution, (c) pronounced corrosion and deposit formation after extended anodic exposure, (d) final surface condition with distinct oxide deposits after further exposure to PS solution [1] [open access].

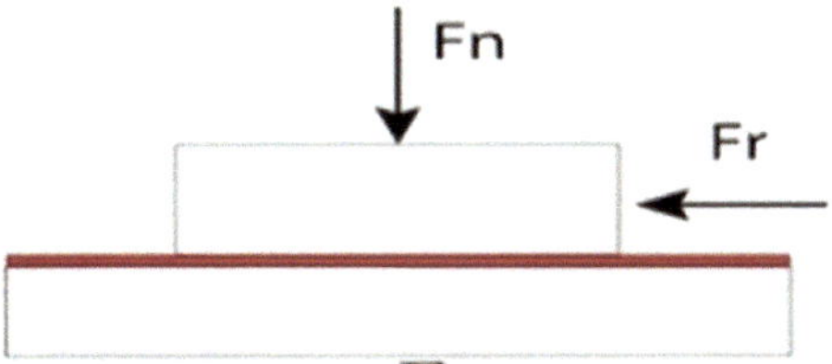

$$\mu = \frac{Fn}{Fr}$$

FIGURE 6.9 This illustrates the tribological system. The red line indicates the addition of lubricant, which reduces the kinetic coefficient of friction. Fn is force applied to sliding surfaces, Fr is friction (motion) force [5] [open access].

6.4 FRICTION AND WEAR

The resistance that occurs when two bodies come into contact is called friction. At the macroscopic level, friction is described by the basic laws of friction. They argue that the relationship between frictional forces and loads is linear. One of the main dimensionless parameters that can be calculated is the coefficient of friction, which is usually denoted by the Greek letter μ. The relationship between the force that pushes two bodies together at the beginning of sliding process and force that causes friction between them is described by the friction coefficient [5]. The materials used when determining the coefficient of friction differ. Ice on steel has a low coefficient of friction, while rubber on concrete has a high coefficient of friction. The amount of friction can be calculated using the following formula for friction coefficient given in Figure 6.9, where Fn is applied normal force, μ is static (μs) or kinetic (μ_k) friction coefficient and Fr is friction force.

6.5 WEAR IN TRIBOLOGY

Wear occurs when the material is lost and surfaces interact irrevocably. Basic physical and chemical reactions in a pair of sliding surfaces [5] lead to wear.

Figure 6.10 shows that after the test, the entire contact area of ceramic ball bearing has showed no signs of damage. However, a black film has formed on the surfaces of ball bearing and raceway of outer ring in several places. The film production process and its effect on the surface friction and wear quality of bearing microcontact area are examined from a microscopic point of view to shed further light on chemical composition of film. We cut the cage and ring of the test bearing nondestructively. They are examined using scanning electron microscopy (SEM), X-ray diffraction (XRD) and other tools in combination with ceramic balls. Shapes of friction partners vary depending on wear mechanisms.

- *Surface fatigue*: This can result in stress on the surface material.
- *Abrasion*: Creates grooves on the surface of the material.
- *Adhesion*: Material transfer.
- *Tribochemical reactions*: Production of particles and residues.

(a) (b)

FIGURE 6.10 After testing, the entire raceway of outer ring of a ceramic ball bearing and rolling elements. (a) Rolling elements; (b) raceway of the outer ring of the bearing [5] [open access].

- *Modern uses of tribology*:
  - Tribology has expanded from its original focus on improving the performance and life of industrial machines to a wide range of applications that impact numerous areas:
    - Traditional tribology studies wear and friction in machine parts such as clutches, brakes, rolling bearings and simple wheels, gears, bearings, etc.
    - The field of nanotribology has become very popular in the last few years. In particular, this concerns nanoscopic tribological phenomena.
    - The field of space tribology deals with tribological systems that function in challenging environments, especially those caused by sudden temperature changes in space.
    - Green tribology studies how to use low-impact technologies to reduce both environmental impact and tribological efficiency losses.
    - The study of lubrication in biological systems such as human hips and knees is known as tribology.
    - Tribotronics is a branch of robotics that creates active tribological devices by connecting mechanical and electrical components.
    - Computational tribology integrates various fields, including fracture mechanics and contact mechanics, to model the behavior of tribological systems, studying friction, wear and lubrication of geological systems such as faults and glaciers. It is based on computational fluid dynamics, mechanics, and computational fluid dynamics.

FIGURE 6.11 Tribology has a major impact on bearing design and maintenance [5] [open access].

6.6 TRIBOLOGY AND LUBRICATION

A significant part of traditional tribology is lubrication. In addition to protecting against corrosion, it helps regulate temperature in various applications and has excellent sealing properties that protect equipment from damage [9]. The main advantage of lubricants is that they significantly reduce costs. By reducing wear and friction, they prevent or reduce surface damage due to adhesion, abrasion, fatigue and tribochemical reactions. When properly selected and used, lubricants extend equipment life and reduce costs for the end-user. Tribology has a significant influence on bearing design and maintenance (Figure 6.11).

By creating a coating that reduces friction between two surfaces, lubricants provide this protection. The lubricant does not necessarily have to be liquid. The lubricant can be a semi-solid substance such as grease, a solid dry material, a liquid such as oil, or water or even polymers and silicones. The main oils used for lubrication are mineral, vegetable and animal oils. From early 1900s the industry has advanced significantly toward offering a wide range of lubricants, and the technology associated with these lubricants has also undergone significant development [4]. The Stribeck curve shows how friction (μ) changes with increasing speed (m/s). The number and density of pits increase with prolonged electrical stress, eventually producing a satin-like surface [5]. As can be seen in Figure 6.12, this change in surface finish is known as icing and is visible at low magnification [6].

These figures are obtained from ongoing work that involves running 2 cm reciprocating tracks for 5400 s using mineral grease. Along with changes in surface quality, physical properties such as surface roughness are expected to change. Under an illuminated microscope, machining grooves and machining damage are undetectable with increasing electrical stress. The surface roughness

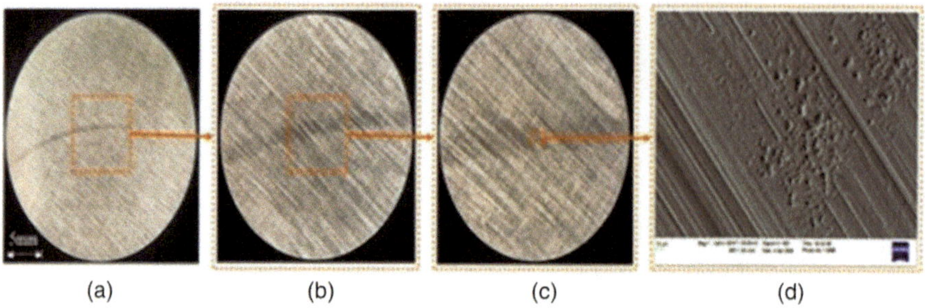

FIGURE 6.12 2 cm matte arc trace from ball-on-disk rolling test: (a) microscopic view at 1× magnification; (b) microscopic view at 3× magnification; (c) microscopic view at 6× magnification. (d) SEM image of electrical pitting resulting from 2 cm reciprocating test for 5400 s using mineral grease [6] [open access].

of the ceramic ball increases significantly, as shown in Figure 6.13. After the friction experiment, the surface of the ball shows grinding marks. The temperature and friction directly affect the appearance of wear marks.

The surface and cross-sectional morphology of the $NbSe_2$-Ti films are shown in Figure 6.14. Each film had flat, smooth surfaces without any obvious defects. The surface of the single-component $NbSe_2$ film in Figure 6.14a, e shows an amorphous structure in cross-section and a loose shape. A distinct columnar crystal structure formed in the film as the sputtering current increased, and the film thickened to become denser. These characteristics are typical of thin-film architectures of transition metal dichalcogenides (TMDs) deposited by radiofrequency sputtering [15]. Although tiny pores were still visible between the aggregated grains, the addition of Ti atoms polished the nanoscale grains in the $NbSe_2$ core and increased the internal density of the film [10].

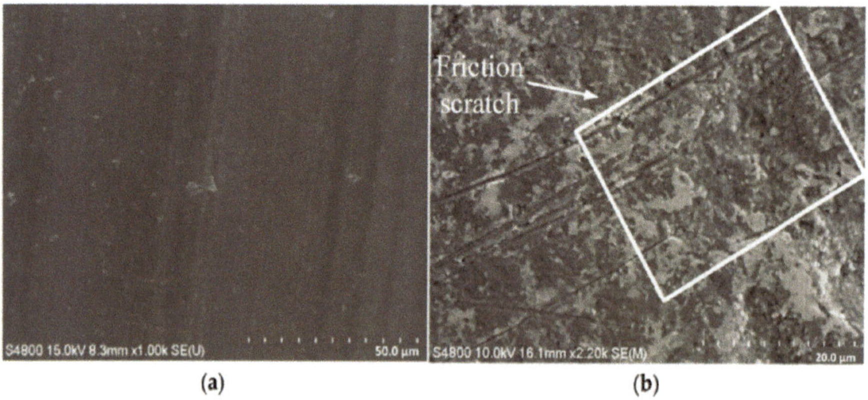

FIGURE 6.13 Surface morphology before and after the experiment. (a) G5 Ceramic ball surface before testing, (b) G5 Ceramic ball surface after testing [14] [open access].

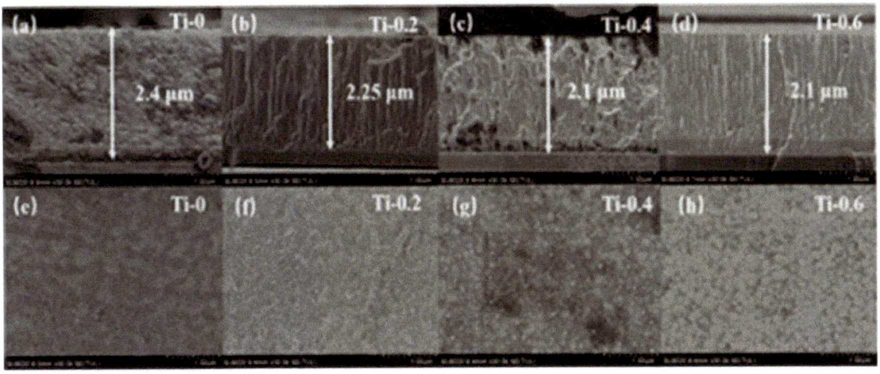

FIGURE 6.14 Cross-section (a–d) and surface (e–h) of $NbSe_2$-Ti films obtained by FESEM [10] [open access].

6.7 BOUNDARY LUBRICATION

Simply put, boundary lubrication is the process of lubricating surfaces with very thin layers of fluid, in which the type of lubrication and surface properties influence the coefficients of friction. Boundary lubrication typically affects bearings, gears, piston rings, pumps and gearboxes under high loads and low speeds. As a result, a thin layer with a thickness of several tens to several nanometers is formed between sliding and rolling objects. Because friction is greatest in this procedure, tribologists consider boundary lubrication to be extremely important. Using lubricants with matched viscosity can help prevent boundary lubrication. High friction and wear occur due to the inability of lubricant to separate materials when the viscosity is too low. On the other hand, using a lubricant with too high a viscosity will shear the internal oil, increase the temperature and cause the layers of oil to slide against each other. However, blurring the boundaries is undoubtedly useful. This has a significant impact on reducing adhesive and chemical wear. Door hinges are one example of low-speed, low-cost applications where boundary lubrication is often used [13].

6.8 MIXED LUBRICATION

Between boundary lubrication and full-diaphragm lubrication (hydrodynamic or hydrodynamic-elastic), mixed lubrication is often considered a transition phase. As the sliding speed increases under boundary lubrication, the friction coefficient decreases sharply. This results in a reduction in friction, to a point known as mixed lubrication, reducing the amount of contact between sliding surfaces. As the name suggests, with mixed lubrication, the sliding of metal on metal is still associated with an abrasive load, and some load also occurs on the lubricant. As the lubricating film thickness increases, the system moves to full-film lubrication, often called flexible hydrodynamic lubrication or hydrodynamic

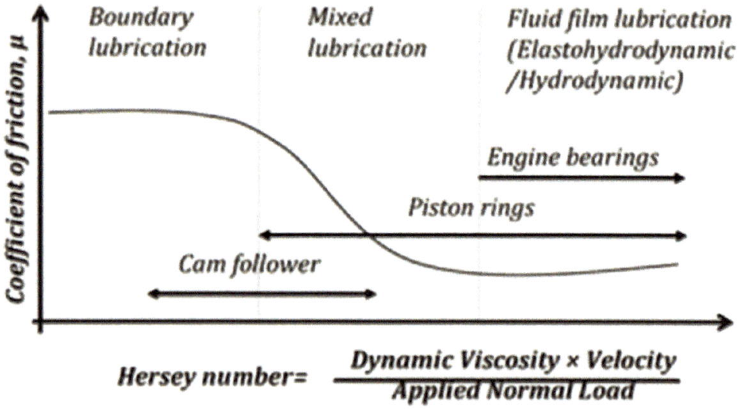

FIGURE 6.15 The Stribeck curve, based on the Hersey number, shows different lubrication regimes [11] [open access].

lubrication [7]. A typical Stribeck curve as a function of Hersey number is shown in Figure 6.15. The curve depicts boundary, mixed and liquid-film lubrication regimes. The same fully formulated engine lubricant is often used in the lubrication systems of important internal combustion engine components found in a typical passenger car, including engine bearings, piston rings and cam followers. On another hand, these engine components are also known to operate in a variety of lubrication regimes, from liquid film to boundary. Therefore, it is important that a properly formulated engine lubricant can operate in a variety of lubrication regimes [11].

6.9 FULL-FILM LUBRICATION

Hydrodynamic lubrication and hydrodynamic elastomeric lubrication are two different lubrication modes of solid-film lubrication.

6.9.1 Hydrodynamic Lubrication

This situation occurs, for example, between a shaft and a bearing, when a working gap is formed and maintained by a full layer of lubricant. The likelihood of corrosion is reduced because the oil lifts the shaft away from the bearing surface during operation. Hydroplaning in a car is similar to hydrodynamic lubrication. In this case, water is a low-viscosity medium. It supports the entire weight of car due to its speed. In order for hydrodynamic lubrication to be studied as intended, a minimum contact pressure between the surfaces and geometric similarity of the sliding parts are required. Friction within the oil itself is the only significant friction that occurs during hydrodynamic lubrication. Therefore, the requirement of this mode is that the oil viscosity maintains its hydrodynamic state under all operating conditions. Sliding materials can come into contact and cause wear and friction if the gap is significantly reduced, which can happen when going from high speed and high load to low speed and high load [8].

6.9.2 Hydrodynamic Elastomeric Lubrication

The action of hydrodynamic lubrication and elastic hydrodynamic lubrication (EHL) is largely comparable. But according to EHL, the surfaces of stabilizers elastically deform under the influence of strong pressure applied to the film. The surface deformation is proportional to the micron thickness of layer separating them. EHL is a major advancement in tribology and in the design and function of machine parts. This mode is often applied to mechanical components such as camshafts, gears and ball bearings. Due to increased viscosity of lubricant and elastic deformation of surface caused by high pressure, the EHL is an improvement over hydrodynamic lubrication. Over time, many standards and test procedures have been created to determine the factors affecting a tribological system. By clearly understanding these factors, lubricant manufacturers can more efficiently formulate their products and use them for specific applications such as pumps, gear systems and bearings [9]. This study discusses the test procedures used to evaluate the suitability of a lubricant for a specific application. To better understand some of the aspects associated with tribological systems, we will also look at various terms that are often used in tribology. This is the section that deals with the main aspects of tribology. Understanding these factors allows lubricant manufacturers to intelligently design and target their products for specific applications, including bearings, gear systems, pumps and more. This discusses the test procedures used to evaluate the suitability of a lubricant for a specific application. In addition, we look at various terms commonly used in tribology that will help you understand the various aspects associated with tribological systems. In this this section, we explore three key aspects of tribology and how it applies to mechanics [16].

6.10 TEST STANDARDS AND PROCEDURES

The lubricant may undergo a series of tests to ensure its quality and ease of use. The study includes how different influencing elements have different effects on the chemical and physical properties of lubricants. The information collected is used to determine the best method for applying the lubricant [17].

6.10.1 Viscosity Test

Depending on the type of product, several measuring instruments are used to calculate oil viscosity. By using a falling ball viscometer or other similar methods, compliance with DIN 51562-1 standards can be achieved. At 40°C the kinematic viscosity is designated as V(ny) [mm^2/s]. To evaluate the drop in viscosity at higher temperatures, for example, 100°C is often important [18].

6.10.2 Flash Point Test

The term "flash point" refers to the lowest temperature at which a sufficient quantity of flammable vapor will ignite upon contact with an ignition source. Using this method, the fire hazard of flammable liquids can be assessed. The most common measurement methods vary depending on the type of product and height of flash point. Two tests can be carried out: open crucible tests or closed crucible tests, also known as closed crucibles (according to DIN ISO 2592) or open crucibles (according to ISO 1523:2002). Since excessive heat generation is usually caused by high friction, this test is very important for rheology [19].

6.10.3 Thread Friction Test

Use a screw test bench to determine thread friction. The force applied when tightening the bolts and nuts is known as the friction coefficient of a threaded joint according to ISO 16047. The material, surface type and thread size affect the test result. Coatings and lubrication can also affect the friction properties of threaded connections. The result will be a standard to measure the impact of different standards [20].

6.10.4 Condensed Water and Salt Spray Test

Understanding how corrosion affects metal components in various applications is one of the most important studies. Condensate and salt spray tests can be used to determine the degree of corrosion protection. The purpose of both tests is to determine how long it will take for a component to show signs of wear. ISO 9227 specifies the salt spray test and DIN 50017 specifies the condensed water test. Moist air condenses on the surface of a wet or dry object in the water condensation test. The lubricated test piece is located in a protected area with special environmental conditions. It takes some time for condensation to appear. These tests allow precise conclusions to be drawn about the behavior and effectiveness of corrosion protection. These tests can also reveal paint defects. Similar settings are used for salt spray experiments. This test simulates a salty environment by exposing coated panels to a special salt jet by ISO 9227. The number of hours required for rust to appear and level of corrosion protection provided by the lubricant are tested [16].

6.11 SLIDING BEARINGS

The thrust bearing was originally designed to transmit the strong axial force created on the shaft of hydraulic units and it is a type of plain bearing (turbines and generators). It is used in applications where roller bearings are impractical due to their limited load capacity and short service life. The operating principle is based on hydrodynamic lubrication, namely the pressure distribution created by a layer of liquid separating a rotating plate from a series of inclined pads, two closely spaced surfaces. The integral of resulting pressure distribution on the rotating plate is equal to transmitted thrust. As long as the oil paint remains, there is no risk of corrosion since it separates the two surfaces, as shown in Figure 6.5, which include the journal bearings that support the crankshaft and connecting rod [2]. The mating surfaces may be subject to sliding wear as a result of the relative movement of the rotating shaft relative to the bearing like slip. Sliding bearings are often lubricated with oil, which creates a liquid film of lubricant [17].

6.12 OIL-LUBRICATED PLAIN BEARINGS

The bearings and guides in simple bearings support precisely defined paths of movement of engine and machine components. The resulting forces are transmitted throughout the entire procedure. To stop wear, lubricant must be used to separate the sliding partners. Figure 6.16 provides a simplified method for predicting the minimum film thickness "hmin" [4].

A diagram of a magazine bearing with a 360° rotation is shown in Figure 6.17. Therefore, the ratio 2 hmin/s can be estimated (often around 0.2) using the graph in Figure 6.17. Typically, hmin in electric motor bearings (Figure 6.16) is less than 1 μm, and the surface roughness of the composite is about 0.35 μm [8]. The plain bearing may be subject to wear even if the film thickness is correct at steady state. In fact, as shown schematically in Figure 6.18a, the bearing and shaft are

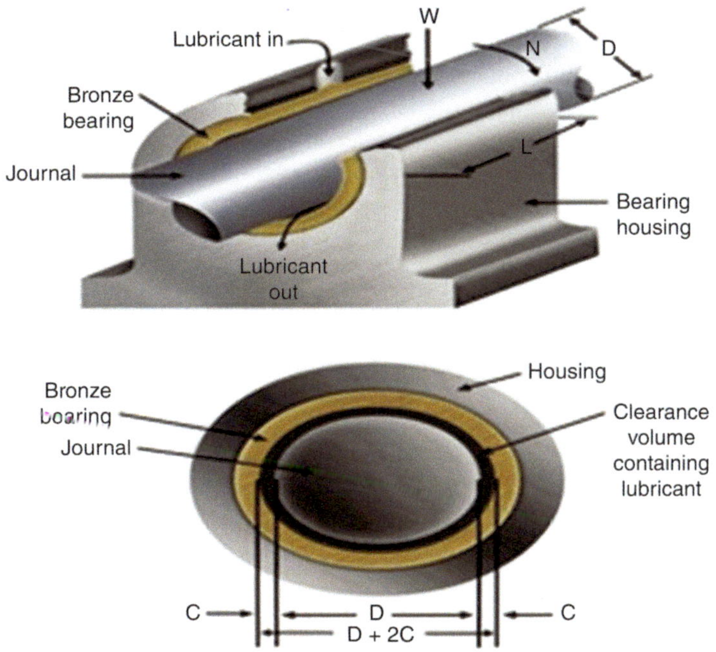

FIGURE 6.16 Journal bearing part names and film distribution [21] [open access].

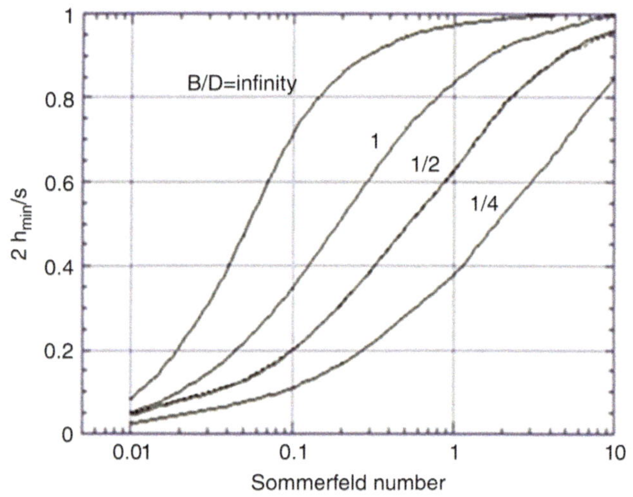

FIGURE 6.17 Using the Raimondi–Boyd design table (adapted from [1]), one can determine the minimal film (for various B/D ratios) for a 360° hydrodynamic [21] [open access].

in contact before rotation begins and oil does not separate the mating surfaces. As shown in Figure 6.18b, the column is rolling along the supporting surface when it begins to move due to the relatively high coefficient of friction at constant equilibrium, with Δ about 30° [9]. As a result, sliding wear due to the clutch may develop at this stage, and the procedure is repeated at every start and stop. When using heavy loads, this wear is very important.

This shift in lubrication terminology from limited lubrication to mixed lubrication and continuous liquid film lubrication ultimately refers to change in film thickness and overall lubrication efficiency over time. After several cycles, the

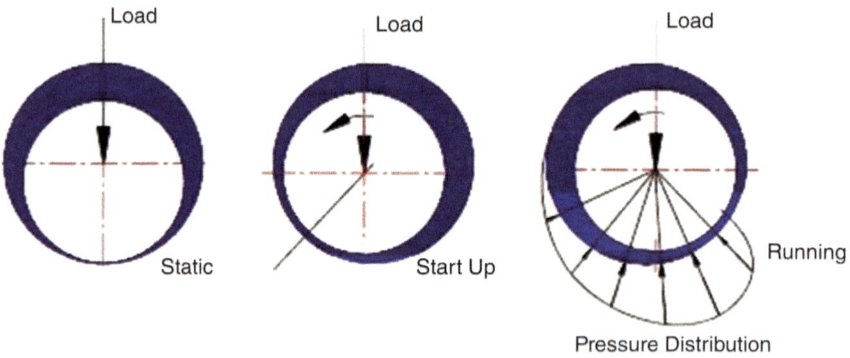

FIGURE 6.18 Stationary condition and transition to rolling motion: limited (a) lubrication and abrasive wear risk, followed by (b) mixed lubrication and (c)sliding wear under start-stop cycles with heavy loads [21] [open access].

amount of oil drawn into passage increases, indicating that as parts wear or surfaces become weathered, a better seal or fit is achieved, allowing for more efficient oil retention. This can result in reduced wear and improved lubrication (Figure 6.18c). However, if the lubricating oil contains contaminated abrasive particles, surface damage due to wear occurs. Wear is influenced by the average particle size relative to hmin and their hardness. If the particle size exceeds hmin, significant corrosive wear may occur. Area 2 in Figure 6.18b is primarily susceptible to abrasive damage. Due to the heat of off-road driving, the lubricant in this area can become very hot, which can promote corrosion of the oil on the bearing surface [21].

If the oil becomes contaminated or oxidized, corrosion can become severe [6]. Then the bearing materials must meet the following requirements:

1. High hardness materials or, moreover, soft materials capable of resisting abrasive wear can be used to achieve high adhesive wear resistance in combination shaft.
2. They consist of solid foreign particles that the lubricant contains, which reduce their abrasive ability and enhance resistance to cavitation erosion or contact fatigue.

Sleeve bearings come in different shapes. For example, they differ depending on the direction of load. The load moves radially in radial bearings towards the axis of rotation. The load in the support bearing moves in a direction parallel to the axis of rotation. Their combination is represented by axial radial bearings. For simple bearings, the type of lubricant is no less important. Examples of common types of lubrication are hydrostatic and hydrodynamic lubrication.

During operation, the lubricating oil must ensure minimal friction losses. As a result, the viscosity of lubricant during use is critical. However, design (e.g., rotational speed) and environmental factors (e.g., temperature and dirt ingress) are only factors that determine the viscosity required for your system. For personalized guidance, please contact our Application Technology Department.

The qualities mentioned here are good adhesion, minimal tendency to foam, no emulsion with water, resistance to aging and corrosion protection – characteristics common to all lubricants used in plain bearings. Solids, oils, greases or gases can be used to lubricate plain bearings. In most cases, our multipurpose grease can be used to lubricate primary bearings. They have proved their worth time and again in a variety of circumstances. Sliding wear is present in dry or unlubricated bearings throughout the entire service period. Furthermore, when dry bearing materials slide over a shaft which is often made of steel, this should result in a fairly low friction coefficient. Checking that bearing wear is sufficiently low forms the basis of design. The PV limit concept and the maximum PV limit concept provide a broad basis for the design process [9].

For internal combustion engines, the piston ring/cylinder liner is also an important component (Figure 6.17), since losses due to wear and friction in this system have a significant impact on fuel and energy consumption, as well as on associated emissions to the environment [2]. This main study is primarily performed on the upper rings of piston assembly, as shown schematically in Figure 6.16. The oil control ring at the bottom restricts the flow of lubricant upward from the crankcase. Sleeve bearings and rolling bearings are the two main types of bearings currently used in equipment. This section focuses on the unique lubrication needs of journals such bearings, often referred to as journal bearings.

A plain bearing consists of a shaft, also known as a journal, and a support element, which may be a composite sleeve, a half-shell containing the shaft, a sleeve surrounding the shaft, or two halves (top and bottom). Sleeve bearings are used at medium to high speeds and at high radial loads (perpendicular to the shaft axis). Typical applications include turbines, large grinding systems, gearboxes, compressors, engine crankshafts and shaft bearing supports [8].

Every plain bearing has some typical design elements. An oil film envelops the shaft and bearing shell of a simple bearing. The shaft is made of premium steel – durable and corrosion-resistant. Depending on the design features of equipment, bearing shell may consist of one or several layers. The lubrication mode under typical operating conditions is a full-fluid hydrodynamic film. When the lubrication between the lubricated surfaces at the point of loading is sufficient to create a liquid wedge separating the sliding surfaces, a hydrodynamic film is formed. Since the lubricated parts do not come in contact with each other, wear and friction are reduced [9].

To express this condition, the formula ZN/P is used, where Z is viscosity, N is speed (rpm) and P is load. The Stribeck curve is the name given to the curve on this graph. This is a classic illustration of how speed, load and friction are related. When visible film loss occurs, resulting in short contact between two surfaces, it is called a mixed film condition. This can occur as a result of short-term changes in load, known as shock loading, which can cause the film to rupture and reverse physical contact. Another possible condition is the lubrication of boundary film. Figure 6.19b, c shows that the wear marks of G5 and G10 are deeper and the friction marks on the surface of the G3 ceramic ball are relatively shallow when the ceramic ball is subjected to the same pressure and speed conditions [10]. This is due to the relatively smooth contact surface and low friction of the high-precision ceramic ball. When the G10 ceramic ball is subjected to high loads, its surface wear characteristics and wear defects increase μm [16, 22].

The piston ring/cylinder liner system has a total roughness (Ra) of approximately 0.2 μm. The surface finish is often improved with a special texture designed to increase oil retention [7]. Sleeve or cervical bearings consist of shaft or cervix that rotates freely within a supporting metal sleeve or housing. These bearings do not have rolling parts. The theory and operation of these bearings can be complex despite their apparent simplicity of design and manufacturing. Now, a brief overview of pins and bushings, dry and semi-lubricated journal bearings, and inclined pad bearings is given. The subsequent text focuses on oil and grease-lubricated plain bearings.

A type of journal bearing known as low-speed pins and bushings occurs when the shaft or bushing rarely makes a full rotation. Metal-to-metal contact

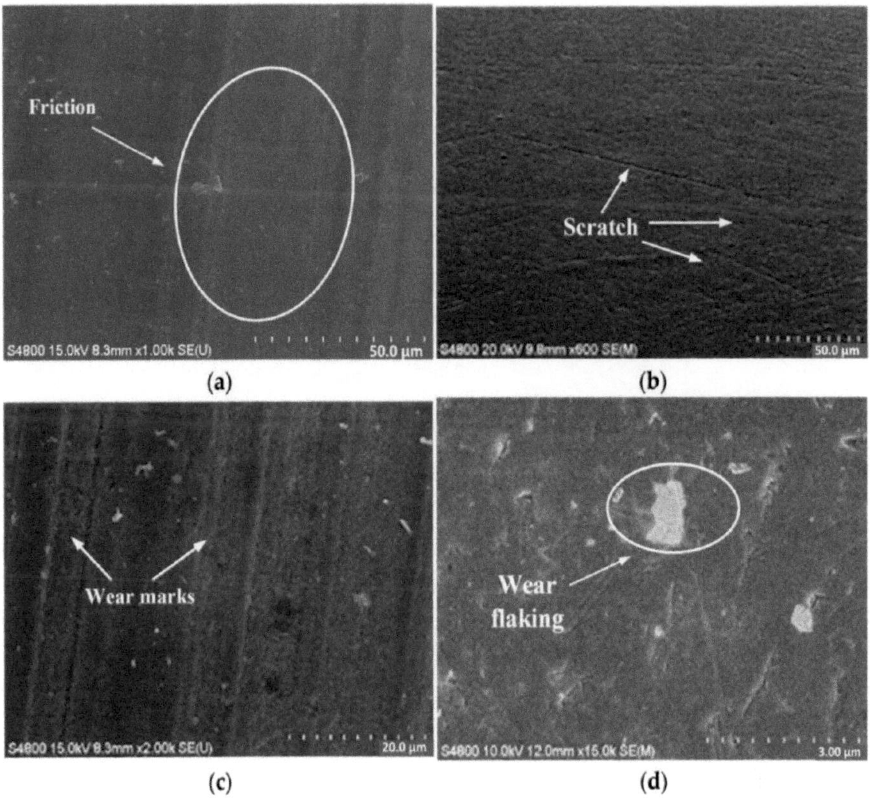

FIGURE 6.19 Surface morphology before and after the experiment. (a) G3 V = 500 r/min, F = 35N, (b) V = 500 r/min, F = 35N, (c) G10 V = 500 r/min, F = 35N, (d) G10 V = 100 r/min, F = 75N, [open access].

does not happen due to low-speed partial rotation that typically occurs before a change in direction. This happens inside the bearing. In extreme lubrication mode, pins and bushings are always on. To support the weight, these bearings are typically lubricated with extreme pressure (EP) grease. The grease contains solid molybdenum disulfide (molybdenum) to improve the load-bearing capacity of the grease [7].

In a dry journal bearing, the shaft rotates inside dry sleeve made from a polymer that can be mixed with hard materials such as nylon, graphite and molybdenum. These bearings are only suitable for low-load and low-speed applications. Partially lubricated plain bearings contain a rotating shaft inside a porous metal material. A bushing is made of aluminum or bronze, sintered with lubricating oil applied into the pores of the porous metal. Low loads, low-to-medium speeds and temperatures up to 100°C (210°F) are the limitations of these bearings.

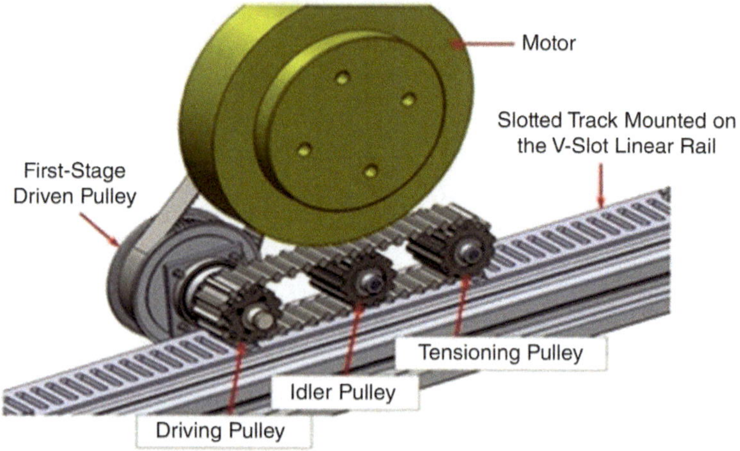

FIGURE 6.20 B-SALD driving component design details (prototype bottom view) [12] [open access].

A rotating bearing shaft with an inclined or rotating support is located inside a housing consisting of curved pads. Each shoe can rotate independently and match the curvature of the rod. The B-SALD system is driven by a disk motor (U8 Lite-KV85, T-Motor, Nanchang, China) that is originally developed for drones to achieve high output torque. A single-stage belt drive is chosen to provide great flexibility in the layout of mechanical components in prototype to simplify the design and transmit rotation to a drive pulley of B-SALD mechanism. The GT2 belt is used to create a 4:1 gear ratio between the 15-tooth drive pulley and 60-tooth driven pulley in this first-stage belt drive, which increases the output torque. As shown in Figure 6.20, the drive pulley of B-SALD mechanism is connected to output pulley of first-stage belt drive. The first-stage belt is located above the motor and B-SALD mechanism, and moving platform is only 121 mm wide, 127 mm long and 50 mm high. This results in an extremely small profile.

The advantages of this design are increased shaft stability and more accurate alignment of support shell with the rotating shaft. Plain bearings, bushings and babbitt bearings are types of journal bearings. The term "bait" describes the layers of softer metals such as copper, tin and lead that make up the metal contact surface of the bearing shell. To protect the stronger steel support housing from the harsher rotating shaft, softer metals are placed on top of it [13].

Insert-type primary support bearings can only support radial loads perpendicular to the shaft, which are typically caused by downward weight or load on shaft. Specially manufactured journal bearings can also support axial or axial loads along the shaft axis. An inclined journal bearing capable of withstanding axial and radial stresses is shown in Figure 6.20. Piston rings, piston skirts and cylinder liners are listed in Table 6.1 along with their surface treatments [16].

TABLE 6.1
Parts and Surface Treatment of Cylinder Liners, Piston Rings and Piston Skirts [16] [Open Access].

	Piston Ring	Piston Skirt Grey	Cylinder
Conventional materials	• Grey cast iron • Malleable/nodular cast iron	Cast iron	Bore/liner Grey cast iron
Recently used materials	• Tool steel • Chromium plating • Thermal sprayed molybdenum on cast iron	Copper and aluminum alloys	Steel; compacted graphite iron; Si-containing aluminum alloy; cast Al-base metal matrix composite
Surface treatments	• Gas nitriding of steel	• Chromium plating on cast iron • Ni-P reinforced with SiC	Chromium plating on cast

6.13 CAM/SLAVE SYSTEM

When a journal bearing is operated at the correct speed, surface area, viscosity and volume of oil, it can withstand very heavy loads. These requirements must be balanced. The viscosity of lubricant must be adjusted according to changes in load or speed. The ZN/P formula shows the complex calculations required to obtain the proper clearance, but there is no simple way to determine the required viscosity for oil-lubricated plain bearings [7].

Once the correct viscosity grade is determined, factors such as corrosion protection, corrosion prevention, oxidation resistance, water/air separation ability, etc. must be taken into account. Sleeve bearings can be used in a wide variety of situations, which is why there is more than one set. The choice depends on the operating conditions and equipment design. Sleeve bearings are usually oil-lubricated, but low-speed equipment may also benefit from lubrication, especially if the equipment is subject to repeated starting and stopping or if the bearings are difficult to access [8]. To maintain hydrodynamic lift and effective lubrication conditions, the type and amount of lubricant depends on the constant replenishment of lubricant mass in the dynamic clearances or voids during bearing rotation. Poorly sealed equipment may require more lubricant and more frequent replacement cycles. In this case, sliding is primarily to blame for wear and friction. Since the speed of rollers can then fluctuate rapidly, some sliding may occur, but rolling motion predominates, so the friction coefficient is orders of magnitude lower than in a pure sliding clutch. In general, when contacting the side and base contours, the lubricant is mixed, but in the nose zone, a boundary lubrication system is achieved, which reduces the service life of system. Therefore, lubricants containing extreme pressure additives are often used [17]. The Hertzian pressure and lubricant film thickness at the front end of cam drive systems of internal combustion engines are about 600 MPa and 0.1–0.15 μm, respectively [4]. On another hand, the surface roughness of nose has changed. Material is removed and redistributed by interaction with surface defects, contributing to an overall reduction in roughness. The surface becomes more uniform as projections wear; this can improve the tribological properties of surface and reduce friction during subsequent contacts [7].

Operating conditions, quality of lubricants and amount of time spent on study affect the amount and frequency of manual (periodic) lubrication. Choosing the right lubricant begins with choosing oil to use. Heavy oil-based greases are used for manual lubrication of simple bearings under severe operating conditions. Once the appropriate viscosity oil has been selected, factors such as soap thickener, oxidation and rust factors, operating stability, pumpability (for automated systems) and durability (EP/AW) are considered. Solid additives such as graphite or molybdenum disulfide can be used for long periods and under extremely high loads. In mixed and boundary lubrication conditions, solid additives mechanically prevent metal-to-metal contact [16].

6.14 ROLLING BEARINGS

Roller bearings consist of two concentric rings containing spherical balls or cylindrical or tapered rollers. Free moving parts allow the rings to rotate while carrying radial load and the inner ring is often attached to the shaft [7]. Each rolling element forms an informal connection to the loop tracks. For application of ANSYS2020R1, as shown in Figure 6.21, the mesh is used to discretize the

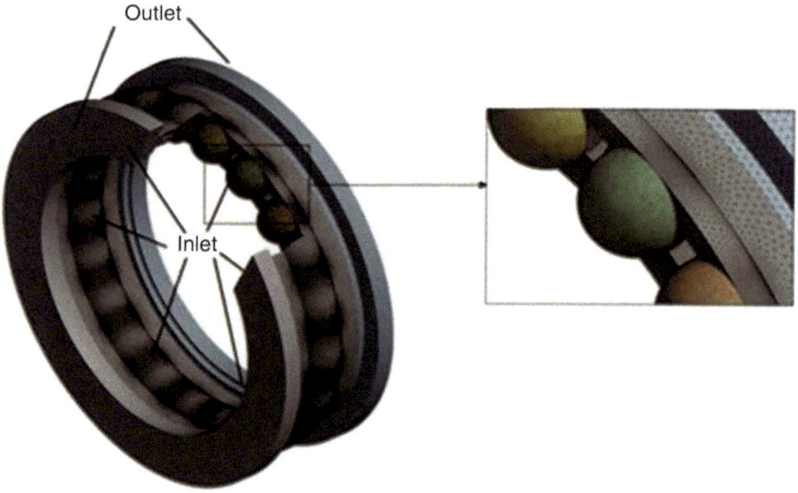

FIGURE 6.21 Computational boundary condition and mesh architecture [7] [Open access].

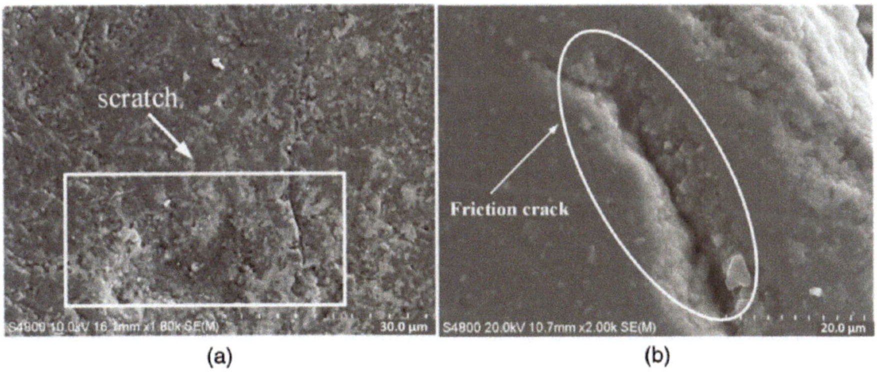

FIGURE 6.22 Surface morphology before and after the experiment (a) G10 V = 500 r/min, F = 75N, (b) G10 V = 1000 r/min, F = 75N [14] [open access].

fluid region of lubricated ball bearing under the raceway. Due to complex internal structure of bearing, the tetrahedral unstructured mesh is used to mesh the entire fluid region. To improve the mesh quality, the clearance in contact region, such as contact region between the balls and cage, balls and inner ring, balls and outer ring, and cage and guide surface is locally cleaned [17]. Contact fatigue can lead to a wear on raceway surfaces and rolling components. High-strength heat-treated steels are often used to manufacture all loaded parts. Because rolling bearings are available in several standardized configurations, a simplified design process is typically used to evaluate bearing life. Figure 6.22a, b shows that the wear marks on the ceramic balls G5 and G10 are deeper and the friction marks on the surface of the ceramic ball G3 are relatively shallow when the ceramic ball is subjected to the same pressure and speed conditions. This is due to the relatively smooth contact surface and low friction of the high-precision ceramic ball. When the ceramic ball G10 is subjected to high loads, its surface wear characteristics and wear defects increase by micrometers. This is due to Cavallaro's [9] formula is more suitable for bottom-lubricated bearings, whereas Parker's [8] formula provides an estimated expression for the average oil volume percentage inside jet-lubricated ball bearings. Considering that it is very similar to Cavallaro [9], the statement given in this study is considered valid in this case [7].

6.15 GEARS

Gears are used to transmit power and rotational motion between shafts. They are important parts of many mechanical tools [23, 24]. There are different types of gears [25]. For example, spur gears have teeth parallel to the axis of rotation of the shaft. Gears are usually lubricated (by splash or oil bath), although they can sometimes operate in dry environments. Figure 6.16 schematically shows the contact of the teeth of a spur gear during contact. Due to the natural flexibility and elasticity of belt, the B-SALD system can reduce shock loads, which is especially useful when interacting with people. For a reliable power/force transmission, it is important to ensure that several teeth are in contact at the same time, however, since the teeth of the timing belt have a comparatively lower force capacity [26]. The B-SALD timing belt has a flat section as shown in Figure 6.23, which allows several teeth on the outside of timing belt to simultaneously engage with the corresponding grooves on the linear track [15].

Spur gears establish linear contact, and Hertzian theory can be used to estimate contact stresses. At the pitch point, the maximum Hertzian pressure is reached [12]. This makes it possible to evaluate the lubrication regime and κ-factor. Given this role, pitch line speed is a direct indicator of lubrication conditions. When calculating hydrodynamic thrust force, speed is important, in particular, low-speed transmission reduces [15].

Tribology is an important science that studies wear, friction and lubrication, which affect the durability and performance of mechanical systems. Its interdisciplinary nature, spanning fields such as computer science and materials science, facilitates careful study and creative solutions. By exploring subdisciplines such as tribology, nanotribology, space tribology and tribotronics, scientists can address specific problems and advance the discipline. Future research and collaboration will advance technologies, enhance commercial applications and expand our understanding of tribological processes in artificial and natural systems.

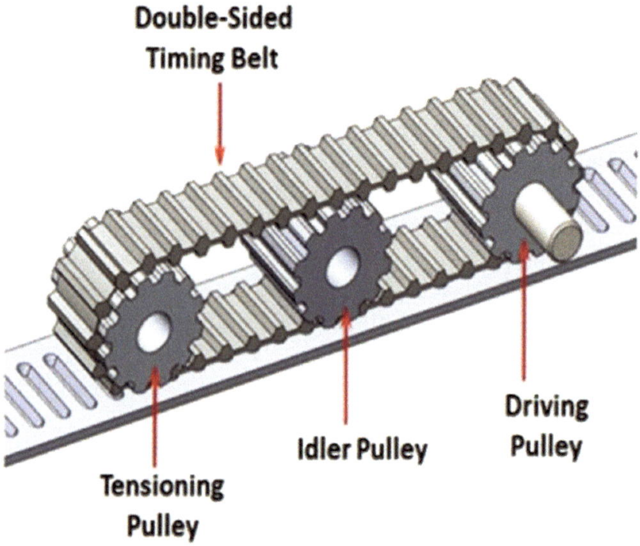

FIGURE 6.23 Schematic showing the working principle of B-SALD [12] [open access].

6.16 PROBLEMS

1. Explain the processes of wear and friction of metals in the metallurgy industry.
2. Explain lubrication processes and tribology. What are the requirements for these operations?
3. What are the techniques used to improve wear and friction resistance?
4. How can the surface of an aluminum alloy be modified to maximize bond strength?
5. How can you reduce the friction between aluminum surfaces moving against each other?

REFERENCES

1. Chen, H., Qin, Z., He, M., Liu, Y., & Wu, Z. (2020). Application of electrochemical atomic force microscopy (EC-AFM) in the corrosion study of metallic materials. *Materials*, *13*(3), 668.
2. Terek, V., Miletić, A., Kovačević, L., Kukuruzović, D., Škorić, B., Panjan, P., & Terek, P. (2020). Surface topography and grain morphology of nanolayer TiAlN/TiSIN coating governed by substrate material and rotation during deposition. *Material Proceed.*, *2*(1), 32.
3. He, B., Zhang, N., Lin, D., Zhang, Y., Dong, F., & Li, D. (2017). The phase evolution and property of $FeCoCrNiAlTi_x$ high-entropy alloying coatings on Q253 via laser cladding. *Coatings*, *7*(10), 157.
4. Chen, Y., Zhang, H., Li, X., Xiao, S., Gu, F., & Shi, Z. (2023). Effects of wear on lubrication performance and vibration signatures of rotor system supported by hydrodynamic bearings. *Lubricants*, *11*(3), 107.
5. Yao, J., Wu, Y., Yang, J., Sun, J., Xia, Z., Tian, J., & Gao, L. (2022). Study on distribution of lubricating oil film in contact micro-zone of full ceramic ball bearings and the influence mechanism on service performance. *Lubricants*, *10*(8), 174.
6. Janik, J.R., Saha, S., Jackson, R.L., & Mills, G. (2024). Exploring the boundaries of electrically induced bearing damage in grease-lubricated rolling contacts. *Lubricants*, *12*(8), 268.
7. Lyu, Y., Li, Y., Li, C., Jiang, L., & Liu, Z. (2024). Oil-air distribution prediction inside ball bearing with under-race lubrication based on numerical simulation. *Applied Sciences*, *14*(9), 3770.
8. Parker, R.J. Comparison of predicted and experimental thermal performance of angular-contact ball bearings; NASA Technical Paper 2275; NASA: Washington, DC, USA, 1984.
9. Cavallaro, G., Nelias, D., & Bon, F. (2005). Analysis of high-speed intershaft cylindrical roller bearing with flexible rings. *Tribol. Lubr. Technol*, *48*, 154–164.
10. Yang, Y., Wang, G., Zhou, X., Gu, X., & Gu, L. (2024). A promising conductive lubricant for space sliding electrical contact: NbSe2-Ti film. *Lubricants*, *12*(2), 44.
11. Chong, W.W.F., Hamdan, S. H., Wong, K. J., & Yusup, S (2019). Modelling transitions in regimes of lubrication for rough surface contact. *Lubricants*, *7*(9), 77.
12. Dooley, M., & Shen, X. (2022, September). A novel self-actuated linear drive for long-range-of-motion electromechanical systems. *Actuators*, *11*(9), 250.

13. Mohan Sankaran, R., Plasma Processing of Nanomaterials in Nanomaterials and Their Application, CRC Press, Taylor and Francis Group, Boca Raton, FL, United States, (2011) 432 pages.
14. Tian, J., Wu, Y., Sun, J., Xia, Z., Ren, K., Wang, H., … & Yao, J. (2022). Thermal dynamic exploration of full-ceramic ball bearings under the self-lubrication condition. *Lubricants*, *10*(9), 213.
15. Zhang, S.D., Liu, Z.W., Wang, Z.M., & Wang, J.Q. (2014). In situ EC-AFM study of the effect of nanocrystals on the passivation and pit initiation in an Al-based metallic glass. *Corros. Sci.*, *83*, 111–123.
16. Pawlus, P., Koszela, W., & Reizer, R. (2022). Surface texturing of cylinder liners: A review. *Materials*, *15*(23), 8629.
17. Baer, D.R., Engelhard, M.H., Johnson, G.E., Laskin, J., Lai, J.F., Mueller, K., Munusamy, P., Thevuthasan, S., Wang, H.F., Washton, N., Elder, A., Baisch, B.L., Karakoti, A., Kuchibhatla, S.V.N.T., & Moon, D. (2013). Surface characterization of nanomaterials and nanoparticles: Important needs and challenging opportunities. *J Vac Sci Technol A*, *31*, 050820.
18. Cho, E.J., Holback, H., Liu, K.C., Abouelmagd, S.A., Park, J., & Yeo, Y. (2013). Nanoparticle characterization: State of the art, challenges, and emerging technologies. *Mol Pharmaceut*, *10*, 2093–2110.
19. Mohanraj, V.J., & Chen, Y. (2006). Nanoparticles-a review. *Tropical Journal of Pharmaceutical Research*, *5*, 561–573.
20. Lopez-Serrano, A., Olivas, R.M., Landaluze, J.S., & Camara, C. (2014). Nanoparticles: A global vision. Characterization, separation, and quantification methods. Potential environmental and health impact. *Anal Methods-Uk*, *6*, 38–56.
21. Someya, T., Mitsui, J., Esaki, J., Saito, S., Kanemitsu, Y., Iwatsubo, T., … & Kanki, H. (2013). Journal-Bearing Databook, Springer Science & Business Media.
22. Rosendahl, S.M., Borondics, F., May, T.E., Pedersen, T.M., & Burgess, I.J. (2011). Synchrotron infrared radiation for electrochemical external reflection spectroscopy: A case study using ferrocyanide. *Anal. Chem.*, *83*, 3632–3639.
23. Kahru, A., & Savolainen, K. (2010). Potential hazard of nanoparticles: From properties to biological and environmental effects. *Toxicology*, *269*, 89–91.
24. Kumar, S., Liu, X., Borondics, F., Xiao, Q.F., Feng, R.F., Goormaghtigh, E., & Nikolajeff, F. (2017). Insights into biochemical alteration in cancer-associated fibroblasts by using novel correlative spectroscopy. *ChemistryOpen*, *6*, 149–157.
25. Kilcoyne, A.L.D., Tyliszczak, T., Steele, W.F., Fakra, S., Hitchcock, P., Franck, K., Anderson, E., Harteneck, B., Rightor, E.G., Mitchell, G.E., Hitchcock, A.P., Yang, L., Warwick, T., & Ade, H. (2003). Interferometer-controlled scanning transmission X-ray microscopes at the advanced light source. *J Synchrotron Radiat.*, *10*, 125–136.
26. Chapman, H.N. (2010). Microscopy: A new phase for X-ray imaging. *Nature*, *467*, 409–410.

7 Surface Treatments

7.1 INTRODUCTION

As wear and friction are surface-related phenomena and wide range of available surface treatment technologies have become increasingly commercially mature, this field is becoming increasingly recognized in the fight against wear. This chapter primarily uses thermochemical machining technology for ferrous metals to show the fundamental concepts of surface technology and its impact on tribology. The fact that surface technologies, such as ceramic coatings and thermochemical treatments, are now applied to a range of engineering components shows how effective surface treatments are in preventing wear. However, surface engineering entails more than simply changing the characteristics of a surface using one of the available surface technologies [1–4]. The basis of surface engineering is designed. Therefore, when components are subjected to complex external loads, multiple surface treatment technologies must be utilized to create the required combination of characteristics to withstand the loads. The study discusses the philosophy of surface design and provides thorough theoretical and practical studies of tribological behavior of low-alloy steel subjected to physical vapor deposition (PVD) coating, plasma nitriding and combined plasma nitriding [1, 5]. It is a natural phenomenon where moving parts and other components of tools or machinery gradually wear out during use, resulting in reduced efficiency, productivity and performance of the product. The need to replace worn machine parts or tools with new ones usually leads to temporarily stopping the machine or the entire technological process. At the same time, the global industry is constantly improving the efficiency and effectiveness of technical processes, as well as machines and vehicles. As a result, working with tools and machine parts becomes more difficult. Therefore, the industry continually strives to develop advanced tools, manufacturing materials and production technologies under certain operating conditions to provide maximum durability at a reasonable unit cost. The development and commercialization of composite materials represent one of the most notable and exciting advances in materials science and manufacturing technology. Many research centers around the world are conducting research and development in the field of metal matrix composite (MCC) production technologies [6–9]. These types of composite materials combine the advantages of a reinforcing material with the high flexibility of metal matrix. In Figure 7.1, the scanning electron microscope (SEM) images illustrate the distinct morphological changes on surfaces of four low-density glass composites (LDGCs) after treatment. SEM scans the HF specimens to show an acicular shape for CAD (Figure 7.1(1B)), Vita Suprinity (Figure 7.1(2B)) and honeycomb

DOI: 10.1201/9781003584735-7

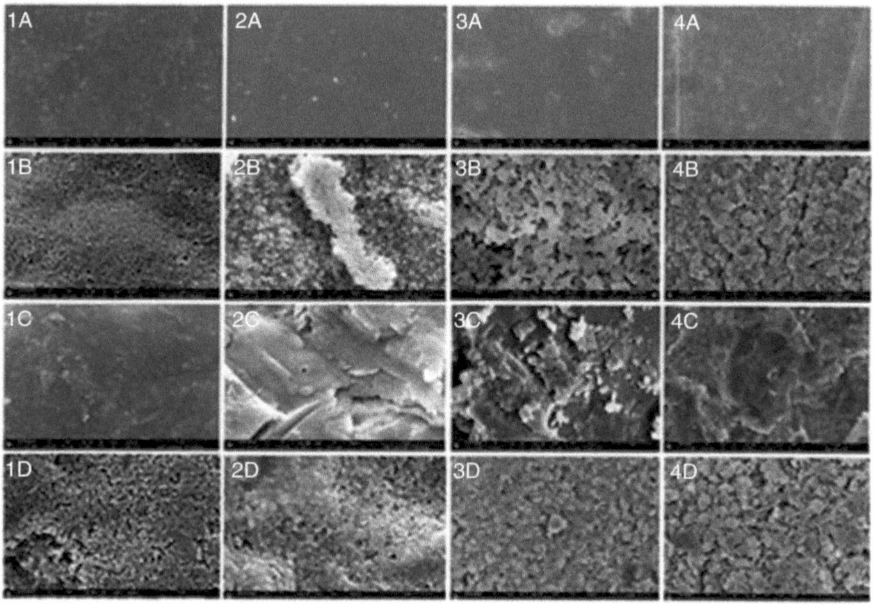

FIGURE 7.1 Surface microstructure after different surface treatments is visualized using four LDGC devices: IPS CAD (1), Vita Suprinity (2), Celtra Duo (3), and the wear resistance of composite coatings is usually increased by dispersed particles or high-hardness phases (4) using scanning electron microscope images. Control (A), HF (B), SB (C) and SB + HF (D) [1] [open access].

shape for Celtra (Figure 7.1(3B, 4B)). The mean roughness values Ra (nm) and standard deviations of the four CAD/CAM LDGCs are extrapolated from atomic force microscope (AFM) digital images. For Vita Suprinity and Celtra Duo, the sandblasting process yielded significantly higher mean roughness values ($p < 0.05$) than the other surface treatments (867.8 nm and 609.6 nm, respectively) [1].

As a result, these composite materials simultaneously have high tensile strength, good fatigue resistance and low specific gravity. The wear resistance of composite coatings is usually increased by dispersed particles or high-hardness phases [7–11]. Another noteworthy development in recent years is the introduction of new materials into manufacturing processes and materials science, in addition to the current widespread use of amorphous materials and nanoparticles [12–15]. However, extensive research is still ongoing, therefore scientists around the world are focusing on production technologies and the use of nanomaterials. The great potential of nanomaterials is determined by their nanometer grain size, which is typically in the range of <100 nanometers. Both metallic and nanostructured composite materials have beneficial properties (e.g., mechanical strength) that are often much higher than conventional materials. Even in the case of known and traditional structural materials, their physical, chemical and mechanical properties, such as increased hardness and strength, increased wear resistance of contact surfaces, increased ductility of brittle materials, etc. can be significantly improved. For example, the tensile strength of copper not only reaches 500 MPa at grain size 50 nm but increases at grain size 8 nm to a value of five times greater than 2500 MPa. By using most advanced coatings and surface engineering methods, it is possible to create coatings from modern metals and composite materials on small or nanometer scales, as well as surface layers with unique properties [16]. According to our findings, using hydrofluoric acid after sandblasting decreased the surface roughness by over 50% (Figure 7.2). Both Vita Suprinity and Celtra Duo were in this category. Sandblasting for max CAD and, however, resulted in a minor increase in roughness [1].

Currently, the following coating technologies are widely used in industry: Arc and plasma surfacing methods provide high efficiency in melting waste filler materials. The disadvantage is obtaining fine-grained structures. This is difficult due to relatively large amount of heat transferred to substrate material. For this reason, these methods are usually used to encapsulate large and rather bulky components. Laser and electron beam cladding provide extremely low heat input, minimal attenuation and limited air heating area. High-velocity oxygen fuel (HVOF) atomization, cold spraying (CS) at supersonic gas jet speeds, flame, arc and plasma spraying are examples of thermal spraying methods that make it possible to create layers with a structure and chemical composition that are

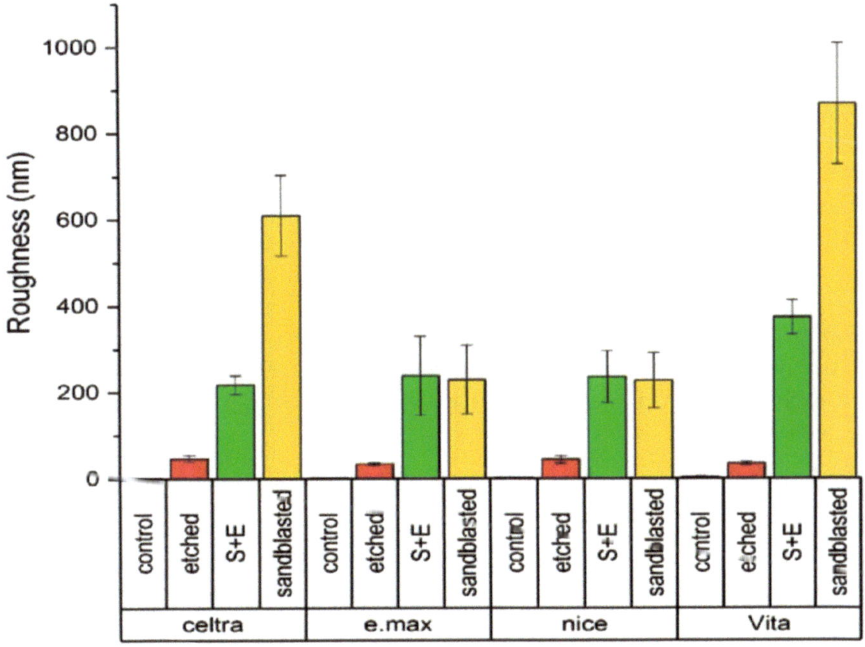

FIGURE 7.2 Mean of different surface roughness on lithium silicate-based glass-ceramic [1] [open access].

impossible using straightening methods. The disadvantage is that coatings are not uniform and often have high porosity, which reduces their usefulness and limits the range of their possible applications. Coating technologies such as PVD and chemical vapor deposition (CVD) enable the production of nanostructured metal and composite coatings. But in this case, the coatings are thin and production process is very labor-intensive and expensive. Structuring of surface layers improves tribological performance and wear resistance using various surface treatment techniques such as heat treatment and thermochemical methodology [15].

In the literature, several methods of surface heat treatment have been studied and described in detail to improve the tribological properties of metals and alloys. The most advanced and interesting are ion implantation, laser and plasma surface treatment. The basic idea of surface thermal treatment methods is to use extremely high rates of heating and cooling to change the microstructure of surface layer without changing its chemical composition. The previous mentioned methods can be used depending on the parts and processing characteristics of thermochemical processing. Thermochemical surface treatment with the enrichment of surface layer with various elements can significantly reduce the friction coefficient, increase wear resistance and corrosion resistance. The most common and widely used thermochemical surface treatments of metals and alloys, especially titanium and its alloys, are oxidation and carburization. The type and intensity of stress, as well as environmental variables such as temperature and corrosive liquids or gases, influence the type and extent of corrosion [17, 18].

Therefore, an individual approach to each working surface is necessary to ensure the durability and reliability of machine or tool components. This includes selecting the right material, designing its microstructure using the best manufacturing or processing technology. Specific properties (e.g., unique mechanical properties or processability) and the modified surface can be made responsible for compensating tribological loads. Selecting the appropriate surface treatment usually saves money by allowing the use of less expensive substrate materials. The higher surface treatment costs required to improve system performance that can be offset by choosing a low-cost substrate [15]. Selecting the best surface treatment option is a complex task as there are many variables to consider and the number of technologies available. The characteristics of machined surfaces under various wear conditions will be described, as well as associated process elements of each treatment to assist the designer in making appropriate surface treatment selections.

7.2 HEAT TREATMENT

Heat treatment is a technique changes the crystalline structure of metals and metal alloys through controlled heating and cooling. Heat treatment can improve a material's strength, ductility, hardness and temperature resistance among other properties, depending on the substance and the treatment method. In addition, surface silylation can produce materials suitable for chromatographic applications. By applying the thermal carburizing process with carbon nitriding to MSS P675, well-distributed metallic and carbon precipitates (carbides) with diameters ranging from about 10 nm to 2 μm are obtained (Figure 7.3a). The surface of specimen is surrounded by a martensitic matrix. Behind the surface, each specimen contained carbides at different levels about 1000 μm radially inward (data not shown). The chemical composition of the carbides and alloying elements that are cleaved from the matrix to form these carbides after heat treatment are determined by EDS analysis of specimen surfaces (Figure 7.3b). However, EDS examination (Figure 7.3a, b) on a wider surface of each steel shows that molybdenum and chromium are mainly cleaved into carbides after heat treatment, raising the possibility of magnetic carbide formation. Due to inconsistent results where large amounts of reaction (by EDS) penetrated both the carbide and surrounding matrix, EDS is performed on the bulk of steel rather than on individual carbides. On another hand, nickel acts as an austenite stabilizer with carbon and iron, promoting the development of non-magnetic austenitic (fcc) structure [5, 7]. Therefore, magnetic force microscopy (MFM) is used to monitor the effects of secondary processing on the surface magnetic properties of these steels.

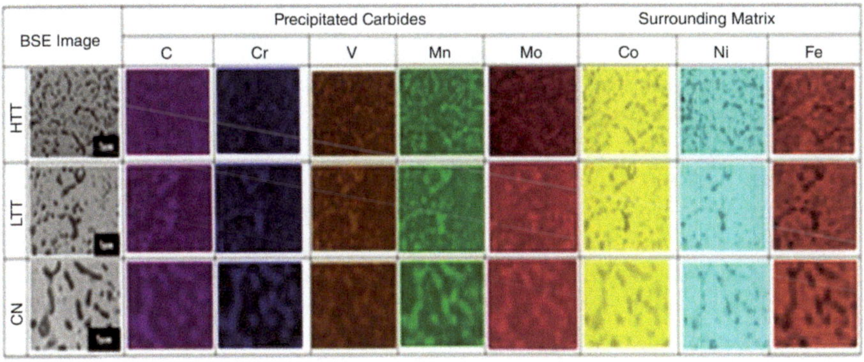

(a)

Samples	C	Cr	V	Mn	Mo	Co	Ni	Fe
HTT	4.1	12.5	0.5	0.8	0.6	5.4	2.12	72.9
LTT	5.3	13.0	0.6	0.7	1.6	5.2	2.4	70.8
CN	6.8	13.4	0.6	0.6	1.8	5.1	2.3	69.0

(b)

FIGURE 7.3 (a) Energy-dispersive X-ray spectroscopy (EDS) composition maps highlighting the major carbide components (middle columns) and bulk matrix (right columns) of HTT, LTT and CN treated specimens (images for each row have the same micron band), compared with backscattered electron (BSE) greyscale images (left column) of three separate P675 treated specimens (carbides appear darker than the surrounding matrix). (b) The weight percentage of elements making up the surface of each steel is determined by EDS analysis (carbides are not identified) [2] [open access].

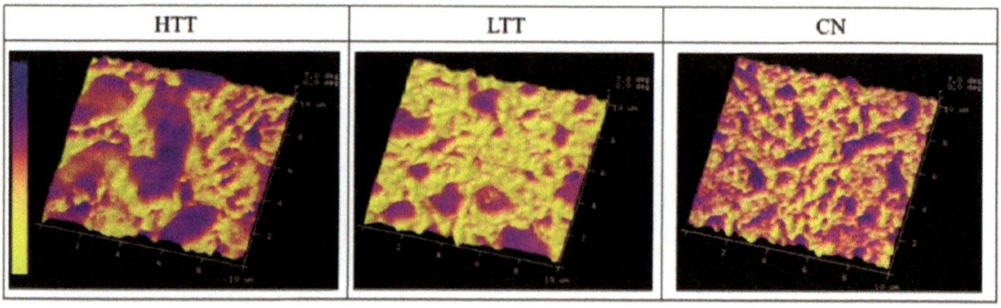

FIGURE 7.4 3D magnetic response maps with changes in height representative of differences in magnetism. Color scale ranges are 7 degrees (0° = yellow, +7° = blue) for magnetic response [2] [open access].

Alternatively, to improve the interaction of silica gel as an inorganic filler in organic polymers, the surface can be functionalized with organic amino, phosphorus, chloro, cyano, acrylate, epoxy and isocyanate groups. To study the interface between the carbides and surrounding matrix, freshly cleaned, polished and dried specimens are imaged ex-situ using scanning probe microscope (SPM) in an inert glovebox. Variations in the magnetic moment projections (surface normal direction) on the steels' surface were mapped using MFM (Figure 7.4) [2].

To confirm whether a substrate will withstand curing, it is important to determine the heat treatment is an integral part of the procedures used in metal fabrication, as it provides materials with the desired mechanical and physical properties without changing the shape of the product. A wide range of heat treatments such as annealing, quenching, hardening and stress relieving can be carried out using our range of furnaces. These high-quality annealing and heat treatment furnaces are precise and efficient due to their low dew point design, climate-saving units and advanced digital microprocessors. Their advanced features ensure maximum productivity and improved material quality, making them ideal for a variety of metal heat treatment processes. These furnaces are designed to meet the unique requirements of modern industry and reliable means of achieving the required metallurgical results. One of the widely used methods for changing the surface properties of silica gels is surface functionalization. Because covalent siloxane bonds can form between the surfactant and silanol modifier, silane coupling agents are often used. It can be applied to the surface of silica to increase its hydrophobicity. Silane contains one to three organic substituents and one to three reactive groups (halide, amide and alkoxide) with which it can react on surfaces. highest temperature the substrate can reach during cure cycle. For example, steel cannot be CVD-coated at temperatures of 1000°C, as this will cause it to lose its mechanical strength. Hard metals are a common base for CVD coatings. Surface functionalization is a useful technique for long-term surfactant-free stabilization of nanoparticles in a nanofluid. Interfacial interactions between liquid molecules and nanoparticles are typically explained by stereostabilization and Deryagin–Landau–Verwey–Overbeek theories.

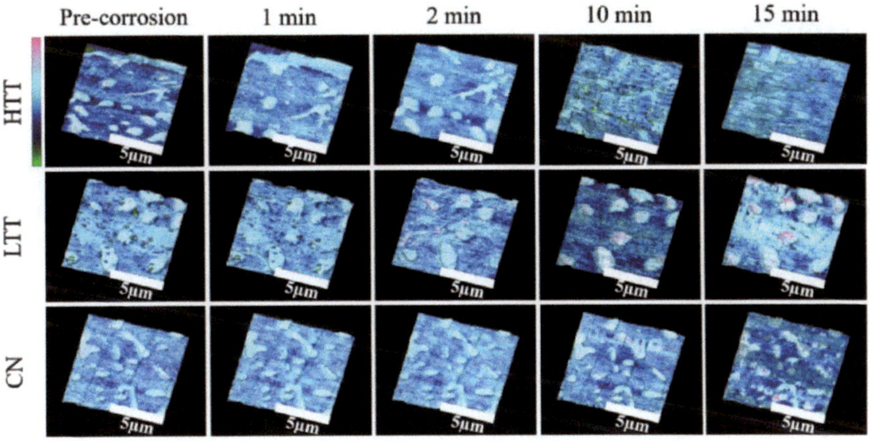

FIGURE 7.5 SKPFM Volta potential maps (green to pink color scale, 400 mV full scale) overlaid on the evolving 3D topography (30 nm full scale) of the three heat-treated MSSs as a function of immersion time in 1 M NaCl solution [2] [open access].

In situ, AFM is used to monitor the topographic changes over time when the specimens are immersed in 0.1 M NaCl solution to track the corrosion progress in real-time. To monitor the surface changes caused by successive prolonged exposure to corrosive conditions, an ex-situ intermittent SKPFM was performed. After immersing the samples in a corrosive brine, VPD maps were produced at cumulative exposure times of 0, 1, 2, 10, and 15 min to a 1 M NaCl solution (Figure 7.5). For the steel, qualitative variations in VPD and carbide matrix appearance (surface topography and shape) over time were observed. The HTT sample showed steady surface particle production and degradation over time, resulting in decreasing variation in surface VPD. On the other hand, the CN sample showed minimal surface topography or VPD change, indicating that the corrosion reaction kinetics were significantly slower even with significant VPD [2–4].

The main types of surface treatments are given in Table 7.1. The following unique factors should be carefully considered during the selection process:

1. Types of materials to be processed using the selected technology;
2. The temperature at which the substrate enters the process;
3. Maximum hardness of the surface layer;
4. Thickness of the hardened surface layer.

Consistent with the almost complete partitioning of corrosion between matrix-exposed and unexposed carbides, CN shows the highest levels, but shallow corrosion is present along the carbide boundaries (Figure 7.5). Between these extremes, LTT appears to operate as follows: initially, there is an accumulation of particles, as in HTT, but gradually these particles disappear, showing

TABLE 7.1
Basic Characteristics of Some Commonly Used Surface Treatments [9] [open Access].

Treatment Category	Treatment	Metals Treated	Processing Temperature (°C)	Maximum Surface Hardness (kg/mm^2)	Typical Thickness (mm)
Treatments for microstructural modification	Surface rolling	Steels, Ti, and Ni alloys	Room temperature	500–600	1–6
	Shot-peening	Steels, Ti, and Ni alloys	Room temperature	500–700	0.2–2
	Flame hardening	Hardenable steels and cast irons	850–1100	500–700	0.1–0.6
	Induction hardening	Hardenable steels and cast irons	850–1100	500–700	0.1–0.6
	Laser hardening	Hardenable steels and cast irons	850–1100	500–700	0.1–0.6
Thermochemical diffusion treatments	Carburizing and carbonitriding	Low carbon steels	800–1100	700–900	0.05–1.5
	Nitriding	Nitriding steels, tool steels (hot working and HSS)	500–600	800–1200	0.025–0.5
	Ion implantation	All	200–600	600–1100	$<10^{-3}$
Chemical conversion coatings	Phosphate coatings	Steels, Al alloys	25–95 (drying at 200 °C)	–	Up to 0.01
	Hard anodizing	Mainly Al alloys	0	>1100	0.04–0.05
Surface coatings	Hard chromium	Most ferrous and non-ferrous alloys	<70	700–1200	0.001–0.5
	Electroless nickel	Most ferrous and non-ferrous alloys	Room temperature (ageing 200–500)	Up to 1000	0.001–0.025
	Physical vapour deposition (PVD)	All metals	150–500	2000–2500 (TiN), 2800–3000 (TiC)	0.001–0.003
	Chemical vapour deposition (CVD)	All metals (limited by process temperature)	800–1000	0.001–0.01	0.001–0.01
	Thermal spray coatings	All metals	<200	700–2000	0.5–1
	Hardfacing by welding	Steels and non-ferrous alloys (>1100 °C melting point)	1200–1600	800–2000	3–10

signs of localized corrosion spreading into the matrix near some CN-like carbides [2, 6]. The stability of the particles is determined by the repulsive force that arises from the overlap of electrical double layers on the particle surface and the attractive attraction that exists between liquid molecules and the particle surface, as per hypothesis. The particles in the nanofluid are stable if the repulsive force is greater than attractive force. There are key differences between the repulsive force types between nanoparticles that are static and electrostatic [10–12]. The results of microhardness tests are obtained on sections obtained by metallographic method.

The effective thickness, defined as the thickness of layer of material with hardness greater than or equal to\specified value (usually fixed at 550 kg/mm^2), determines the thickness of the treatment. The roughness experiences the greatest contact load during adhesion, tertiary oxidation and abrasive wear. Therefore, it is important to select processing methods that reduce the damage caused by these systems and significantly increasing the hardness of outer layer of workpiece. For example, it is possible to increase the frictional resistance of cutting tool that is subject to adhesive (and possibly abrasive) wear during operation by applying a ceramic coating to high-speed steel made by PVD method. It is necessary to select a surface treatment that can provide a sufficiently high surface hardness (as well as suitable fracture hardness in the event of abrasive wear under high loads). It is necessary to select the treatment that can provide sufficient effective thickness.

In general, a minimum depth of 2 μm should provide high rigidity. The possibility of permanent deformation existing in machined areas is a unique factor in the field of surface engineering. By their nature, residual stresses can be either tensile or compressive. Tensile stresses can accelerate tribological wear, while compressive residual stresses usually take precedence. Common stabilized nanofluids by surface modification include coating of Fe_3O_4 nanoparticles with oleic acid, nitrogen activation of TiO_2 nanoparticles [16], functionalization of ZnO nanoparticles with polymethacrylic acid [10] and grafting of silanes onto SiO_2 nanoparticles [11]. And the placement of water jobs clusters on CN [12] and diamond plasma treatment [14]. Since amino groups are present on the particle surface, the addition of molecules with amino functional groups usually stabilizes the nanoparticles in the liquid and increases their adsorption capacity [17]. Two additional factors related to coatings should also be carefully considered:

1. The coating's adherence to substrate
2. The substrate's capacity to hold coating in place.

In case of ceramic coatings, this difference should generally not exceed 800 kg/mm^2. Increasing the coating thickness is an additional option. There are different ways to perform roofing study. The initial goal of developing strategies such as ligand exchange or encapsulation is to transfer organically synthesized nanoparticles into water. In contrast, electrostatic packing and in situ binding are commonly used to create nanoparticles. This is a summary of popular surfactant processing methods, along with examples of typical compounds or surfactants used in each method. These tactics are organized according to the basic principles of substitution reaction, non-covalent bonding and covalent bonding. The physicochemical interactions that occur upon contact and determine adhesion are significantly affected by the presence of defects and contaminants. In chemical conjugation, various conjugation methods can be used, such as (i) direct conjugation of nanoparticles through reactive linkers such as isothiocyanates, maleimides and succinimidyl esters; (ii) click chemistry; and (iii) through an intermediate link [13].

It is important to remember that neither the intended function of biomolecule nor the intended function of magnetic nanoparticles should be compromised throughout the operation process. Ensuring that the magnetic nanoparticles maintain their magnetic properties while preserving the biological activity of biomolecule requires a careful selection of operating conditions and procedures. For applications such as targeted drug delivery, imaging and biosensing, where both elements must cooperate to achieve maximum performance, this balance is essential.

To better highlight the surface morphological differences, the three-dimensional (3D) AFM images, which are topographic maps of the scanned areas are presented in Park Systems' "Enhanced ColorTM" view mode. In "classic" view mode, 2D AFM images are displayed with a gradient of one color (brown in this case) representing the depth on the z-scale. Each sample was subjected to time-dependent linear analysis of specific carbide particles (Figure 7.9), which validated the qualitative results derived from the images of Figure 7.8. The surface contrast of the carbides increased up to 50 nm, according to HTT, and the height of the surrounding matrix changed slightly but uniformly in tandem. Initially (44 min), the height changes across the carbide/matrix interface for LTT revealed a deep attack of about 100 nm adjacent to the carbides (Figure 7.6). Then the height of the carbides increased with longer times (112 min), and the apparent attack depth of the nearby matrix region also contracted. These changes are probably related to the formation and accumulation of insoluble corrosion products. It needs constant temperature and gradual cooling, even if the high temperature helps the adhering paraffin molecules to spread by tearing the surface diffusion barrier. In addition, the objective of the study is the closed-type temperature-controlled micro-evaporation with the best uniform temperature. The topography and adhesion degree of the coating that evaporated for 8 min is shown in Figure 7.7a, b, while the coating that evaporated for 25 min is shown in Figure 7.6c, d. The cross-section of the white line in the topography is shown in Figure 7.6e. The cross-section of the white line in the adhesion state is shown in Figure 7.7a, b, f [4].

A common morphological feature common to all the surfaces photographed is the presence of large accumulations of material or hills (sometimes called "bumps") with protruding shapes, accompanied by tiny particles at surface level. The dark areas in 2D AFM images or the holes and depressions in 3D images (which are more logical as an aspect), pits or even surface cavities can also be seen. From a quantitative point of view, each of these changes results in a change

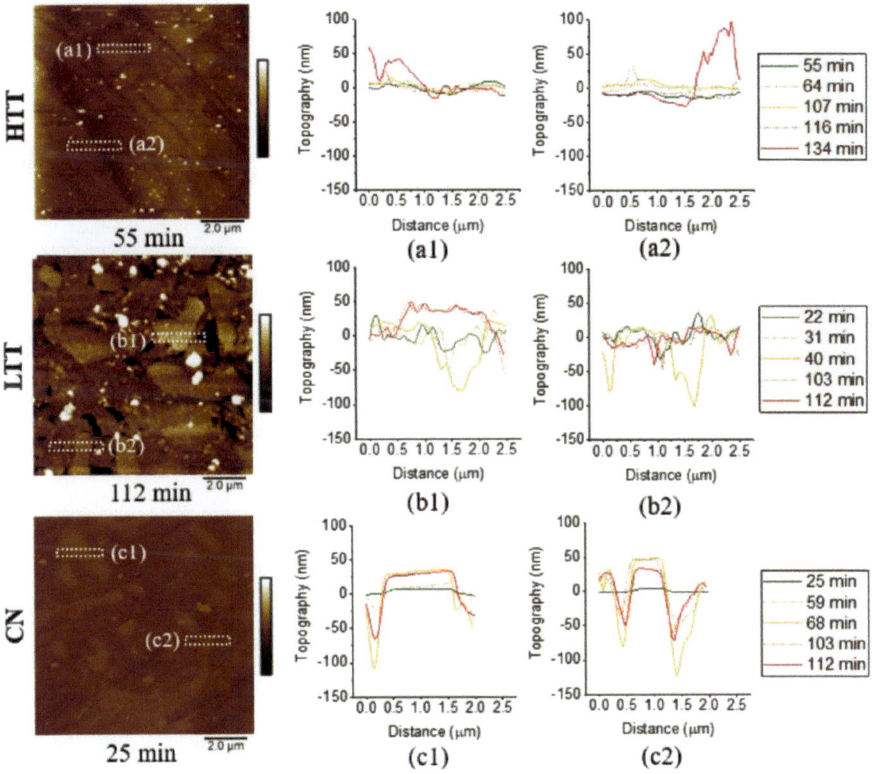

FIGURE 7.6 Topography maps (a, b, c), 160 nm full scale, exposure time indicated below corresponding map) for each of the three heat-treated MSSs with height profiles across selected carbide-matrix interfaces shown as a function of exposure time to 0.1M NaCl solution (a1–c2). Location of each profile is indicated by the corresponding box in the exemplary topography maps presented at left for each of the three heat-treated steels [2] [open access].

in the values corresponding to amplitude parameters listed in Table 7.1, especially when it comes to roughness behavior [4].

Using pulsed laser deposition, bio-derived hydroxyapatite (HA) doped lithium carbonate (LiC) and lithium phosphate (LiP) coatings are fabricated on additively manufactured Ti6Al4V substrates. We have carefully evaluated and compared the effects of post-deposition heat treatment (TT400–TT700) on the mechanical, physical and chemical properties of the synthesized coatings with those of no-heat-treated (NTT) coatings. The highest post-deposition heat treatment temperature (i.e., 700°C) in a water vapor environment yielded the best overall properties for both LiC and LiP synthesized coatings in terms of structure, morphology, composition, wetting and mechanical properties. WC-6 wt% Co backing pads (11.5 mm × 11.5 mm × 3.5 mm) are used. The pads (Figure 7.8a) are typically used as "seats" for indexable cutting tools. They are positioned between the tool

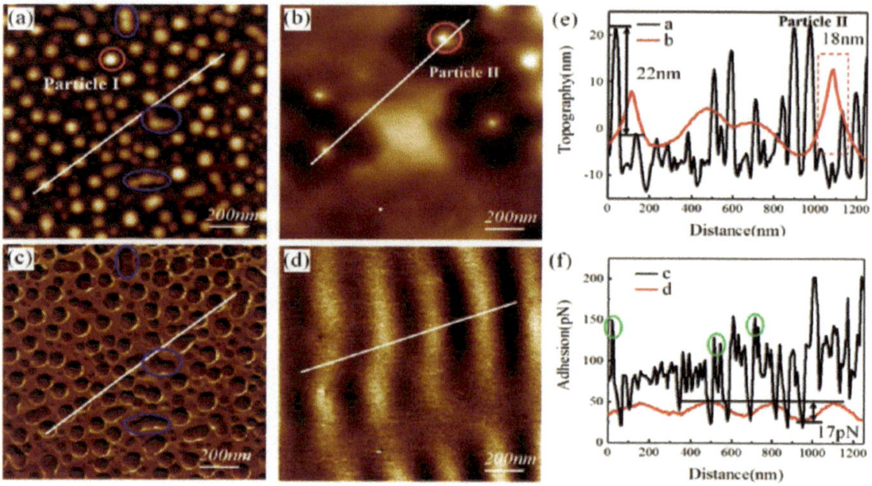

FIGURE 7.7 Paraffin coating produced by closed-type precise temperature-controlled evaporation for 8 min and 25 min: (a) topography with evaporation for 8 min; (b) topography with evaporation for 25 min; (c) adhesion corresponding to (a); (d) adhesion corresponding to (b); (e) cross section of the topography corresponding to (a, b); and (f) cross section of the adhesion corresponding to (c) and (d). The blue circled areas are gathered paraffin particles [19] [open access].

holder (Figure 7.8b) and cutting inserts to protect it in the event of cutting insert breakage. Traditionally, the pads are made from the same material as cutting inserts. Three different pad samples are made: the first is machined (M), known as received; the second is diamond polished (DP) and the third is sandblasted (GB). The beveled surface gaskets are obtained directly from a commercial supplier; at our request, the supplier ground the standard gaskets (CS 120200) to make them 200 μm thinner to match the 200 μm surface coating [5].

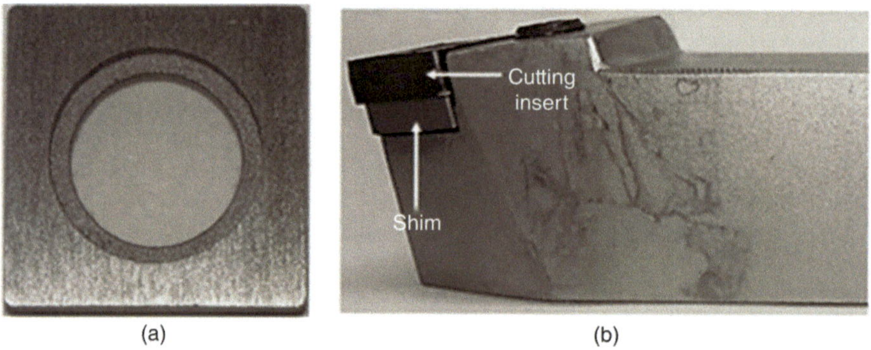

FIGURE 7.8 Shims located between the cutting insert and tool holder are (a) normal shims and (b) spacers [5] [open access].

7.3 USING A LASER, LASER SURFACE TREATMENT CHANGES SURFACE CHARACTERISTICS

In the modern context, surface treatment is an important topic. Surface treatment helps to develop better components with a desired bulk and surface qualities, and saves important materials. Considering the limited resources available, it is practical and cost-effective to use traditional materials for large-scale production coating to meet modern needs. Many methods are available to coat and modify substrate surfaces, including thermal spraying, laser surface treatment, PVD, CVD and electrodeposition. In terms of workpiece material, shape, adhesion, thermal deformation, each method has its limitations. Coatings with low adhesion strength are produced by CVD, physical deposition, or electrodeposition methods. The repeatability and controllability of electrodeposition are particularly challenging. Although thermal spraying has outstanding repeatability, controllability adhesion strength, it poses a serious problem with the heat-affected zone in critical applications. Laser surface treatment (LST) is a solution to all these problems. With LST, any shape can be used, as well as materials with excellent adhesion and a smaller heat-affected zone [1–4]. In addition, due to excellent repeatability and controllability, LST-processed products can be easily customized. Laser energy is used as a heat source in LST procedures. Impact peening, engraving, selective laser melting, sintering, glazing, cladding and surface alloying are some examples of LST processes. Laser material processing procedures according to the surface temperature achieved, where Tm and Tv denote the melting and evaporation temperatures, respectively [3]. To study the wettability, friction characteristics and antibacterial behavior, three laser-etched surface samples are selected based on their morphology and homogeneity (see Figure 7.9).

One method to improve wear resistance is laser hardening. The surface to be modified is exposed to a high-energy laser beam during the laser hardening process. In laser hardening, the high cooling rate hardens the components, resulting in formation of martensitic structure. These acicular phases of martensitic structures provide exceptional hardness and ultimately increased wear resistance [5–7]. A pulsed laser beam impacts the substrate during laser impact peening, creating a strong plasma on the substrate surface. The rapid expansion of laser plasma creates a shock wave that passes through the substrate material and causes plastic deformation. This results in the creation of residual compressive stresses near the exposed surface. This method has been used by some researchers to create compressive stresses on the substrate surface [8–10]. Laser impact peening and laser surface texturing are comparable processes. Small pits and compressive stresses are introduced into surfaces using laser ablation or laser interference technology to improve lubrication, friction control and wear resistance [11–13]. When applying mineral-based coatings in a solid state, that is, without melting the material in significant quantities, a friction surface (FR) is usually used. This makes them particularly interesting for materials that are considered difficult or even impossible to weld [16]. It is initially established that the repair or restoration of worn or damaged components is an acceptable use [16]. Additional technically

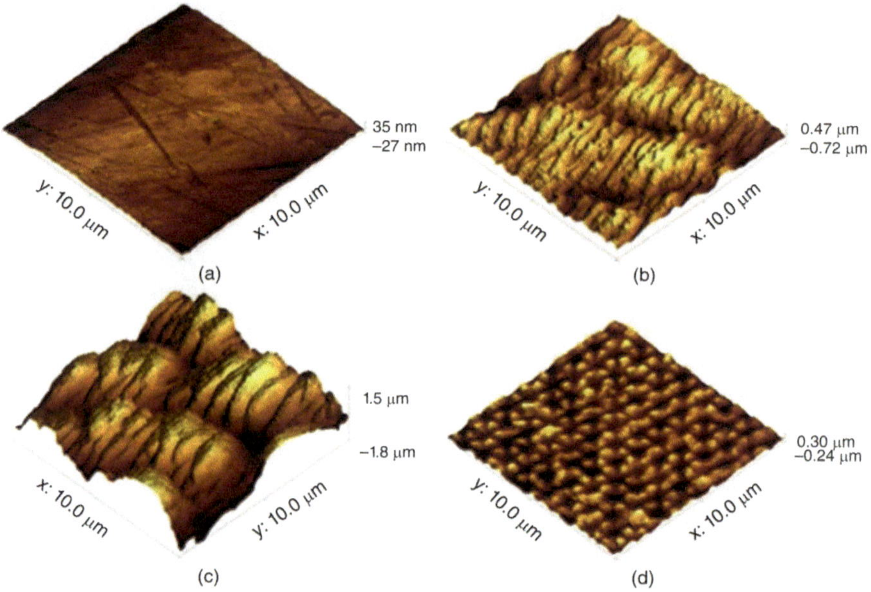

FIGURE 7.9 AFM microscopic images of CrCoMo surface are shown in untreated surface (a), LSFL (b), grooves with LSFL layer (c), and triangular nanopillars (d). (a) Polished CrCoMo sample under AFM microscope. (b) AFM microscopic image of LSFL. (c) AFM microscopic image of LSFL and grooves. (d) AFM microscopic image with triangular nanopillars [3].

advantageous aspects of this procedure include the ability to deposit multiple layers of different materials on top of each other to form functional coatings, transition layers with a gradient of desired properties or additive layer manufacturing techniques [17]. Compounds can be formed at the throat by adding powder to them during deposition, although the complex and poorly understood movement of the material ultimately causes uneven particle distribution [18].

This process uses a primer made from the coating material. The screw rotates about its longitudinal axis as its face is pressed against the substrate to lock into place using a predetermined axial force. The temperature of material raises due to frictional heating upon contact, causing the surface of screw to adhere to substrate. The screw now rotates by plastic flow in a shear plane within the thermally softened material at the tip of screw, rather than at interface with a stationary substrate. This elastomeric material is often referred to as a shear zone or "semi-fluid layer" [16]. At this stage, the substrate is fixed on machine table, which is moved to provide lateral movement. A layer equal to the diameter of nail is left along the path, as part of soft material that is deposited from the tip of nail onto the substrate. A small area at the boundary of each coating layer suffers from insufficient adhesion because soft material along the outer edge of pin cannot fully transfer the axial force of forging to substrate. In addition, some of the elastic material around the edge of screw is pushed up and out of contact area, creating an axial flash force [14]. Previous studies have shown that substrate surface properties, material interactions and process factors influence metal-to-metal bond strength in fiber-reinforced composites [15].

7.3.1 Procedures Using Machines

Rolling a solid cylinder over a surface with an applied load capable of causing full plastic contact produces plastic deformation by surface rolling, which is, however, not often used in tribology. During shot peening, a stream of tiny steel, ceramic or glass spheres with diameter 0.2–2 mm is blown onto the surface being treated at speed 40–120 m/second. The desired level of surface hardening can be achieved by varying the intensity of te treatment. To improve the slip or contact fatigue resistance of powder metallurgy sintered parts, shot peening and surface rolling are sometimes used to close the surface pores of these parts. However, for re-sintering procedure to be truly effective, the surface of pores with which plastic deformation has come into close contact must be welded to match the processing progress. Otherwise, blocked pores can still act as cracks and contribute to wear due to contact fatigue.

In most cases, surface heat treatment is applied to hardenable steel containing a sufficient amount of carbon (0.3–0.5%) to promote martensitic transformation and cause surface hardening. Previous studies [18, 20] have shown that the best AFM scan size for distinguishing between malignant and normal cells is 10 × 10 μm squared. Examples of normal height and adhesion images of cancerous and precancerous cells are shown in Figure 7.10. The main conclusion drawn

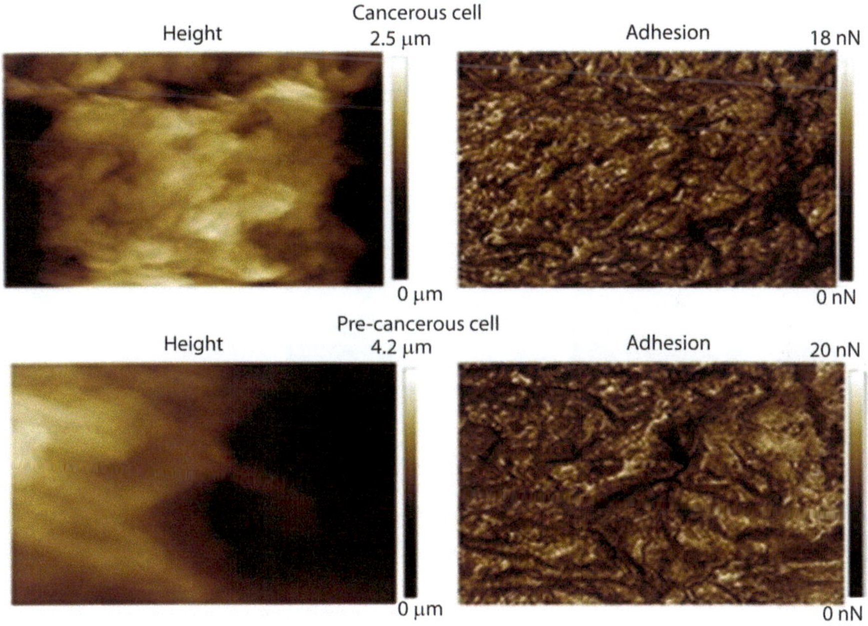

FIGURE 7.10 AFM square images of malignant and cancer cells measuring 10 × 10 μm [21] [open access].

from these images is how difficult to distinguish between different cell shapes based on their appearance alone. Investigation of fractal geometry properties of cell surfaces and their adhesion maps has proposed as a quantitative method for distinguishing between normal and malignant cells [18].

The effect of stress can be understood by examining the graph. As the modulus decreases, the residual stresses can be relieved by plastic deformations during contact, and so this assumption loses its significance. However, this is not the complete picture. The presence of residual compressive stresses, which may be presented in outer layer, also contributes to the high resistance to contact stresses in case-hardened cast irons and steels [7]. On the other hand, NaCl particles were found at the grain boundaries and caused the carbide to separate from the matrix and subsequent grain separation, indicating that the LTT and CN carbide boundaries were targeted (Figure 7.10). LTT showed some generalized attacks, as evidenced by the roughness of the entire surface caused by the deposition of wear products, as well as some attacks along the carbide boundaries. Compared with LTT, CN showed significantly more localized attack at the carbide boundaries (dotted oval in the right panel of Figure 7.11), and the CN carbides were surrounded by narrow "valleys" about 0.5 μm wide [9].

Hardened parts undergo geometric deformations after processing due to thermal cycling and phase changes, requiring a final grinding process. This must be done correctly so that no residual tensile stress remains on the surface. Now let us look at corrosion damage caused by tribo-oxidation and adhesion. It can be said that hardened steels and surface heat-treated steels behave very similarly. Due to highly rough deformations and subsequent local heating, these elastic stresses are weakened during sliding. Corrosive wear can be justified using the same logic. Surface hardening alone will not solve the problem of corrosion of

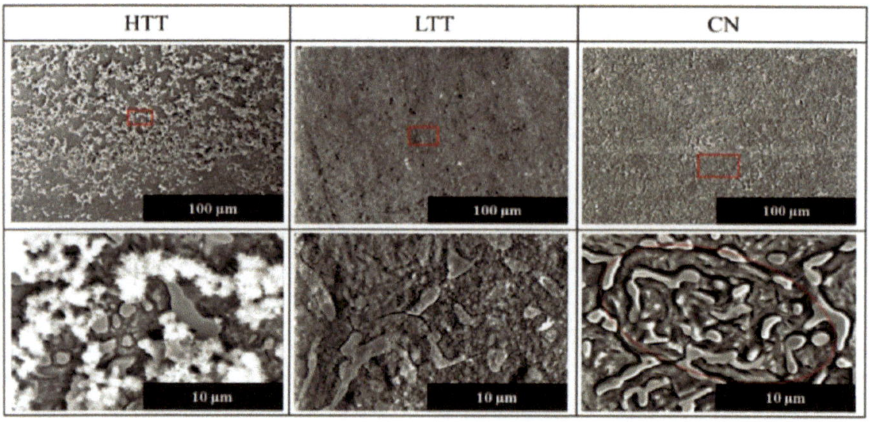

FIGURE 7.11 SE SEM images of the sample surfaces following in situ AFM testing. Red squares in the images in the top panels indicate areas of magnified images. Dotted red oval area in magnified CN image indicates the "line of attack" [2].

steel by particulates such as silica inclusions, given the maximum surface hardness that can be achieved. However, this may be a reasonable solution if stress wear is minimal [12].

7.4 THERMOCHEMICAL DIFFUSION TREATMENT

Often, the metal materials that are traditionally used do not meet the ideal specifications required for the purposes for which they are intended. Therefore, to achieve reliable performance and long service life of metal components, it is necessary to modify their surface properties. Among the various surface treatment methods, plasma-assisted thermochemical diffusion treatment has attracted great interest in modifying surface characteristics due to its ease of use, availability and minimal environmental impact. The improved surface properties in this case are a result of interaction between the metal surface and gaseous species generated in the treatment environment. The carburizing and nitriding are two main methods of thermochemical diffusion of steel.

7.4.1 Carbonation Process

It is then heated with oil so that it undergoes a martensitic transition and acquires high surface hardness. After hardening, steel parts are hardened at temperature, 170–180°C. To ensure good fracture toughness in the core, it is recommended to treat steels with a carbon content of less than 0.4%. A typical microhardness profile after carburization. As with surface heat treatment, carburizing is a good way to protect against damage caused by oxidation, tribo- and adhesives and low-pressure abrasive wear. However, carburization works best on mechanical parts that suffer from contact fatigue damage due to large effective thickness, high surface hardness.

The mechanical energy input promotes phase dissolution at process temperatures and the subsequent cooling rate is often very high. As a result, coatings containing martensitic, nonequilibrium precipitates, or supersaturated solid solution phases are formed [17]. Insoluble phases also crumble or become spherical as a result of mechanical processing [16]. These changes in microstructure have been shown to improve the wear resistance of material and result in FS-treated materials exhibiting less wear compared to a conventional balanced microstructure [14]. An important element in production of aluminum (Al) alloys is silicon (Si). Silicon alloys are often used as casting alloys for engine blocks or pistons because silicon facilitates the casting of aluminum alloys [15]. In addition, silicone increases wear resistance and corrosion resistance [16]. Si alloys are strengthened by addition of magnesium (Mg), which promotes deposit formation. Higher levels of Mg can cause increased oxidation of products due to its strong affinity for oxygen (O), which can cause brittleness of materials due to pronounced roughness [16]. In aluminum alloys, copper (Cu) promotes the formation of additional cohesive precipitates, which increases the yield strength. Among other things, traces of metallic iron (Fe) contribute to the precipitation of poorly soluble mineral phases containing silicon and aluminum. Manganese (Mn) supplements counteract these effects. More brittle material results from small concentrations of other elements such as zinc (Zn) and chromium (Cr) in aluminum alloys, which promote the formation of complex hard phases between the metals [20] described in the previous paragraph. If $\tau'_{max} < \tau_{end}$ over the entire depth. There are two potentially dangerous circumstances:

1. When the τ'_{max} profile is too high relative to the surface endurance limit or surface hardness. In this case, the damage is caused by cracking when a crack forms in the subsurface area too early.
2. In cases where the effective is too small compared to the profile $\tau'_{max,}$ the crack near the body/core interface initiates prematurely. This type of damage is known as hull crushing and is a type of spalling.

For the situation without residual stresses, local shear is a function of hydrostatic stress as a determined criterion; for case with residual stresses, the results correspond to technologies in Figure 7.12. Strong and uniform contact is demonstrated by FS pin samples used in this study; This connection is not discussed further here. The FS process is based on high-temperature shear deformation, which treats weld deposits thermomechanically. As a result of dynamic recrystallization, small equiaxed grains are often formed in coatings [1].

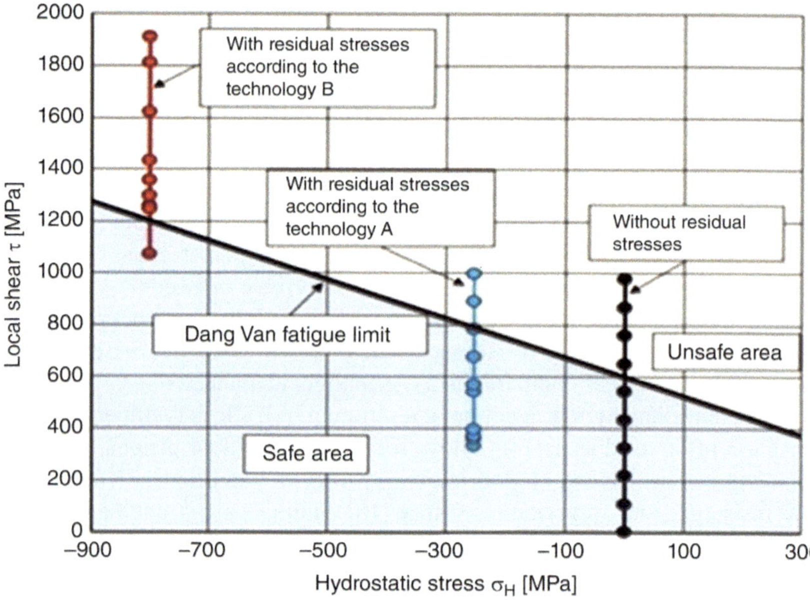

FIGURE 7.12 Local shear as a function of hydrostatic stress in situation without residual stresses and in case with residual stresses, determined for technologies A and B respectively using the Dang–Wan criterion [14] [open access].

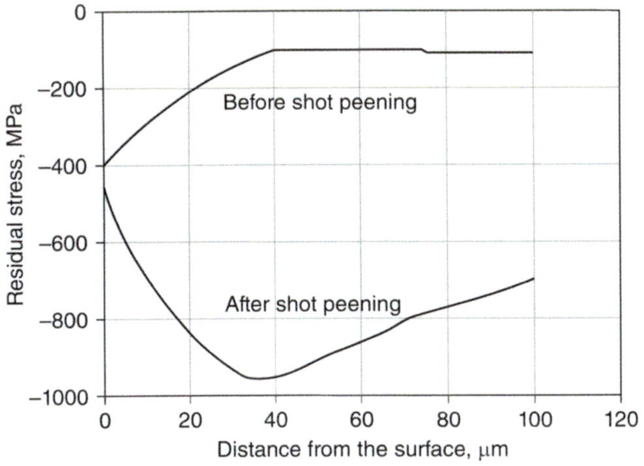

FIGURE 7.13 Carburized 16MnCr5 steel's residual stress evolution before and after shot peening [9].

Due to the elastoplastic contact between the spheres and surfaces as shown in Figure 7.13. Geometric distortions may occur in carburized parts; however, machining is a solution that can restore the correct final geometry. These procedures, which require the removal of surface layers, require special care [12].

In addition, higher retained austenite content will result from higher surface carbon content. Conflicting results have been published in the literature regarding the effect of retained austenite on contact fatigue resistance [11]. The amount of retained austenite can be quite significant during carbonitridation or enrichment of surface with carbon and nitrogen. In this way, hardness and more importantly, compressive residual stresses can be increased.

7.4.2 Nitriding

Nitriding can be carried out in any of four states of matter - solid, liquid, gas, or plasma - in the same way as carburization. While gas and liquid nitriding are not environmentally friendly and often result in thick and brittle composite coatings, batch nitriding is unreliable or repeatable. In addition, plasma nitriding can successfully solve the surface oxide problem encountered during conventional nitriding procedures on stainless steel. Plasma nitriding is environmentally friendly and can alleviate or eliminate the abovementioned problems. Nitriding causes less distortion and deformation than carburizing because it does not have a hardening mechanism and operates at relatively low temperatures; 400–570°C. This is a critical for some precision wear components [22–30].

Cyclic potentiometric polarization (CPP) scans were performed on each sample to investigate the effects of different heat treatments and to clarify the corrosion and passivation behavior of the samples. The resulting polarization curves

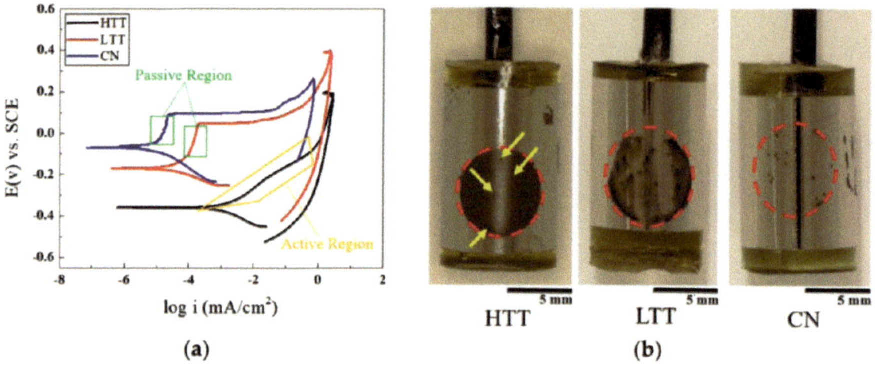

FIGURE 7.14 (a) Cyclic potentiodynamic polarization (CPP) scans (0.01 M NaCl electrolyte) for all three surface treatment samples. Passive regions for LTT and CN are indicated by green squares. (b) Images of the samples post-testing (after the area masking tape was removed) with dotted red circles indicating the test location on each sample surface. All samples display some isolated pitting. However, due to the difficulty in clearly seeing the pits on the HTT sample (which, in contrast to the other samples, underwent generalized corrosion attack), yellow arrows indicate the location of the pits present on the HTT sample [2] [open access].

and macroscopic images of the sample surfaces after electrochemical testing are shown in Figure 7.14. According to the test, HTT had the lowest OCP (-400 mV), followed by CN (−80 mV) and LTT (-200 mV). Previous research that classified the corrosion resistance (i.e., the corrosion rate determined by electrochemical methods) of identical steels is consistent with our classification [7, 9]. Typical passive behavior was demonstrated by the LTT and CN samples, which showed a rapid voltage shift upon a small increase in current density (Fig. 11a, green boxed areas). Breakdown voltages were obtained at 40 mV and 95 mV, respectively, for the LTT and CN samples. In contrast, the linear increase of current density over the sweep voltage range indicates active corrosion behavior in the HTT sample. However, pits were first observed on the HTT surface (-200 mV), but did not expand when the anode overvoltage was increased [15].

The PN specimen has showed significantly lower current densities in anodic region and higher corrosion potential when exposed to same no-slip conditions. Compared with uncorroded region, the corroded region has showed a slight color change (see Figure 7.14b). Therefore, the normalized specimen in test solution is in an active state in the anodic region, and γ'-Fe4N layer that nitriding produced on the surface is beneficial in reducing the decay rate of the mild steel [31–34].

As a result, the normalized specimen in the test solution is in an active state in anodic region, and the dissolution rate of mild steel is effectively reduced by γ'-Fe4N layer that nitriding produced on the surface. The corrosion potentials of both specimens were cathodically shifted under tribocorrosion (sliding) conditions, and current densities in anodic region increased (Figure 7.15a) [35].

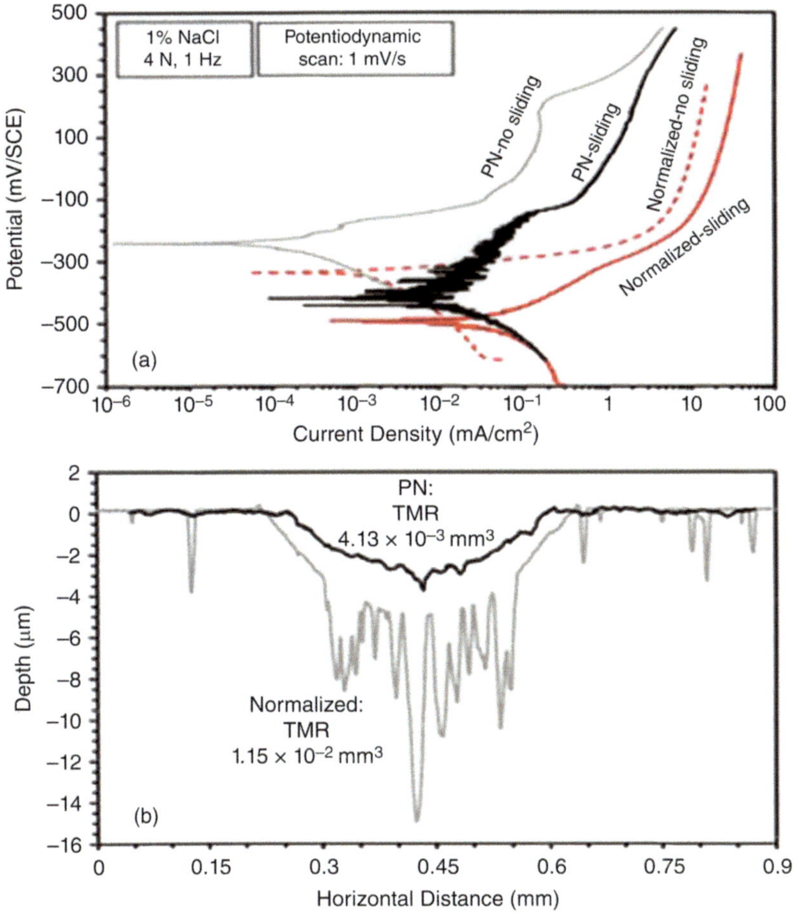

FIGURE 7.15 (a) Surface profiles measured from sliding tracks created during anodic polarization and (b) anodic polarization curves measured both with and without sliding [15] [open access].

PeakForce tapping mode used here revealed differences in the elastic modulus distribution of carbides and steel matrices in conjunction with topography (see example in Figure 7.16) The carbides exhibited a larger relative elastic modulus than the matrix, as shown in the CN image in Figure 7.16, suggesting that the potential sites for microcrack formation and fracture may be at the interface between carbides and matrix where the local variation in elastic modulus was greatest [36–39].

For example, in case of AISI M2 tool steel is subjected to gas nitriding at a temperature of 570°C for 8 hours, Figure 7.17 illustrates the experimentally demonstrated close [17]. The improved sliding wear resistance is particularly suitable for nitriding treatments. If the white coating during dry sliding has a small

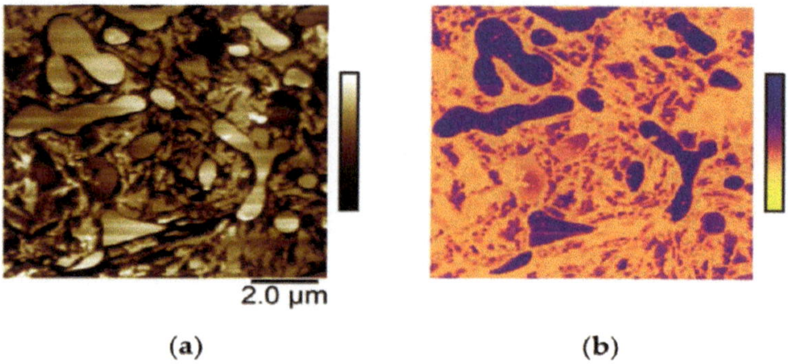

FIGURE 7.16 (a) CN topography (160 nm full scale), and (b) DMT Modulus (1.5 GPa full scale). Images are representative of 103–112 min submersion in 0.1M NaCl solution [2] [open access].

thickness and a sufficiently high contact pressure during the running-in stage, it is eliminated. So, the diffusion layer is what provides wear resistance.

Tribo-oxidation is the main cause of wear, and a very low specific wear rate, 5×10^{-15} m^2/N. In addition, the diffusion layer has a high degree of heat resistance, and local temperature increases caused by frictional heating softens slightly compared to carburized or through-hardened steels. The critical transition temperature from light to severe wear is raised by this characteristic. This quenched and tempered steels, as shown in wear map as illustrated in Figure 7.17 [11]. The white coating usually does not need to be removed before serving, since the hardness of iron nitrides is comparable to the hardness of the diffusion layer. The residual compressive stresses of the composite layer help to seal any potential cracks in the surface. Such a rise leads to a noticeable decrease in κ-factor, and cleaning of

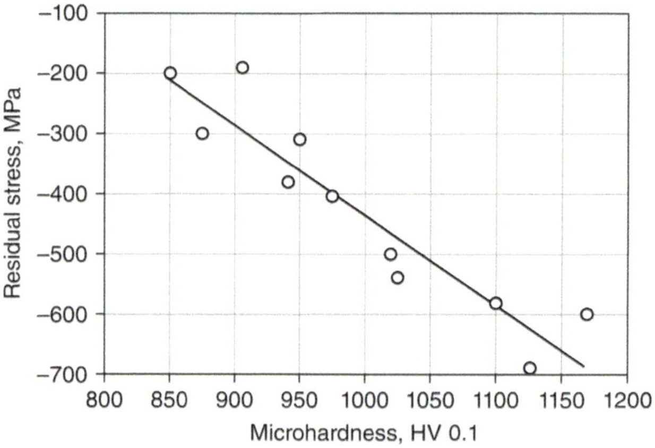

FIGURE 7.17 Dependence of microhardness (in the diffusion layer) on residual stresses in an experiment using nitrided tool steel AISI M2 [9].

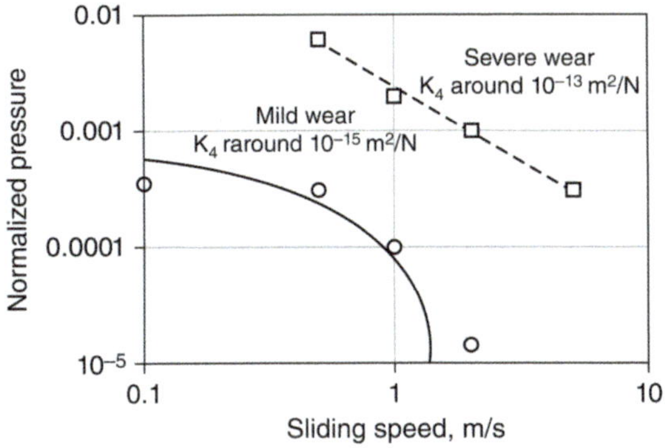

FIGURE 7.18 Data presents a map of wear of nitrided steels during dry sliding on a hardened steel counterbody. Comparatively, the boundary between normalized, quenched and tempered steels between moderate and intense wear is also fixed [9].

large defects during running-in only slightly reduces these losses [16]. In addition, nitriding is well suited for tribological applications involving abrasive wear, especially low-stress abrasion. This is due to outer areas of nitrided layer can become extremely hard, up to 1250 kg/mm^2. For example, Figure 7.18 shows the exact wear rates of untreated steel (H: 391 kg/mm^2), nitrided steel and chromium steel, which are determined using a modified DSRW testing apparatus [12]. The test used low-stress abrasive machining to simulate the wear process occurs inside a glass-filled injection molding cylinder at three different temperatures. Optimal results at all temperatures are demonstrated by nitrided steel, which has a surface hardness of 1284 kg/mm^2, in contrast to 1125 kg/mm^2 for chromium steel.

Nitrided steels have a carbon content of 0.2–4% and a sufficient amount of nitride-forming alloying elements (Cr, Mo) to ensure the necessary precipitation hardening process. 720 M24 (En40B) is a standard nitride steel with a Cr content 3%. High-strength corrosion steel is used to achieve the best quality parts used in automobiles and Formula 1 racing engines. In motorsports there may be some overlap in using carburizing and nitriding (for applications such as crankshafts, gears, etc.). However, since the maximum thickness of nitrided body is often less than 0.6 mm versus ~4 mm for carburized parts, the load-bearing capacity of carburized parts is significantly higher than nitrided parts. Conversely, nitriding can provide much higher corrosion resistance. However, due to the relatively small effective thickness, it is less likely to maintain high contact pressure in well-lubricated environments. It is important to note that the Λ-factor tends to decrease with increased roughness after nitriding. As you can see, some polishing during break-in somewhat mitigates this drop. The thick white coating remains in place and increases the resistance of the nitrided material to pitting corrosion on the sections [11]. For untreated, nitrided, and chrome-plated steel, Figure 7.19 demonstrates the typical specific wear rates for low-stress abrasive.

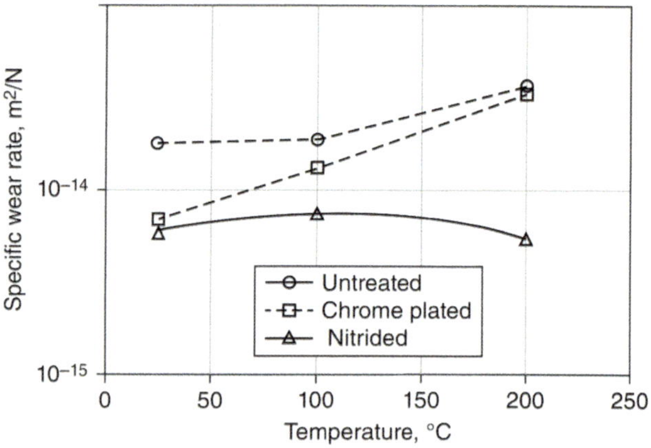

FIGURE 7.19 Untreated, nitride and chrome-plated steel, the average specific wear rates for low-stress abrasive wear are taken from [9].

7.4.3 Alternative Medical Interventions

Another diffusion treatment of steels is boriding. This entails the enrichment of surface with boron. Creating a sufficiently thick composite layer of iron borides ensures maximum surface hardness, which ranges from 1550 to 2000 kg/mm^2. Toyota diffusion (TD) technology produces surface layers with hardness levels ranging from 2500 to 3800 kg/mm^2, which provides an excellent abrasion resistance. This method is usually used to process tool steels. It creates extremely hard vanadium and titanium carbides that are passed through a salt bath. Ion implantation is a surface treatment that produces the thinnest case thickness. High-energy ions are involved, which, depending on implantation parameters such as accelerating voltage and type of ions, strike surface. Various ions such as phosphorus, carbon, nitrogen and boron ions are commonly used for therapy. Components typically undergo an annealing process after implantation to activate certain diffusion processes and restore the crystalline structure [12].

7.5 ION IMPLANTATION IS PROCESS OF CHANGING ITS CHEMICAL AND PHYSICAL COMPOSITION

A great deal of research is being carried out in the field of materials science worldwide on methods to change the electrical, chemical or physical properties of a solid. Ion implantation, a materials science technique in which ions of a material are accelerated in an electric field and slammed into a solid, is the process discussed. By transferring their energy and momentum to the electrons and atomic nuclei of target material, the ions alter the elemental composition of the target, resulting in several chemical and physical changes in the target. This results in structural changes in which cascades of energetic impacts can damage or even destroy the crystalline structure of the target. A brief overview of various aspects

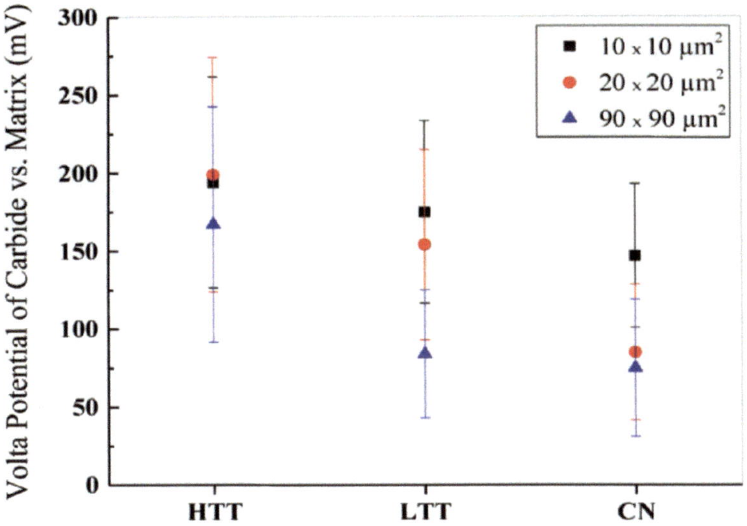

FIGURE 7.20 Plot of measured VPDs (with standard deviation error bars) of carbide precipitates versus the surrounding matrix for the three P675 surface-treated steels as a function of scan area [2] [open access].

of ion implantation procedure has been provided. The chamber body serves as the anode and connected to ground potential. Following the methodology described above, the digital VPD data were calculated and aggregated for comparison (see Figure 7.20, where error bars represent one standard deviation). Depending on the scan size examined, the measured VPDs of the carbides ranged from 60 to 200 mV higher than the steel matrix, with the smallest difference in CN and the largest difference in HTT. Regardless of the scan size, the relative sizes of the VPDs of the carbides and matrix remained constant, indicating that even the smallest imaging areas (10 × 10 μm²) were large enough to represent the sample and provide the best spatial resolution of VPD variations [40].

The power supply, which serves as the cathode plate, is connected to the sample holder. Real-time monitoring of voltage and current throughout the processing is possible thanks to the equipment connected to an oscilloscope and a computer. After the vacuum system is purged to a base pressure of less than 2.0 103 Pa, nitrogen is refilled with a gas flow rate of 30 cm³/min to achieve a working pressure, 0.3 Pa. Finally, nitrogen ion implantations are performed at room temperature, with an acceleration energy, 20 keV, at the same high dose, 1.215 × 1020 ions/cm² and several high dose rates, 2.60 × 1017, 5.18 × 1017, 1.04 × 1017 ions/cm² h and 7.85 × 1017 ions. We can adjust the dose rate by varying experimental parameters such as pulse width and frequency while maintaining the same nitrogen ion implantation dose [8]. Figure 7.21 shows the microscopic images of corrosion scars on plasma-treated 8Cr4Mo4V steel at different dose levels. Both the original sample and sample enhanced with a dose level, 1.04 × 1018 ions/cm².h

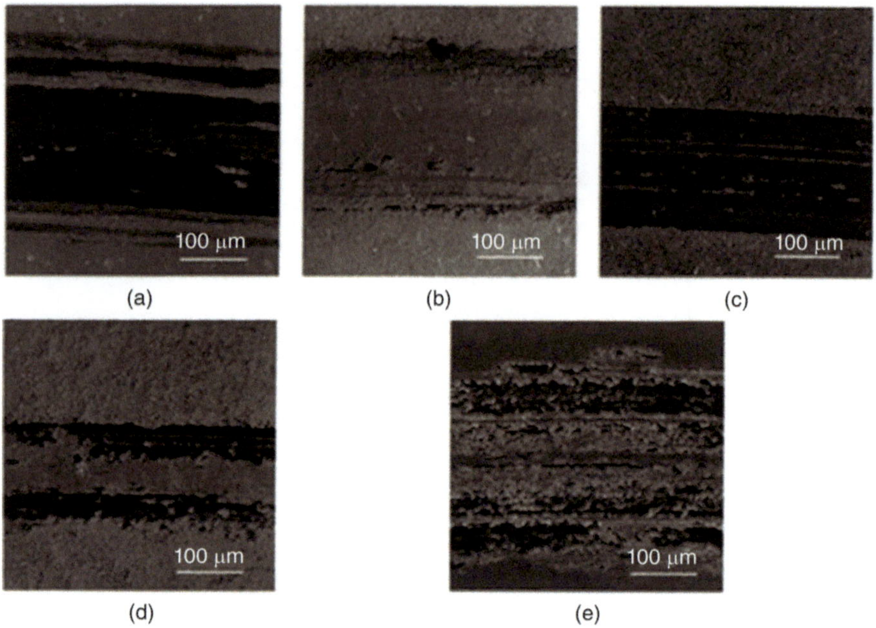

FIGURE 7.21 Corrosion morphology of obtained samples, (a–e) corrosion behavior of samples treated with different doses. The corresponding values are 2.60 × 1017 ions/cm^2.h, 5.18 × 1017 ions/cm^2.h, 7.85 × 1017 ions/cm^2.h and 1.04 × 1018 ions/cm^2.h [40] [open access].

exhibit significant corrosion as shown in Figure 7.21a, with deep scratches and visible debris in their corrosion scars. This indicates a large number of debris that is removed during the corrosion test. On another hand, the samples implanted with ion implantation at a dose level exhibit more limited and smooth corrosion scars with corrosion rates ranging from 2.60 × 1017 to 7.85 × 1017 ions/cm^2.h [40].

High-molecular-weight polymers are attractive options due to their adaptable composition, shapes and properties. Many different sectors including biomedical, electrical, structural and optical have used them in the past to create medical implants, microelectronics, automotive parts, anti-reflective coatings and optical sensors [1–4]. These applications stem from the intrinsic and unique qualities of polymers which include their corrosion resistance, light weight, affordability, moldability, optical clarity, weatherability and ease of availability [5, 6]. Changes in the chemical and structural characteristics of polymers can be easily implemented to meet functional requirements and improve such applications. Modification techniques such as gamma irradiation, ultraviolet (UV) irradiation, neutron, electron and ion implantation can be useful in improving various properties of polymers. Various polymers including polycarbonate (PC), polypropylene (PP), polyallyl diglycol carbonate (PADC), low-density polyethylene (LDPE), polyvinyl alcohol (PVA), polytetrafluoroethylene (PTFE), polyvinyl chloride

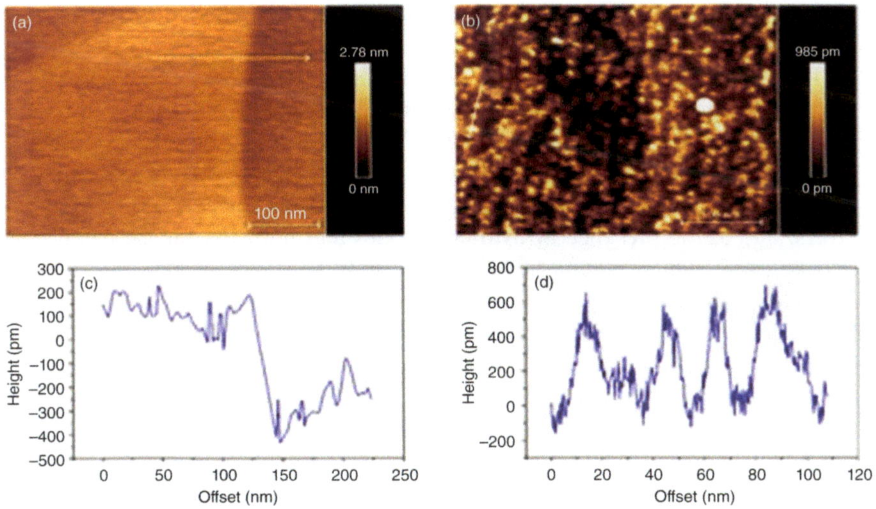

FIGURE 7.22 AFM images (a) HOPG surface without radiation; (b) surface exposed to 23 MeV I radiation during frictional incidence; (c) height profile across the step surface; and (d) height profile of the ion track shown in (b) [11] [open access].

(PVC) and polymethyl methacrylate (PMMA) exhibit remarkable chemical and physical properties after surface modification [7]. This works how radiation affects the surface morphology of highly ordered pyrolytic graphite (HOPG). The ion tracks generated by irradiation with 23 MeV beam at scratch angle of 1° concerning, the surface under ambient conditions is observed by atomic force microscopy (AFM) as shown in Figure 7.22. Compared with the unirradiated region (Figure 7.22a), which has a fairly smooth surface, the irradiated region (Figure 7.22b) shows many ion tracks. The surface height profiles are depicted in Figures 7.22c and 7.22d, respectively, for unirradiated and irradiated HOPG. The location of the profiles is indicated by white arrow in Figure 7.22b. It is noteworthy that the irradiated HOPG shows significant waviness in terrace range, about 100 picometers, while the flat surface of unirradiated HOPG shows wavy tracks and an increasing in surface waviness, 500 picometers. These results demonstrate how radiation significantly affects the surface properties of HOPG, which has implications for its using in a variety of applications.

7.5.1 Anodizing

An electrochemical conversion process known as anodic oxidation is used to coat a metal surface with a hard and durable oxide layer. Its main application is aluminum alloys (although some Al-Cu and Al-Si alloys do not oxidize well). Alloys of titanium, magnesium and zinc. You can purchase decorative coatings (a few micrometers are enough) for architectural coatings. Because paint is created by dissolving the surface of metal; a layer of 50 microns will raise the surface level by 25 microns. Anodizing of aluminum alloys is usually carried out in a sulfuric acid bath.

The "hard oxidation" method, which produces protective coatings, is characterized by thorough mixing of solution and precise control of process parameters, for example, a cold electrolyte (<10°C). The moderately porous layers consist of a mixture of hydrated alumina and alumina. The pore size ranges from 200 to 350 nm and is located perpendicular to the metal surface. The hardness of layers, due to the presence of hydrated oxides and porosity, ranges from 350 to 600 kg/mm^2, which is significantly lower than hardness of alumina, which is 1600-2600 kg/mm^2. Depending on the process parameters, these layers have a fairly high surface roughness (typical value Ra = 2.5 μm). Grinding can be used to reduce it. Hard anodic coatings are primarily used in applications where resistance to wear and corrosion at low loads is required. Pneumatic cylinder tubes, valve spools, automotive brake calipers and pistons and pump parts (operating in aggressive environments) are just a few examples [16].

Aluminum alloys are subjected to significant abrasive wear due to their very low hardness (typically 100 to 180 kg/mm^2). After hard anodizing, the specific corrosion rate at low voltage (measured by the Dynamic Surface Roughness Widening (DSRW) method takes values, 10^{-13} m^2/N, which may be sufficient for various applications. Several factors influence resistance to particulate corrosion. Due to the brittle behavior caused by residual porosity, the wear rate increases with increasing impact angle [23]. Untreated ductile alloys generally exhibit greater resistance than oxidized alloys in most cases. The dry sliding of machined parts is affected by residual porosity. Partial corrosion rates are around 10^{-14} m^2/N. In contact with abrasive materials the porous layer can be coated with a solid lubricant such as MoS_2 or PTFE (polytetrafluoroethylene), or impregnated with oil to reduce wear and friction of alloy being processed, as well as antibody. Solid lubricants can be applied in several ways; as coating components or topcoats. PTFE is the most widely used solid lubricant and can reduce counterbody and workpiece wear by 50% [22]. Microarc oxidation (MAO), often called plasma electrolytic oxidation (PAO), produces extremely durable coatings. Unlike hard anodizing, the deposition process gives superior results, for example creating surface layers with a density, 1400 to 1800 kg/mm2. Hardness is typically highest in the inner non-porous portion, which is formed by the melting and solidification processes of aluminum alloy, and lowest in the outer porous portion of the layer.

7.6 PHYSICAL VAPOR DEPOSITION

Atoms collide with the substrate during PVD, creating a thin layer on top of substrate. It focuses on critical elements in material engineering such as layer thickness, grain boundary, homogeneity, film composition based on stoichiometric concentrations of precursors, which study together to create noble metal-based nanosystems with the desired dimensions. A key component of PVD method, the annealing process (precursor/substrate pre-annealing and post-annealing after deposition), can provide a significant control over the previous described parameters [25, 26, 28, 29]. The antibacterial efficacy of silver nanoparticles (Ag-PCL mats) against Gram-negative bacteria (*Escherichia coli*) is investigated using this synthetic method, and the results have showed that the material is more than 99.9% effective within 4 hours [30]. Among the most efficient and promising manufacturing processes are lasers, which have contributed to the development of technology in many different industries, especially those where material processing is the main focus. The advantages of laser processing including selectivity, non-contact operation and the potential for full process automation. Since non-contact laser processing provides a clean processing area, it allows for remote control of laser beam through transparent protective barriers, in a gas atmosphere, vacuum, or underwater. The ability of laser beam to be concentrated to a very small size – even fractions of a micrometer – is crucial. This allows the beam to be selectively directed to carefully selected areas of the material, such as hard-to-reach areas, areas subject to mechanical stress, etc., without having to worry about how the heat input may affect nearby areas, surrounding features, or deformation of the part. For discussion, it can be divided into metal coating, thin coating and thick coating.

7.6.1 Metallic Paint

Deposits of hard chromium and chemical nickel are mineral deposits that tribologists are interested in studying. To make both, you can use almost any metal.

7.6.1.1 Hard Conversion Coating

A metal object is immersed in a chemical solution to form a conversion layer, with or without an electrical charge on the object. A coating layer is formed on the surface of substrate as a result of interaction of material with a chemical solution. A coating is not just a layer applied to surface of component. As a result, the coating and base material have no noticeable mechanical interaction.

Documented corrosion rates are very low, averaging about 10–16 $m^2/N \times 5$ [13]. In dry conditions, the equivalent friction coefficient is 0.6, which is typical for a steel-steel pair. Aging of coating due to adhesion and corrosion of steel surface may not be sufficient to cause moderate oxidative corrosion. Reported corrosion rates are typically in the range, 41014 m^2/N [26]. However, in this case the equivalent friction coefficients are 0.4 or less, which can be useful in various applications. One study for example, found that chromium coating containing 1.13% silicon carbide nanoparticles has a specific corrosion rate, 1.4×10^{-14} m^2/n [27]. A first-order flattening filter was used on all topographic images to remove individual line-to-line displacements and sample tilt. To account for accumulated debris, a second-order flattening of the HTT images at 116 and 135 min was required. A thresholding technique using a user-defined cutoff voltage based on the voltaic voltage distribution shown in the associated data histogram (512 bins) was used to scale the results from the SKPFM mapping (see example image in Figure 7.23). The mean voltaic voltage (and associated standard deviation) for each of the two surface phases—matrix and carbides, whose identities were verified by SEM/EDS

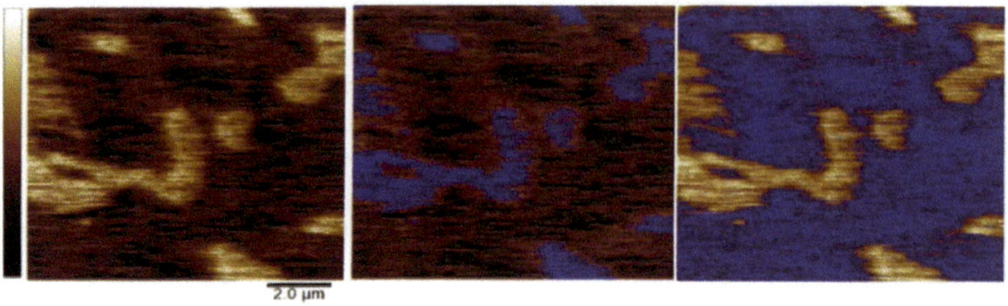

(a) (b) (c)

FIGURE 7.23 Representative 10 × 10 µm^2 scanning Kelvin probe microscopy (SKPFM) images of P675- high-temperature temper (HTT). Dark brown corresponds to the softer matrix phase, which is lower in height following polishing than the harder, lighter brown carbides. Images show (a) the original Volta potential image (600 mV full-scale range) and subsequent implementation of thresholding cutoffs (blue) to calculate average Volta potential differences (VPDs) for the (b) matrix and (c) carbides [2] [open access].

characterization—was calculated from the resulting thresholding data [52]. A representative SKPFM voltaic voltage map of HTT P675 is shown in Figure 7.23a. While the light brown regions in Figure 7.23c correspond to carbides (with the matrix excluded and indicated by the dark blue regions), Figure 7.23b shows the matrix in dark brown with the carbides omitted (data in blue). For each SKPFM image, the average VPD between the matrix and carbides was determined using this technique [3].

Figure 7.23 illustrates atomic force microscopy images and scanning Kelvin force microscopy (SKPFM) voltage maps taken from a coated sample in areas of uniform corrosion. Within the corrosion filaments, intermetallic compounds appear as dark spots as shown in the VPD images. Images obtained after 2 hours, 4 hours and 15 minutes show that the filaments contain bright anodic regions and dark cathodic regions. These results highlight the characteristic electrochemical behaviors that arise during filamentary corrosion and offer suggestions for techniques to mitigate corrosion in coated materials. The filament moves in a jumpy manner towards the cathodic intermetallic compounds shown in the photographs as shown in Figure 7.23a. Figure 7.23b shows that while the filament activity appears to be decreased, head appears to be moving forward and slowly toward intermetallic compound indicated by the red circles in Figure 7.23b.

Figure 7.24 Effect of scanning speed on LFM images adjusted according to the color range. (d-f) Raw LFM data of graphene at different scanning speeds (scale bar: 1 nm); (a-c) Raw LFM data of MoSe2 at different scanning speeds (calibrated scale bar: 1 nm). Each green arrow indicates a zigzag pattern [3].

Examples of chemical conversion coatings include phosphate coatings on steel or zinc, chromate coatings on aluminum, cadmium, copper, magnesium and silver, and oxide coatings on copper, iron, steel and zinc alloys. Electrochemical conversion coating, known as anodizing, converts the surface of some materials,

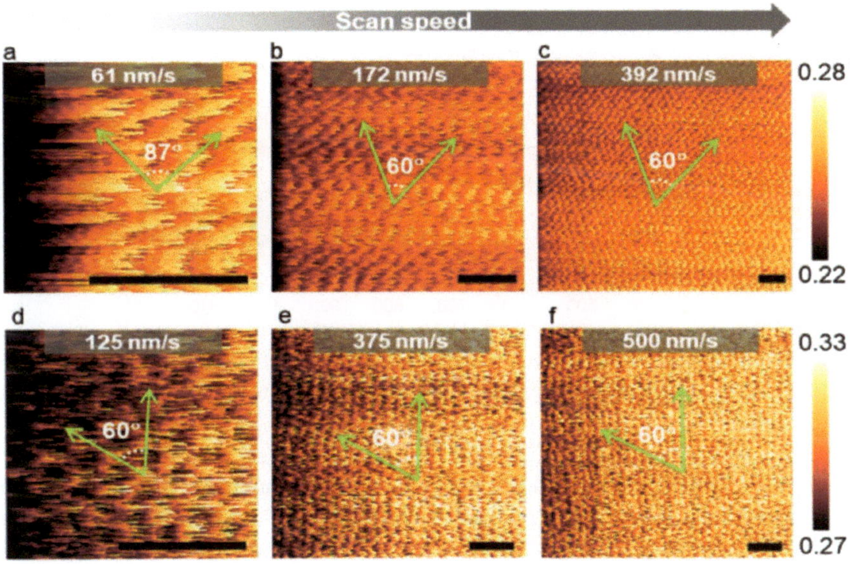

FIGURE 7.24 The effect of the scan speed on LFM lattice images (color range adjusted). (a–c) LFM raw data of $MoSe_2$ at various scan speeds (calibrated scale bar: 1 nm); (d–f) LFM raw data of graphene at various scan speeds (scale bar: 1 nm). All green arrows are aligned along the zigzag direction [2] [open access].

including titanium, magnesium and zinc alloys, into metal oxide. This procedure involves immersing a component that acts as an anode in an electrical circuit of acid solution, then conducting electricity through it. Common applications include dies, rolling mills, hydraulic rods, valves and piston rings. These coatings are used in restoration procedures (when corrosion thickness is less than 250 microns). Hard chrome deposits are generally highly resistant to wear and slip, especially when the coated part is in constant contact with and sliding against a steel object.

7.6.1.2 Electroplated Nickel Plating

Hard chrome plating is known to have negative impacts on human health and environment. Alternatively, electroless nickel (EN) coatings and thermal spray are viable alternatives. At the catalytic surface, nickel ions are chemically reduced to form these coatings. Regardless of the geometry, the coating has a very constant thickness since no electrical current is used [9].

Figure 7.25 illustrates the sliding tests as carried out using Ni-P coating of 20 μm thickness approximately at various heat treatment settings to obtain values ranging from 650 to 1000 kg/mm². As hardness increases, the specific wear rate decreases until it approaches a peak value, 1000 kg/mm², which is normal for hard chrome plating. Adhesion/abrasion caused wear and, in each case, the friction coefficient is around 0.3. As shown in Figure 7.20, electroless nickel coatings

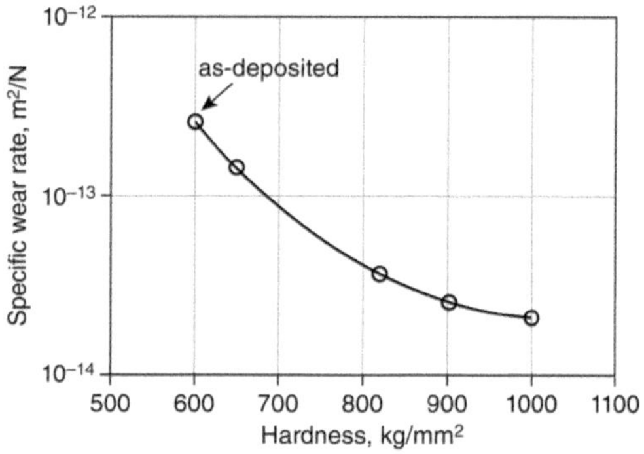

FIGURE 7.25 Impact of hardness on Ni-P coatings ability to withstand wear during dry sliding [9].

often exhibit much less wear at low loads than hard chromium coatings [17]. When dealing with abrasives, it is important to optimize both the cure cycle and phosphorus (or boron) content to prevent excessive brittleness of the coating, which can lead to increase wear due to brittle contact. Three grades of EN coating; alloy coating, composite coating and pure metal coating. Ni-P and Ni-B binary alloy coatings are the most commonly used types. They have strong corrosion resistance and practically amorphous in precipitated state. However, wear resistance is relatively low, although hardness in Ni-P coatings can reach 600 kg/mm^2 and 650–750 kg/mm^2, respectively. Heat treatment at 300–400°C, transforming the amorphous layer into a fine-crystalline nickel matrix containing deposits of Ni_3P or Ni_3B, can provide higher hardness values. The maximum hardness can reach 1000–1200 kg/mm^2, which is equivalent to hardness of hard chromium plating [18]. When EN coatings are properly heat treated, their sliding wear resistance. For example, the study has examined how post-heat treatment affects residual stress in DLD 316 L stainless steel. According to results, the residual tensile stress of specimen decreases sharply after 2-hour heat treatment at 400°C. After heat treating the specimens for 2 hours at 400°C, the residual tensile stress decreases to 356.29 MPa (a decrease of 53.7%) from 769.27 MPa for the untreated specimens [22]. The yield strength and UTS values are constant at same temperature, however, the elongation increases to 40.2% from 26.48%. The yield strength and ultimate tensile strength (UTS) stands for specimen decreasing as heat treatment temperature continuously increases. The microstructure evolution in Figure 7.26 shows the specimens under different conditions after heating [12].

Figures 7.22a and b [38, 39] illustrate how irregular residual stress that builds up during the manufacturing process can lead to the failure of selective laser melting (SLM) components. Figure 7.27d shows the effect of loading on crack initiation in presence of residual stress, while Figure 7.27c [28] shows how residual stress causes the component to deform and detach from the base plate. For thin-walled components, residual stress during manufacturing without an adjacent base makes structural deformation a common source of failure [28]. The fatigue life of components

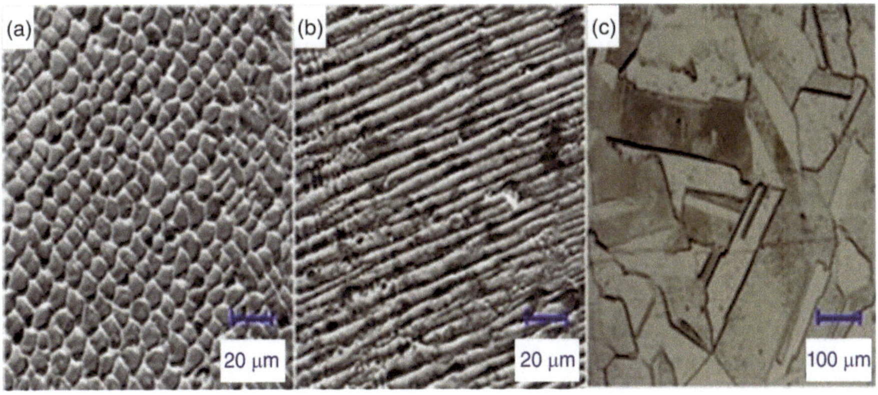

FIGURE 7.26 Microstructure Evolution of DLD 316L Stainless Steel Under Different Post-Heating Conditions: Initial State (a), After 2 Hours at 400°C (b), and After 2 Hours at 1150°C (c) [12] [open access].

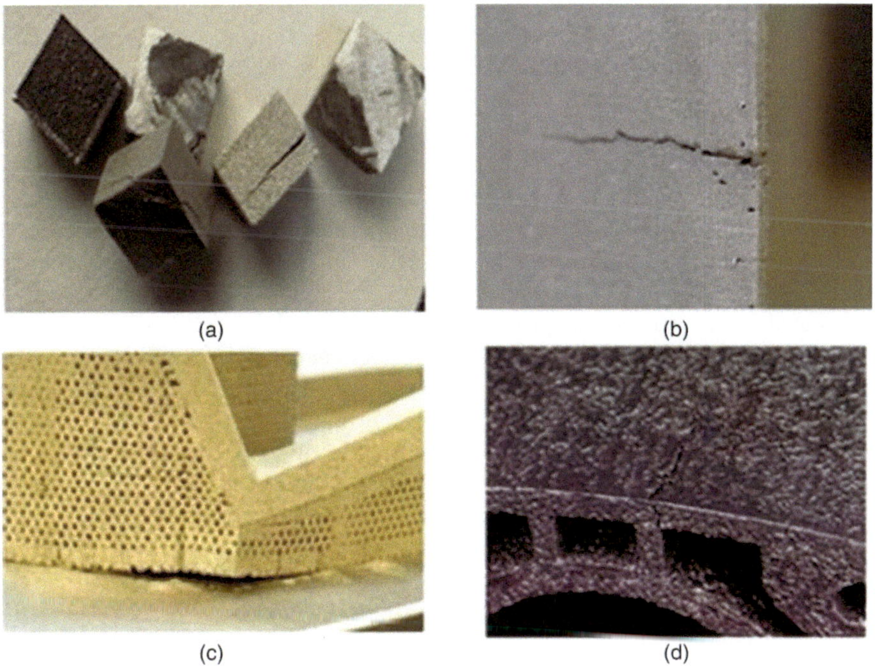

FIGURE 7.27 Processes: (a) cracking of the Inconel-718 alloy part, (b) cracking of the Ti-6Al-4V alloy part, (c) deformation and (d) separation from the base plate [12] [open access].

is limited by this residual stress, which also compromises dimensional accuracy and leads to an inadequate fit of components during assembly [8]. These results highlight the vital need to understand and reduce the effects of residual stress to improve the performance and reliability of parts manufactured using SLM.

7.6.2 Lightweight Cases

Many industries, including automotive, aerospace, electronics, nanobiotechnology and biomedical engineering have found applications for coated materials, especially ceramic coatings. Choosing the wrong technique or characterization method can lead to unreliable pavement functional evaluation results. Several general suggestions have made regarding the basic methods for determining pavement characteristics. There is not much research describing the qualities required for each application and accurate methods for characterizing coatings that are still in their infancy. As a result, this chapter provides an overview of scant research conducted to date, especially on physical, chemical, mechanical and biological methods for characterizing coatings. Extensive study of physical, chemical, mechanical and biological characterization methods required in three major applications – automotive, biological and electronics manufacturers. The fatigue life of a part can be increased by using thin films, since with mixed or boundary lubrication, fatigue cracks initiate predominantly on the surface. In general, the application of thin coatings should be assessed individually based on the involved costs [17]. The deterioration and damage of old steel structures have led to significant financial losses and safety risks [10–12]. One of the main causes of damage is corrosion of steel members, which can lead to the loss of material from surface of steel members and reduce the thickness, rigidity and load-bearing capacity of steel structure. Figure 7.28 shows how the experiment is carried out. Artificial salt water with a concentration, 3.5% is sprayed onto the surface using spray equipment, which resulted in a salt deposition, 28.0 ± 2.8 g/m2. Each cycle of the repeated drying and wetting procedure required eight hours, which included three hours of dryness, 60°C, 35%RH, and three hours of humidity, 40°C, 95%RH. The transition time between the two conditions is one hour each. This is done alternately in 8 cycles (3 days) and 11 cycles (4 days).

Thirty-six (4 × 3 × 3) corrosion samples are divided into three groups and three samples of each type of steel in each group. The duration of tests carried out by three groups is 28 days, 84 days and 168 days, respectively. To further achieve more effective and efficient maintenance management, it is worthwhile to conduct a detailed analysis at modeling and predicting the corrosion of steel structures. Traditional corrosion assessment is labor-intensive and lacks precise standards. However, the target material and research conditions on this topic are limited, and there is no generalization. So, we propose GAN-based machine learning model to improve the dataset and model and predict the weathering of SMA400AW and SMA490AW weathering steel and corrosion of SM400A and SM490A carbon steel in four environments: Figures 7.29, 7.30, 7.31 and 7.32 correspond to four different types of steel plates used in this study [37].

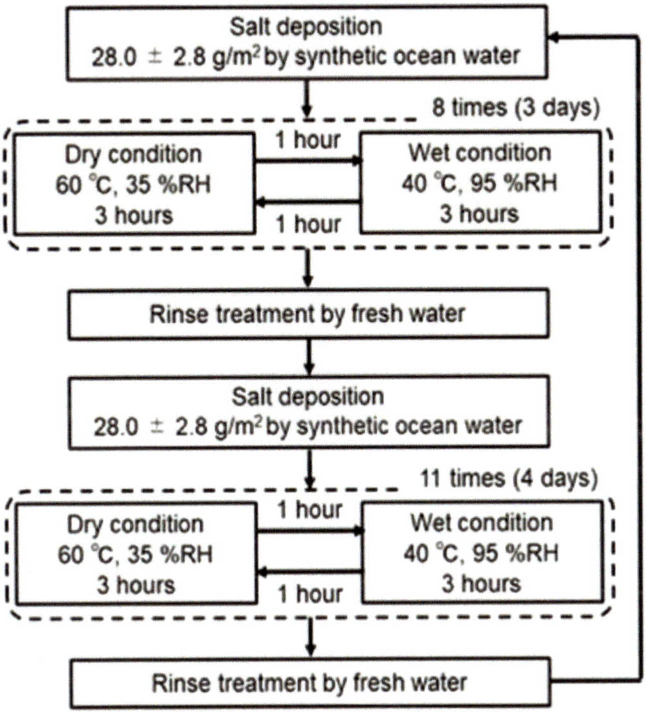

FIGURE 7.28 ISO 16539 sea water spray experiment protocol [37] [open access].

Using the third stage corrosion of SM400A type steel plate as an example, Figure 7.29 illustrates the process. Figure 7.29b depicts the simulated steel plate, while Figure 7.29a demonstrates the actual steel plate. The colors indicate how far the rust has progressed. Its minor corrosion activity is indicated by its yellow color. Rust is worse when it is redder and darker in color. Figures 7.29a,b show a comparable pattern of corrosion distribution, which means that the corrosion on the observed area can be correctly simulated by our technology. The outstanding performance of our approach in modeling or predicting corrosion on three other types of steel plates is further illustrated by Figures 7.30, 7.31 and 7.32. The corrosion of steel plates in the third stage of ISO16539 experiment and corresponding simulated corrosion obtained by [37].

7.6.2.1 Coating Techniques

This method can be applied to a variety of oxides, nitrides and carbides, as well as a variety of coatings [14]. TiN and TiC are the most widely used CVD coatings. Since chemical bonds are formed at the interface due to high temperature of process, adhesion to the substrate is excellent. Plasma combustion around the components in plasma CVD activates the gaseous reactants. Polycrystalline diamonds are deposited in this way [15–18, 20]. PVD processes operate at sub-atmospheric

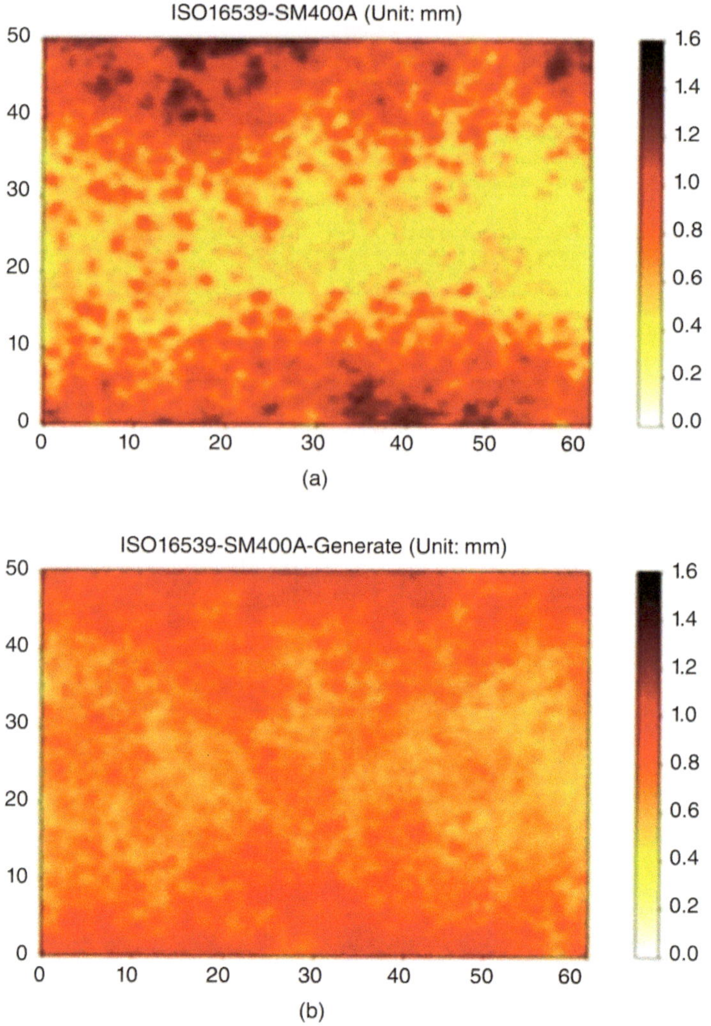

FIGURE 7.29 Associated simulated corrosion and third-stage corrosion on SM400A ISO16539-type steel plate. The three stages of corrosion are observed on two different steel plates: (a) the SM400A ISO16539 type steel plate; and (b) the simulated SM400A type steel plate [37] [open access].

pressures and require lower coating temperatures (500°C or lower). It can be applied to polymers, as well as any metal to create coatings. The main PVD methods are:

- When a metal (target) is exposed to physical environment such as an electron beam, metal vaporizes and condenses on workpiece. Reactive gases such as oxygen, nitrogen or hydrocarbons react. For example, to create TiN, a target of nitrogen and titanium is used.

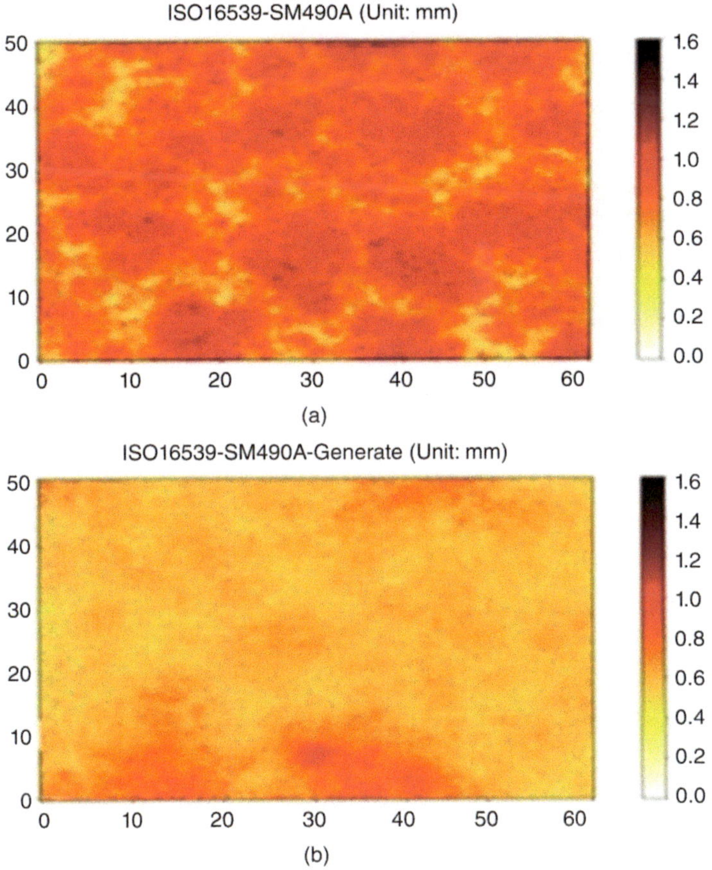

FIGURE 7.30 Corresponding simulated corrosion and corrosion on steel plate type SM490A ISO16539, stage three. (a) corresponding simulated corrosion; (b) corrosion on steel plate type SM490A ISO16539, stage three [37].

- Ions of noble gases, usually argon, are directed toward the target, causing the released ions to be attracted to the workpiece. Reactive gases are used.
- Vehicles are allowed to rain. In general, speckled films adhere more easily to substrates than sprayed coatings.
- The atoms enter the argon plasma after the target metal evaporates. The ionized part of the atoms is attracted to the workpiece. Thermal collisions accelerate others towards the workpiece. In this case, reactive gases are present, allowing the compounds to precipitate. This method is often used in different contexts.

The inability of PVD method to cover cavities deeper than their width is one of its disadvantages.

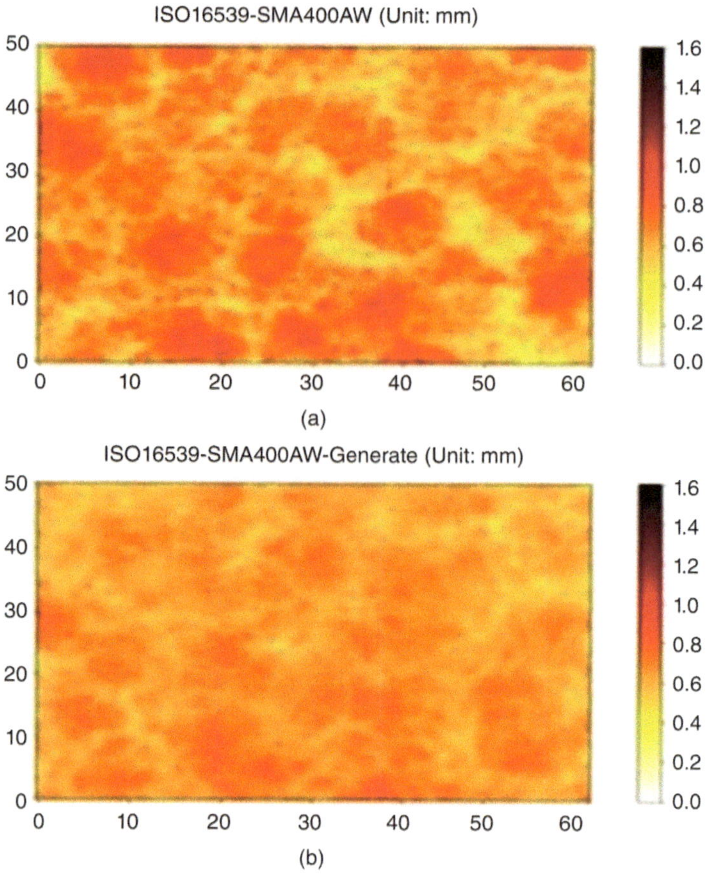

FIGURE 7.31 Third stage corrosion on SMA400AW ISO16539 type steel plate and corresponding corrosion simulation. (a) Third stage corrosion on SMA400AW ISO16539 type steel plate; (b) third stage corrosion on SMA400AW type imitation steel plate [37] [open access].

7.6.2.2 Main Types of Thin Films

Here are some additional notes regarding primary coating systems:

- *Vase*: His general breeding characteristics are excellent. It is often used to coat stamping and cutting tools. In comparison, their operating conditions are limited by low oxidation resistance at high temperatures.
- *TCN*: Due to its higher hardness than TiN, it resists corrosion better, especially corrosion under light loads. Improved adhesion to base carbides makes TiC ideal for hard sandwich coatings on hard metals such as TiC/Al_2O_3/TiN coatings.
- *Vase*: First of all, its thermal resistance is higher than TiN, but it is also harder. It exhibits excellent oxidation stability.

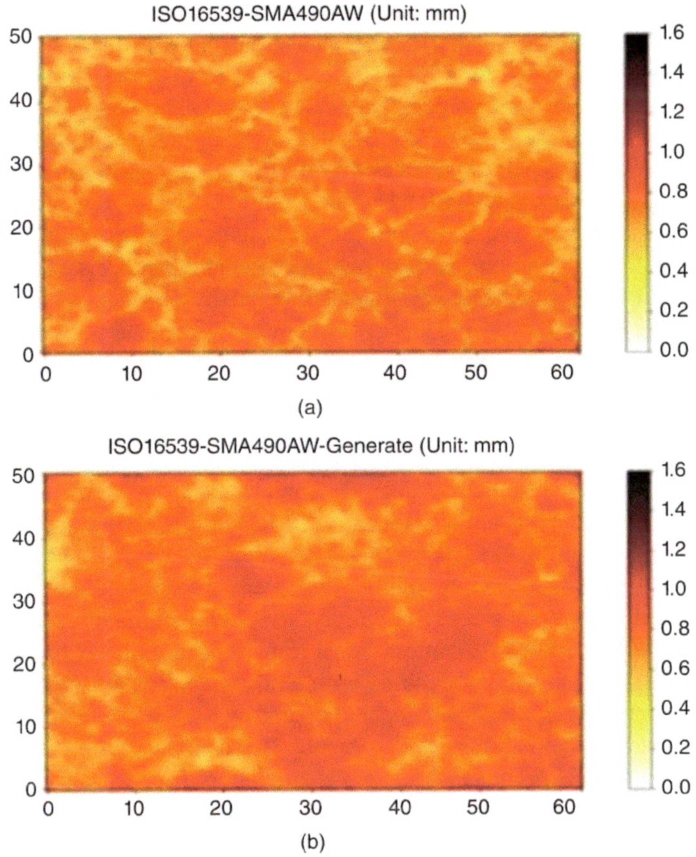

FIGURE 7.32 Corresponding simulated corrosion and corrosion on SMA490AW ISO16539 type steel plate, stage three. (a) Corresponding simulated corrosion is shown, (b) simulated corrosion on the SMA490AW ISO16539 type steel plate, stage three is shown [37] [open access].

- *Karn*: It is often used where high wear resistance and corrosion resistance are required.
- *Solid particles*: They consist of layers of amorphous carbon and solid carbon nanoparticles that alternate, giving them a lamellar structure. While amorphous carbon creates a solid lubricant, solid particles make the coating harder.

Residual stresses from the manufacturing process are found in thin coatings [25]. The two main factors that contribute to this are usually thermal and internal stress. The mechanisms of coating growth are determined by internal stresses. They are usually compressive, and as process temperature increases, their density also increases. according to a study of the coefficients of thermal expansion.

In general, coating properties and process conditions have a significant influence on residual stresses. Experimental measurements of residual stresses in various coatings [26]. When polycrystalline diamond (PCD) coatings are applied to WC-Co that have not been properly pretreated, they exhibit poor adhesion, making the problem particularly serious [27–31]. The primary factor influencing the selection of materials for a particular friction application. Many substances [32–37] have been discovered that suppress all harmful mechanisms These are discussed briefly in the previous chapter. As described in this chapter, there are many situations in which the use of a particular surface treatment is more appropriate or practical. A review of materials and surface treatments specifically designed to combat various corrosive agents. Adhesion and tertiary oxidation are grouped because their material selection criteria are comparable, while corrosion and contact stress are divided into two categories according to the factors discussed in the discussion of corrosion processes. For more detailed information, it is recommended to familiarize yourself with the contents of the above points, specialized literature (some of which are included in the bibliography), catalogs of manufacturers, and their experience [24].

This process converts molecular oxygen into atomic oxygen and ozone through UV stimulation of surface molecules. Converted into harmless volatile compounds such as H_2O, CO, CO_2, and N_2, which are absorbed by the surface, these powerful oxidizing agents destroy organic surface contaminants produced by UV rays. In this case, inorganic contaminants transform into highly oxidized states, which are easily removed by washing with a liquid, such as ultrapure water. This procedure requires minimal start-up and operating costs and can be performed at room temperature and atmospheric pressure. Because they do not use hazardous chemicals or toxic solvents, they are environmentally friendly and do not require expensive treatment or disposal. Common uses include cleaning metals, cleaning blocks that serve as standards, cleaning electrical and semiconductor components, cleaning optical materials and CN, cleaning probe tips, cleaning polymers and biomaterials, decontaminating incubators contaminated with microbes, and ashes of biological materials. Samples for analysis of trace elements and radiological disinfection. Here are some examples of ingredients [21, 38–40].

The disciplines of tribology and surface engineering have been thoroughly reviewed in this chapter, with particular emphasis on the critical issue of wear resistance. We have developed a thorough understanding of the development, application and performance of surface treatment methods and coating systems through a thorough analysis of these approaches. This knowledge, which is based on both theoretical foundations and practical evidence, opens the door to future developments in improving the performance and wear resistance of materials in a variety of applications [41].

7.7 PROBLEMS

1. Explain the physical vapor deposition (PVD).
2. Explain how the ion implantation is process of changing its chemical and physical composition.

3. Should you be responsible for determining the surface treatment technology? How will you improve the engineered surface and reduce the failure rate of associated equipment?
4. List four sectors in which coating methods are widely used, along with one use case for each.
5. List six advantages of surface coating as a primary method of treating and enhancing surfaces.

REFERENCES

1. Bebsh, M., Haimeur, A., & França, R. (2021). The effect of different surface treatments on the micromorphology and the roughness of four dental CAD/CAM lithium silicate-based glass-ceramics. *Ceramics*, *4*(3), 467–475.
2. Kvryan, A., Efaw, C.M., Higginbotham, K.A., Maryon, O.O., Davis, P.H., Graugnard, E., … & Hurley, M.F. (2019). Corrosion initiation and propagation on carburized martensitic stainless steel surfaces studied via advanced scanning probe microscopy. *Materials*, *12*(6), 940.
3. van der Poel, S.H., Mezera, M., Römer, G.W.R., de Vries, E.G., & Matthews, D.T. (2019). Fabricating laser-induced periodic surface structures on medical grade cobalt–chrome–molybdenum: Tribological, wetting and leaching properties. *Lubricants*, *7*(8), 70.
4. Duta, L., Stan, G.E., Popescu-Pelin, G., Zgura, I., Anastasescu, M., & Oktar, F.N. (2022). Influence of post-deposition thermal treatments on the morpho-structural, and bonding strength characteristics of lithium-doped biological-derived hydroxyapatite coatings. *Coatings*, *12*(12), 1883.
5. Rashid, M.M.U., Tomkowski, R., & Archenti, A. (2022). Effect of surface pretreatment on the adhesion between HiPIMS thick Cu: CuCNx coating and WC-Co shim. *Coatings*, *12*(10), 1484.
6. Dazzi, R., Prazeres, F., Glotin, E., & Ortega, J.M. (2007). Analysis of nanochemical mapping performed by an AFM-based ("AFMIR") acousto-optic technique. *Ultramicroscopy*, *107*, 1194–1200.
7. Bostwick, E., Rotenberg, J., Avila, M.C., & Asensio, J. (2012). Zooming in on electronic structure: NanoARPES at SOLEIL and ALS. *Synchrotron Radiation News*, *25*, 19–25.
8. Avila, M.C., & Asensio, J. (2014). First NanoARPES user facility available at SOLEIL: An innovative and powerful tool for studying advanced materials. *Synchrotron Radiation News*, *27*, 24–30.
9. Al-Samarai, R.A., & Al-Douri, Y. (2024). Friction and Wear in Metals, Springer.
10. Vogt, P., De Padova, P., Quaresima, C., Avila, J., Frantzeskakis, E., Asensio, M.C., Resta, A., Ealet, B., & Le Lay, G. (2012). Silicene: Compelling experimental evidence for graphenelike two-dimensional silicon. *Phys. Rev. Lett.*, *108*, 155501.
11. Tomić Luketić, K., Karlušić, M., Gajović, A., Fazinić, S., O'Connell, J.H., Pielić, B., & Kralj, M. (2021). Investigation of ion irradiation effects in silicon and graphite produced by 23 MeV I beam. *Materials*, *14*(8), 1904.
12. Bastola, N., Jahan, M.P., Rangasamy, N., & Rakurty, C.S. (2023). A review of the residual stress generation in metal additive manufacturing: Analysis of cause, measurement, effects, and prevention. *Micromachines*, *14*(7), 1480.
13. International Union for Vacuum Science, Technique and Applications (IUVSTA). Available at: http://www.iuvsta.org/.

14. Mozetic, M., Ostrikov, K., Ruzic, D.N., Curreli, D., Cvelbar, U., Vesel, A., Primc, G., Leisch, M., Jousten, K., Malyshev, O.B., Hendricks, J.H., Kover, L., Tagliaferro, A., Conde, O., Silvestre, A.J., Giapintzakis, J., Buljan, M., Radic, N., Drazic, G., Bernstorff, S., Biederman, H., Kylian, O., Hanus, J., Milosevic, S., Galtayries, A., Dietrich, P., Unger, W., Lehocky, M., Sedlarik, V., Stana-Kleinschek, K., Drmota-Petric, A., Pireaux, J.J., Rogers, J.W., & Anderle, M. (2014). Recent advances in vacuum sciences and applications. *J. Phys. D Appl. Phys.*, *47*, 153001.
15. Sun, Y., & Bailey, R. (2023). Tribocorrosion behavior of γ'-Fe4N nitride layer formed on mild steel by plasma nitriding in chloride-containing solution. *Lubricants*, *11*(7), 281.
16. M. Rajkiewicz, W. Tyszkiewicz, Z. Wertejuk(Eds.) (2013). Chemistry and Physics of Complex Materials: Concepts and Applications, Apple Academic Press. ISBN: 9781926895604.
17. Baer, D.R., Engelhard, M.H., Johnson, G.E., Laskin, J., Lai, J.F., Mueller, K., Munusamy, P., Thevuthasan, S., Wang, H.F., Washton, N., Elder, A., Baisch, B.L., Karakoti, A., Kuchibhatla, S.V.N.T., & Moon, D. (2013). Surface characterization of nanomaterials and nanoparticles: Important needs and challenging opportunities. *J. Vac. Sci. Technol. A*, *31*, 050820.
18. Cho, E.J., Holback, H., Liu, K.C., Abouelmagd, S.A., Park, J., & Yeo, Y. (2013). Nanoparticle characterization: State of the art, challenges, and emerging technologies. *Mol. Pharmaceut.*, *10*, 2093–2110.
19. Wei, J., Ma, Z., Wen, H., Guo, H., Tang, J., Liu, J., ... & Sugawara, Y. (2020). Adhesion effect on the hyperfine frequency shift of an alkali metal vapor cell with paraffin coating using peak-force tapping AFM. *Coatings*, *10*(1), 84.
20. Li, P., Gong, Y., Liang, C., Yang, Y., & Cai, M. (2019). Effect of post-heat treatment on residual stress and tensile strength of hybrid additive and subtractive manufacturing. *Int. J. Adv. Manuf. Technol.*, *103*, 2579–2592.
21. Kim, Y., & Lee, S. (2023). Applications of electron microscopy in surface analysis of biomaterials. *Materials*, *12*(3), 1192.
22. Lopez-Serrano, R.M., Olivas, J.S., Landaluze, C., & Camara, C. (2014). Nanoparticles: A global vision. Characterization, separation, and quantification methods. Potential environmental and health impact. *Anal. Methods-UK*, *6*, 38–56.
23. Kahru, K., & Savolainen, K. (2010). Potential hazard of nanoparticles: From properties to biological and environmental effects. *Toxicology*, *269*, 89–91.
24. Rosendahl, S.M., Borondics, F., May, T.E., Pedersen, T.M., & Burgess, I.J. (2011). Synchrotron infrared radiation for electrochemical external reflection spectroscopy: A case study using ferrocyanide. *Anal. Chem*, *83*, 3632–3639.
25. Kumar, S., Liu, X., Borondics, F., Xiao, Q.F., Feng, R.F., Goormaghtigh, E., & Nikolajeff, F. (2017). Insights into biochemical alteration in cancer-associated fibroblasts by using novel correlative spectroscopy. *Chemistryopen*, *6*, 149–157.
26. Kilcoyne, A.L.D., Tyliszczak, T., Steele, W.F., Fakra, S., Hitchcock, P., Franck, K., Anderson, E., Harteneck, B., Rightor, E.G., Mitchell, G.E., Hitchcock, A.P., Yang, L., Warwick, T., & Ade, H. (2003). Interferometer-controlled scanning transmission x-ray microscopes at the advanced light source. *J. Synchrotron Radiat*, *10*, 125–136.
27. Chapman, H.N. (2010). MICROSCOPY: A new phase for X-ray imaging. *Nature*, *467*, 409–410.
28. Bauer, E. (2014). Surface microscopy with low energy electrons. Springer, New York.
29. Kumar, C. S. S. R. (2015). Surface science tools for nanomaterials characterization. Springer, New York. Bozz.

30. Bozzola, J.J., & Russell, L.D. (2002). Electron Microscopy: Principles and Techniques for Biologists, Jones & Bartlett Publishers.
31. Benyahia, L., Lajaunie, L., Sablier, M., Gagnaire, F., & Martouzet, L. (2023). Surface characterization and materials analysis. *Materials Characterization*, *1*, 34–44.
32. Parchen, S., & Heuer, S. (2023). Nanotechnology and applications in surface engineering. *Surface Engineering*, *1*, 56–67.
33. Azarov, A., & Dubitsky, M. (2022). Surface Analysis and Imaging of Nanomaterials, Springer Nature.
34. Petrakova, D., Timoshenko, V., & Radziuk, A. (2021). Surface characterization of carbon materials. *J. Surface Sci.*, *6*(1), 12–19.
35. Wang, S., Huo, J., & Zhang, Y. (2020). Atomic force microscopy for surface characterization in nanotechnology. *Nature Nanotechnol.*, *15*(9), 810–818.
36. Goncalves, J., Ferreira, T., & Figueiredo, H. (2021). Surface properties of biomaterials. *J. Material Sci.*, *45*(8), 142–148.
37. Li, X., & Zhang, H. (2018). Scanning electron microscopy in materials science: An overview. *J. Microscopy*, *23*, 351–362.
38. Davidson, J., & Chang, P. (2022). The impact of scanning probe microscopy techniques in materials characterization. *J. Nanotechnol.*, *30*(2), 243–255.
39. Smith, A., & Jones, R. (2017). Atomic force microscopy: Principles and applications. *J. Vacuum Sci. Technol. A*, *35*(6), 060602.
40. Zhang, Z., & Wei, L. (2023). Surface analysis techniques in material science. *Materials Manufact. Process.*, *25*(4), 232–245.
41. Wang, Z., & Lin, C. (2022). Surface analysis by X-ray spectroscopy. *J. Microscopy*, *12*, 105–110.

8 Surface Engineering and Manufacturing

8.1 INTRODUCTION

Surface engineering, sometimes referred to as advanced surface engineering, is an important field of study essential to many different industrial processes. It involves methods for modifying the surfaces of solid materials to improve their quality or impart them new properties. Over the past few decades, much attention has been paid to manufacturing equipment and protecting industrial components and manufacturing equipment from excessive wear, oxidation, corrosion, and friction, which are factors that cause surface deterioration. In addition to incurring significant manufacturing costs due to the need to replace and dispose of components and tools, degradation is expected to result in economic losses in industrial economies equivalent to 4% of GDP. Surface engineering developments aim to overcome these barriers to improve the durability and functionality of materials, ultimately supporting more environment friendly manufacturing methods. These qualities amount to over USD280 billion per year for the United States. In addition, other studies have estimated that in Germany alone, the manufacturing industry spends up to USD1–2 billion per year [1] on petroleum-based lubricants to avoid wear. Surface protection costs resulting from the production of galvanic process residues, such as hexavalent chromium, are additional. The development of micro- and nanofabrication (MNF) technologies has brought new and exciting challenges to surface engineering. Surface modification processes will play an important role in enabling the industrialization of several technologies, including microforming, micromachining and micronanotexturing. In addition, unique functional surfaces must be discovered by new and emerging technologies to bring to market elements that can outperform those currently available in the market that use traditional concepts. Some examples include (1) biomaterials that require sophisticated surface biofunctionalization techniques, or (2) fabrication of functional membranes and other functional coatings for H_2 fuel cells for sustainable energy production [2].

Given the critical role played by the surface condition, many surface modification techniques have been developed and implemented to date [3], resulting in a variety of surface topographies and properties, each with unique applications and characteristics tailored to a specific body part. These techniques can be categorized into groups based on many criteria, including the type of technique and topographic size. Although the surface modification processes are briefly depicted in Figure 8.1 based on the type of technique, the purpose is to discuss them in terms of the structure size [4].

 DOI: 10.1201/9781003584735-8

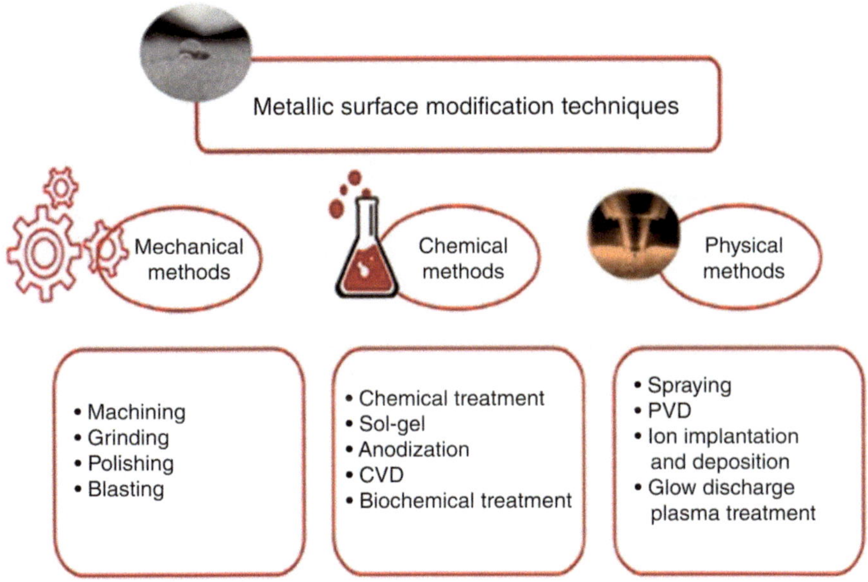

FIGURE 8.1 Review of surface modification methods [4] [open access].

Additional top-down techniques include block copolymer lithography, nanolithography and scanning probe lithography [5]. The other side of top-down nanomanufacturing is bottom-up. Bottom-up nanofabrication, or self-assembly, techniques assemble using small physical or chemical forces. Modern technologies facilitate the development of a range of materials such as coatings, monoliths, powders, dispersions, polymer nanocomposites (PNCs), prototypes and devices from the bottom up. Due to their adaptability, hybrid technologies such as viral templates for lithium batteries can be developed. Nanomaterials meet the requirements for higher quality and improved performance at the nanoscale and are essential for many applications such as oil sponges, catalysts, health products, electronic storage media, solid-state sensors and vehicle cleaning coatings. With annual sales in the billions of dollars, the nanomaterials industry is significant and reflects the growing importance and demand for these materials in many industries.

However, producing end-user products using nanoparticles requires careful planning and research. Bottom-up nanostructure construction requires a comprehensive understanding of individual molecular structures, self-assembly and dynamic behavior [6, 7]. As the passivation period increases, the width of nanoscale features, i.e., scaly spots on sample surface, increases, which mean the growth of passive film. The sample surface is covered with a thick passivation layer when the grooves left by surface treatment are not visible. The calculation of the surface roughness of topographic profile has showed that the passive film is relatively uniform since it is modest and fluctuated little. As shown in Figure 8.2, the roughness increases and passive film became nonuniform when the potential reaches 0.7 V saturated Calomel electrode (SCE) [1].

Investigation of the mechanism behind the initial stages of localized corrosion at the cutting edges of adhesively bonded Zn (Z) and Zn-Al-Mg (ZM) galvanized steel substrates. Under anodic polarization in the corrosive solution, in situ AFM equipment was used to measure the initially localized corrosion topography of Z and ZM specimens. As shown in Figure 8.3, the zinc layer was mainly responsible

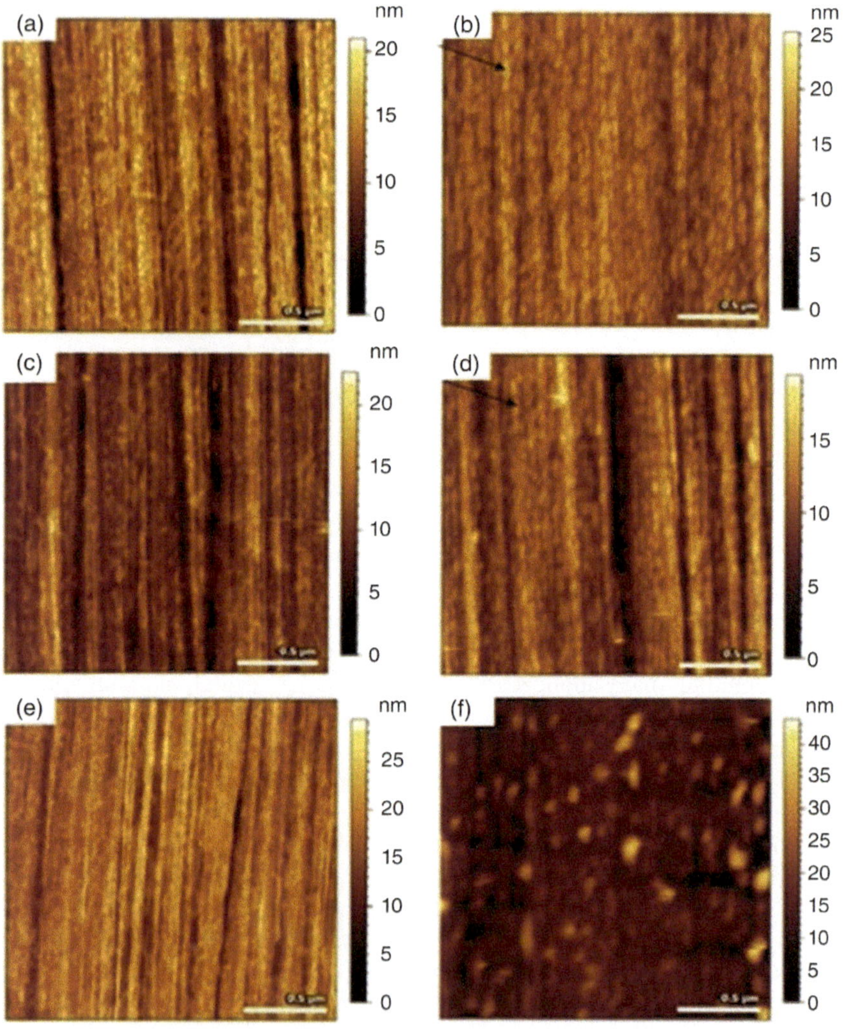

FIGURE 8.2 AFM images of steel sample polarized for (a) 0.5 minute and (b) 60 minute, (c) 0.5 minute and (d) 60 minute, and (e) 0.5 minute and (f) 60 minute at −0.1 V (SCE). The black arrow points to scaly regions [1] [open access].

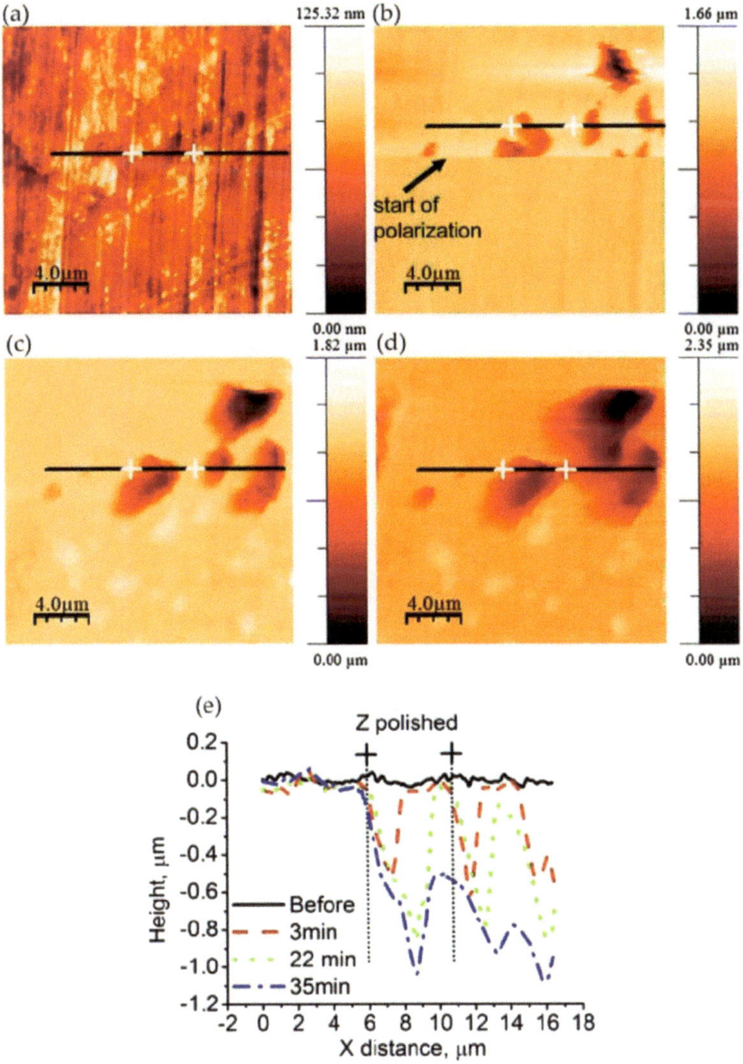

FIGURE 8.3 AFM topography maps made on cutting edge of adhesively bonded Zn (Z) substrate before immersion (a) and during 24 min (b), 1 h 5 min (c), and 2 h (d) of immersion in 0.001 M NaCl. (e) Evolution of topography across the black line profiles drawn in the same zones on AFM maps. Black arrows indicate local pits on the zinc layer [1] [open access].

AFM images of an annealed $Al_{88}Ni_8Ce_4$ sample immersed in 0.01 mol/L NaCl at open circuit potential (OCP) for three minutes in touch (TM) mode are shown in Figure 8.4. Figure 8.4a shows the surface topography, highlighting the height deviations. Image While Figure 8.4b identifies structural or mechanical variations, and Figure 8.4d magnifies the microstructural changes, Figure 8.4c provides a 3D phase view giving a stereoscopic view of the crystalline distribution. To better understand the corrosion resistance and surface interactions in aqueous solutions, the study evaluates the sample stability and surface behavior in chloride environments.

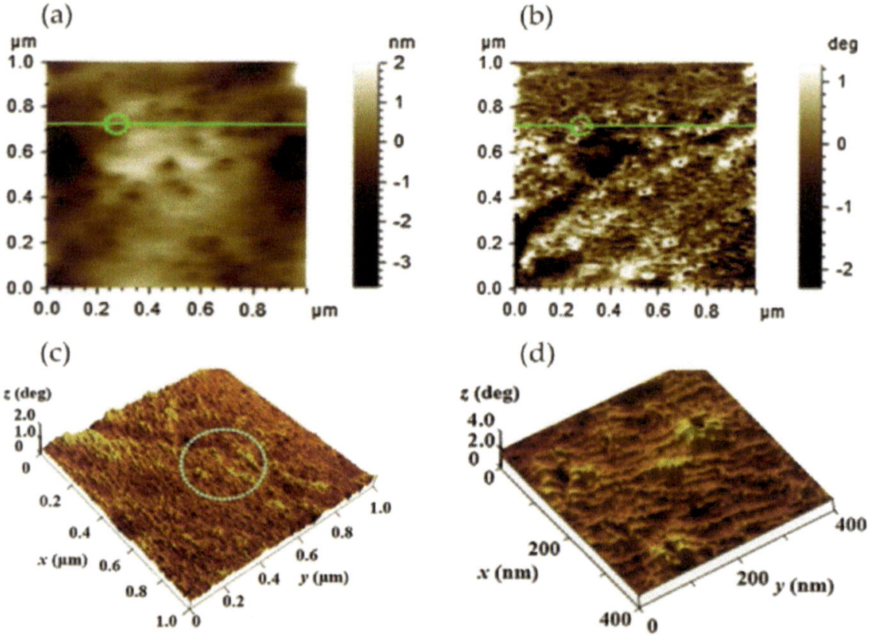

FIGURE 8.4 In situ tapping mode (TM) AFM images of the annealed $Al_{88}Ni_8Ce_4$ amorphous nanocrystalline sample at open-circuit potential (OCP) held for 3 min in 0.01 mol/L NaCl solution. Topography image (a) and phase image (b). Three-dimensional (3D) images of phase (c) and the (d) local enlarged phase image of the marked circle in (c) [1] [open access].

for the initial corrosion pits at the Z cutting edge; neither adhesive/zinc nor zinc/steel showed any preferential corrosion [1].

The negative film was formed by in situ EC-AFM in the Al-Ni-Ce metallic glass. Open circuit potential (OCP) conditions were used for in situ EC-AFM measurements and the voltage was introduced at the drip potential (Epit). Using TM-AFM, they were able to identify the α-Al nanophase from the amorphous matrix. The phase image shows that small α-Al islands, which showed a negative phase transition for the matrix, were randomly scattered across the surface. Figure 8.5 shows how the α-Al island is surrounded by bright ring spots in the slightly magnified phase image. Furthermore, in situ, EC-AFM was used to monitor the nucleation process and growth of the passive film in real-time. It was found that the presence of α-Al nanoparticles changed the nucleation mechanism of the passive film in the amorphous matrix from instantaneous nucleation to gradual nucleation. The formation of the corrosion product Al(OH)3 occurred early in the formation of the passivation film due to the galvanic coupling between the α-Al nanophase (anode position) and the surrounding amorphous matrix (cathode position). This product was mixed into the passive film, changing its local structure and composition [1].

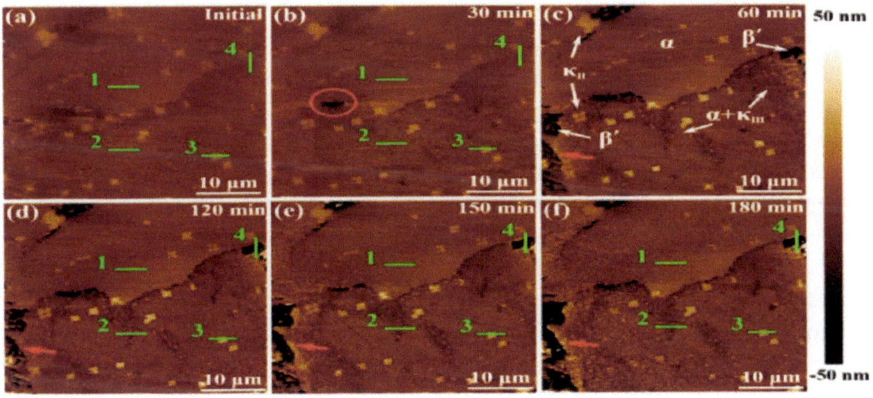

FIGURE 8.5 In situ topography images of nickel–aluminum bronze (NAB) specimen surface after exposure to 3.5 wt% NaCl solution at different times: (a) initial; (b) 30 min; (c) 60 min; (d) 120 min; (e) 150 min; (f) 180 min. Site 1 corresponds to the α phase, site 2 corresponds to the $\alpha + \kappa_{III}$ eutectoid structure, and site 3 corresponds to the κ_{II} phase, while site 4 corresponds to the β' phase [1] [open access].

In the field of corrosion electrochemistry, electrochemical atomic force microscopy (EC-AFM) cannot conduct in-situ microscopic corrosion (passivation) studies in real time. In addition, its high resolution provides detailed information on submicroscopic corrosion phenomena including passivation and activation, especially in the early stages of corrosion. We can better understand the corrosion process and its mechanism by quantitatively analyzing the changes in the surface topography of target sample using the information on changes in surface height with applied voltage or time. Functional supramolecular objects observed in living organisms are examples of structures that can be created reliably, accurately, and with atomic precision through bottom-up assembly. Bottom-up methods include wet chemical processes such as co-precipitation, microemulsions, and gelatinization. One approach is a top-down strategy that starts with large, uniform pieces of material and using powerful lithography and etching techniques to create the desired nanostructures. Three general categories can be used to group nanofabrication methods; physical processes often decompose some bulk material (precursor) into smaller components in a top-down manner. The scientists are more interested in developing advanced post-synthetic methods for characterization, functionalization, synthesis, purification and modification. The creation of nanomaterials is possible using gas phase, liquid phase and solid phase methods. They can be created directly as molecules or distributed among liquids, glass and gas. Liquid phase methods are relatively inexpensive, easy to use and versatile. Low temperatures, polar and nonpolar solvents are used. Researchers are still making efforts to improve the synthesis process through a better understanding of nucleation and crystal formation processes. Since this sector is still young, it requires a lot of knowledge and improvisation. Nanomaterials must be functionalized to realize their full potential. Nanomaterials can be easily made hydrophobic, hydrophilic, conductive and corrosion resistant simply by changing their surface [8–10].

The production of biological nanoparticles is a promising and greener approach that allows synthesis under aqueous conditions due to low energy requirements and low cost. The inclusion of inorganic nanoparticles in the coating also increases the metal's resistance to scratches and stickiness. In clinical studies, multifunctional nanoparticles have improved targeted drug delivery and active drug uptake into cells. These multifunctional nanoparticles are produced by conjugation with a specific chemical. Simple functionalization or modifications of biomolecules, surfactants, polymers and copolymers can also produce these multifunctional nanoparticles. As a result, NPs with improved optical, thermal, electrical and mechanical properties are uniformly distributed throughout the polymer matrix.

8.2 INVESTIGATION OF DLC COATING DEGRADATION USING TRIBOEMISSION IMAGING AND CHARGE MEASUREMENTS

To protect components subject to sliding and rolling contact, coatings are widely used in various industrial and commercial applications, both with and without liquid lubricants [1]. Hard coatings and other surface modification techniques improve the resistance of rolling elements (i.e., bearings) to wear, corrosion and friction [4, 11]. For example, thin diamond-like carbon (DLC) coatings provide low friction, wear resistance and protective surfaces for a wide range of industrial applications, including razor blades, magnetic hard disks, invasive and implantable medical devices, and microelectromechanical systems [5–7, 12–15]. However, residual stresses and chemical reactions can occur at material–material interface due to mismatches in mechanical and electrical properties between the substrate and coating, leading to the layer's structural degradation. To increase the lifetime of widely used devices, damage assessment, the electrical state of film/coating and substrate, and ability to track coating failure are becoming critical tools. Recently, various techniques have been used to study coating failure. These include scanning electron microscopes (SEMs) and thermographs to detect damage progression [8, 16], AFM to map crack progression in nickel films on polyimide substrates [17] and ultrasonic force microscopy to measure delamination of glass films on polyethylene terephthalate substrates [18]. In addition to these qualitative methods, real-time in-situ acoustic emission-based methods have been developed to study the interface properties of coating/film systems by obtaining quantitative stress/strain information [2, 9, 10]. The surface topography of the sample in situ using an atomic force microscope is shown in Figure 8.5. The corrosion behavior of the α phase varied with location. The α phase far from the Kaya phase did not corrode, mostly due to the formation of numerous micro galvanic cells in the eutectoid, while the lamellar α phase within the α + Kaya III eutectoid structure was severely corroded. Because a dense, hard protective layer was formed quickly, the Kaya II and Kaya III phases exhibited great corrosion resistance. Since the Kaya II phase was based on Fe3Al intermetallic compounds and included more iron, it is noteworthy that the layer in this phase was thicker [11].

These results demonstrate the potential of this method to track the location of coating failure and to assess how emission events have changed over time. However, the requirement to test the microchannel plates in a vacuum is a challenge. Subsequent charge measurements are performed [17]. The wear scar is captured on video using a conventional optical microscope in combination with air charge measurements. The average color intensity data collected at contact exit allowed a local assessment of immediate coating integrity on sample surface during sliding. Since steel is more reflective than DLC coating, the increasing in intensity indicates the influence of metal substrate. Figure 8.6a illustrates a graph of observed surface charge and average intensity at contact exit as a function of time. This figure shows how the surface becomes positively charged as the tip scratches the coating without failure. However, once the coating deteriorates (and the failure point is defined), as shown in Figure 8.6b (shown by the dotted line), the metal substrate is exposed and surface charge begins to decrease to negative values, which is consistent with the previous results [11].

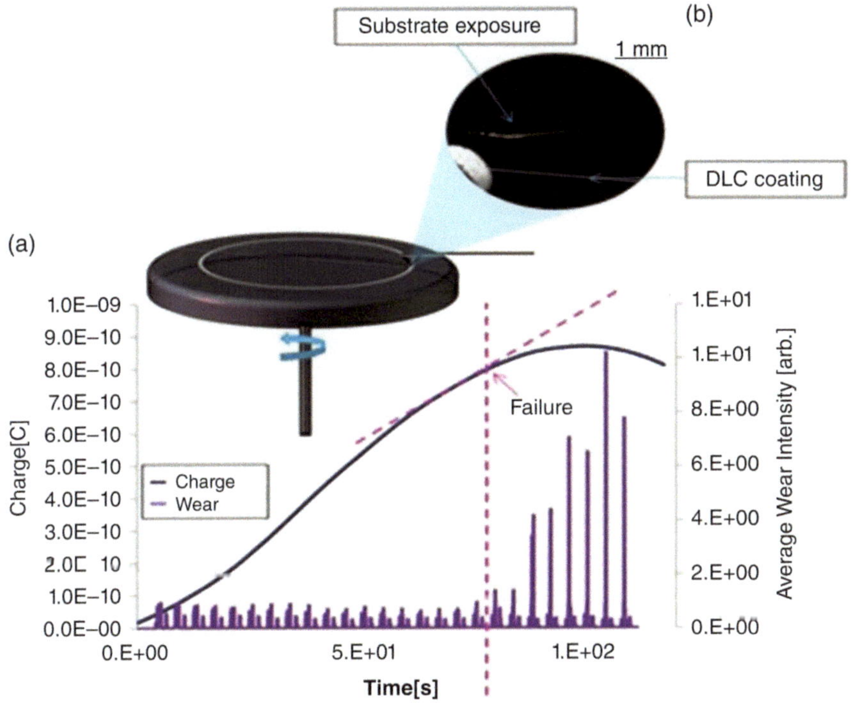

FIGURE 8.6 (a) surface charge and average intensity of worn track over time; (b) tip exit pattern of worn track with 5× magnification [11] [open access].

8.3 PHASE DEVELOPMENT AND CHARACTERISTICS OF HIGH-ENTROPY FeCoCrNiAlTix ALLOY COATINGS ON Q253 USING LASER CLADDING

High entropy alloy (HEA) is a new design philosophy that has just emerged and is turning traditional alloy design thinking upside down. Multiple primary elements can be incorporated into an alloy system using the HEA technique without creating complex phases such as intermetallic compounds (IMCs), which can lead to embrittlement and complicate processing and applications. Rather, simple solid solution formation is promoted by high entropy. This has led to a significant increasing in interest in metallic materials research field of HEA. It is important to understand how alloying elements affect the composition and functionality of HEA to advance their practical applications.

In this work, laser cladding is used to tune the titanium content of FeCoCrNiAlTix HEA coatings (x = 0, 0.25, 0.5, 0.75, and 1 at.%) on Q235 steel. X-ray diffraction (XRD), SEM, microhardness testing and friction wear testing are used in the analysis process to study the constituent phases of coatings, microstructure, hardness and wear resistance. According to results, the coating exhibits a eutectic microstructure with face-centered cubic (FCC) and body-centered cubic (BCC) 1 phases for x = 0. Less eutectic structures are presented with increased titanium levels. The Fe–Cr and (Fe, Ni) phases respectively indicate FCC and BCC1 systems, which are incorporated into FeCoCrNiAl coating, according to XRD study. These findings expand our knowledge of HEA behavior and its possible using in real production conditions. Figure 8.7 shows the results

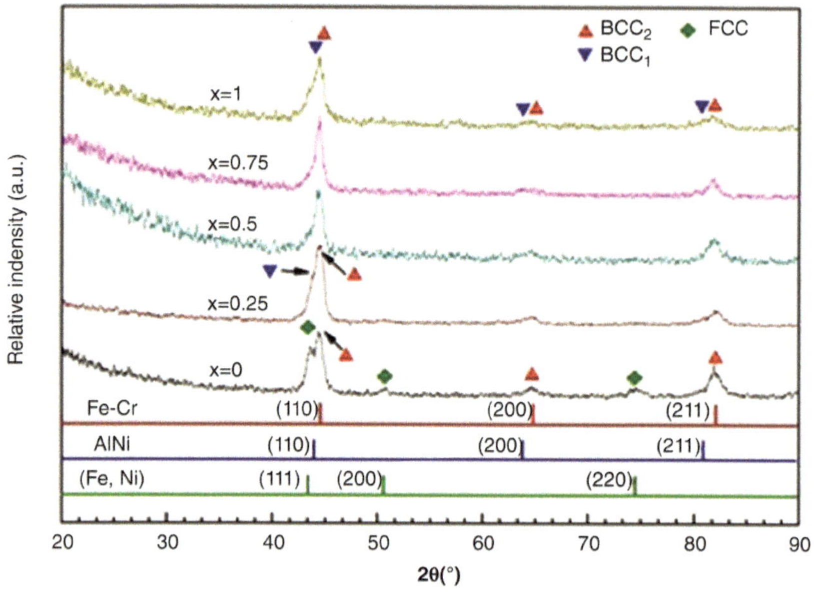

FIGURE 8.7 XRD scheme of FeCoCrNiAlTix high entropy alloy (HEA) coatings [12] [open access].

of X-ray diffraction (XRD) analysis of FeCoCrNiAlTix high-entropy alloy coatings. The analysis is based on determining the crystal structure and phase of the coating based on the locations of diffraction peaks. The main points that can be observed from XRD. The peaks appear at certain diffraction angles (2θ) and are related to the interatomic spacing (d-spacing) in the crystal phase. The peaks are prominent at specific locations indicating the main crystal phase of the alloy. The figure provides a comprehensive overview of the effect of titanium addition on the crystal structure of the high-entropy alloy, with an analysis of the phases formed and their stability. This analysis helps in identifying potential applications of these coatings based on their mechanical or thermal properties [12].

But in $FeCoCrNiAlTi_{0.25}$ HEA coating, the FCC system disappears and AlNi material forms BCC. Ti is a stabilizing element of BCC, which suppresses the formation of FCC. BCC2 is the identification of AlNi. In the experiment where AlCoCrFeNiTix is deposited by arc melting and injected into a water-cooled copper mold [8]. The cross-sectional SEM images of HEA-coated FeCoCrNiAlTix alloy. At x = 0, there are two different phases in microstructure. When examining the EDS data, it is obvious that there is a significant similarity between the components of zone A and zone B. As a result, zone B is a basic hole. Here, the only eutectic structures are fcc and bcc1. The inclusion of Ti causes a sharp change in the structural morphology. The α-eutectic phase and β-eutectic phase constitute the structure at x = 0.25. The β-eutectic phase is a typical lamellar phase, but the α-eutectic phase has a chrysanthemum-shaped microstructure. This demonstrates how the weak ΔH prefers to occupy the adhesion site. In numerous FCC and BCC alloys, which in small quantities correspond to certain composition formulas, the cluster formula has been verified [8, 10]. The morphology of the FeCoCrNiAlTix HEA coating after wear is shown in Figure 8.8.

Based on EDS results, it can be concluded that α phase has high Ni and Al content, while the β phase shows high Fe and Cr content. When compared with previous XRD results, it is clear that α phase corresponds to BCC1 and β phase to eutectic phases BCC1 and BCC2, respectively. Since the enthalpies of mixing are known to show an ordering or clustering tendency, the segregation of Ni, Al, Fe and Cr into individual phases can be related to the enthalpies of mixing of atomic pairs among these six elements. The atomic interactions between the solute components and base solvent are represented by the enthalpies of mixing. The solute with a negative ΔH tendency easily occupies the cluster center [12]. The microscopic morphology of wear surface of FeCoCrNiAlTix HEA coatings is shown in Figure 8.8. The addition of Ti tends to provide a smooth surface. During wear, the drag force on contact surface tends to correlate with the abrasion rate [19]. A higher abrasion rate is usually caused by a higher drag force. The changes in the friction coefficient curve of FeCoCrNiAlTix coating with the wear test period are shown in Figure 8.9

In Figure 8.10, the mass loss is proportional to friction coefficient. Based on the surface morphology, no glare pits or other morphological features of adhesive wear are visible. The worn surface has debris and grooves. As a result, abrasive wear is the wear mechanism. x = 0 is where the grooves are deep and there is a significant amount of detritus FeCoCrNiAlTix coatings consist of two types of BCC phases. The number of tactic structures decreases as continuous increasing

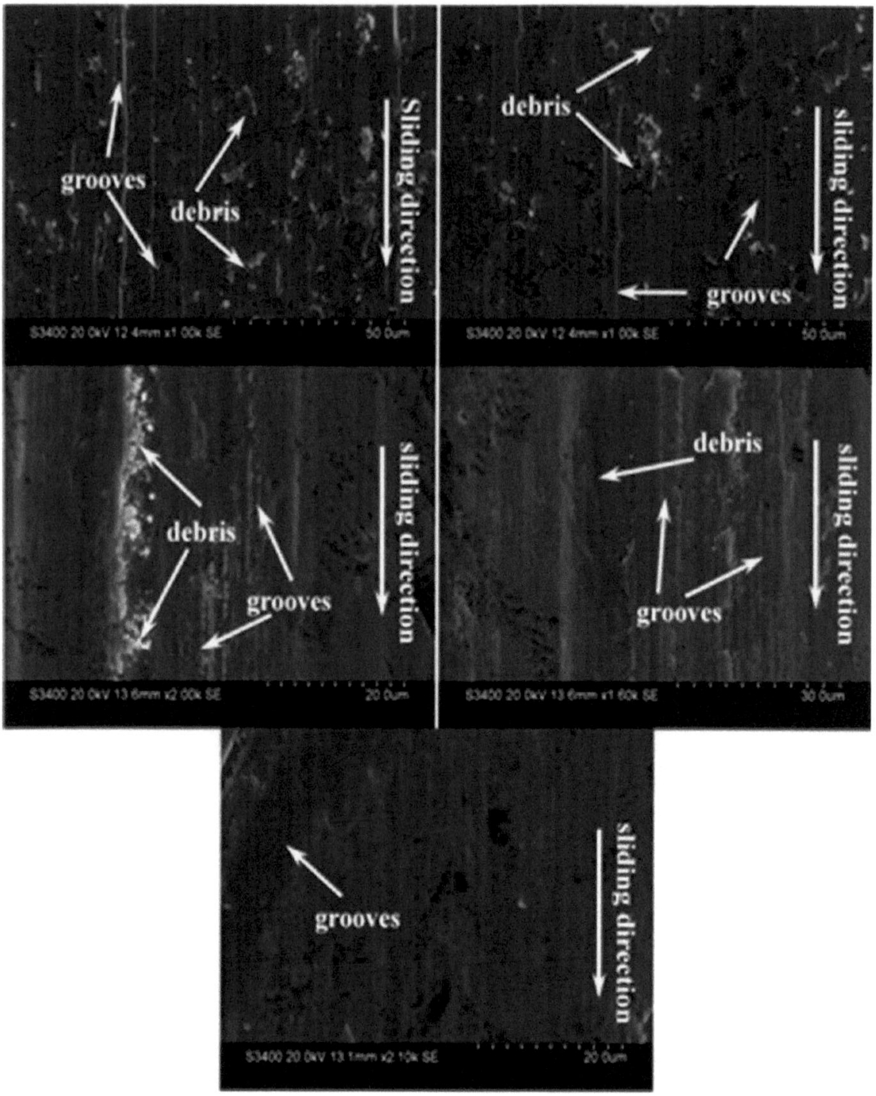

FIGURE 8.8 FeCoCrNiAlTix HEA coatings' morphology following wear [12] [open access].

in Ti concentration, as does the growth of tactic phase and precipitation of Ti-rich second phase. Between x = 0 and x = 0.25, the tactic points BCC1 and BCC2 are located. With the increasing of Ti concentration in FeCoCrNiAlTix coatings, the solution strength effect and second phase strengthening effect increases. In addition, the medium microhardness increases. The wear resistance increases and friction coefficient of FeCoCrNiAlTix coatings decreases with increasing of Ti concentration. Otherside, HEA coatings at x = 1 show the best wear resistance [12].

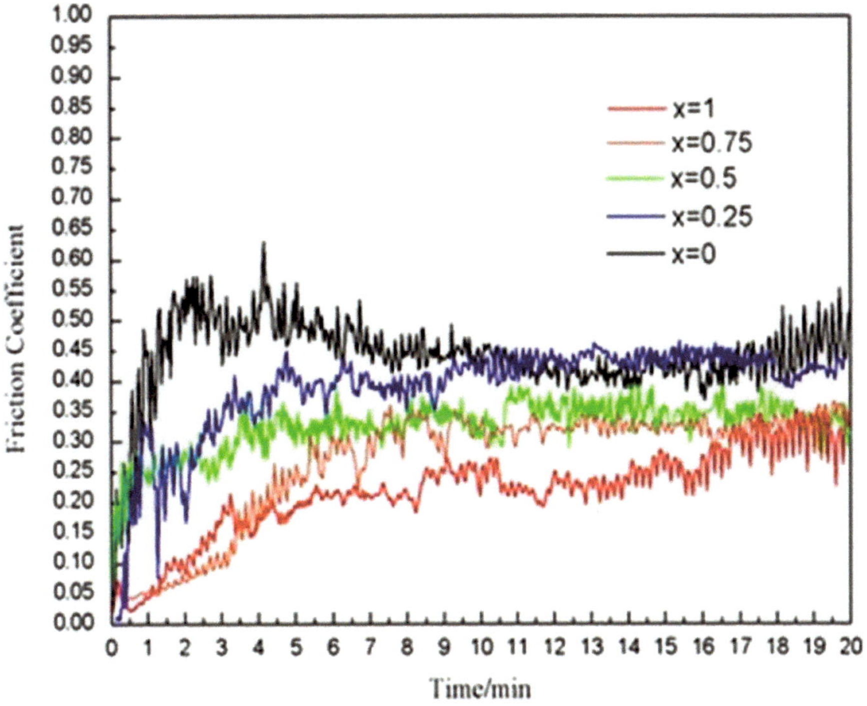

FIGURE 8.9 Coating friction coefficient [12] [open access].

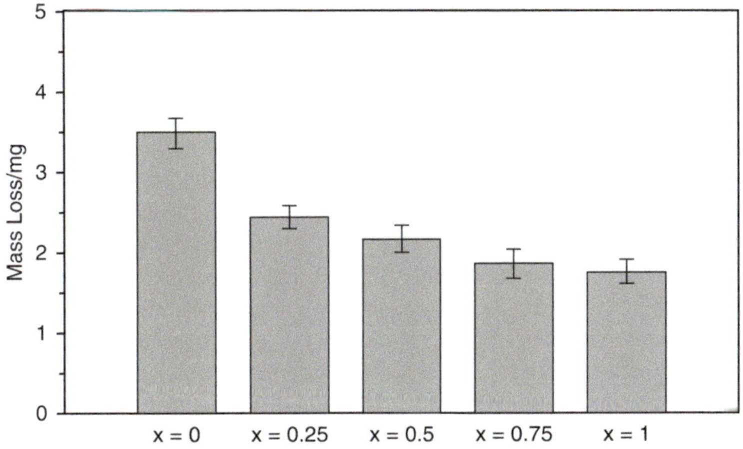

FIGURE 8.10 FeCoCrNiAlTix coating mass loss [12] [open access].

8.4 EFFECT OF IMPACT WEAR ON STRUCTURAL STEEL WITH YIELD STRENGTH OF 235 MPa IN DIFFERENT LIQUIDS

Maintaining the navigability of waterways or creating additional land in freshwater or saltwater bodies requires dredging. Pipelines are needed to transport the sedimented sands or other solids mixed with the water during the dredging process. The interaction of erosion, abrasion and corrosion causes pipe deterioration, resulting in costly pipeline replacement and maintenance [1, 4, 11]. For decades, researchers have studied and documented pipeline deterioration [5–7, 12–15]. Three determinants of deterioration are compiled by Truscott [17] after reviewing research conducted over 20 years before 1972; materials, flow regime and slurry characteristics. The main slurry characteristics are hardness, size, sharpness, specific gravity and particle concentration. How the dispersion and velocity of particles are ultimately determined by flow, which in turn determines the deterioration process. The main characteristics of materials are their hardness, microstructure and chemical composition [5].

Pulp pipeline wear due to erosion is caused by two mechanisms; particle impact and abrasion, the latter is caused by sliding abrasive wear. Some devices have been developed to study the mechanism of pulp pipeline wear [16, 18]. These devices are designed to study the influence of various parameters such as particle size, flow velocity and flow concentration. The wear of pipes used for slurry transportation results in high maintenance and replacement costs. The interaction of abrasion, corrosion, impact and other factors is the wear mechanism. In this study, we investigate the influence of wear mechanism and wear rate. The results show that electrochemically generated surface modification dominates the wear process when the impact force is low. This results in a decreasing in wear rate of corrosive environment compared to non-corrosive environment. In contrast, the wear process changes significantly when the impact force is high. Plastic deformation is significant in this range. In the high-impact regime, the effect of corrosion can be ignored. Figure 8.11 compares the wear scars of three fluids at different impact heights [5].

It is obvious that as the impact height of each fluid increases, the depth of scar increases. At 3.3 mm, the impact height is the same for all fluids, which is consistent with the comparison of wear rates. The total volume loss due to wear consists of two parts; deformation and material removal. When the impact height exceeds 1 mm, the deformation part plays a major role in the total wear.

The effects of impact on the wear of S235 structural steel are investigated in a range of fluids. Due to the beneficial interaction between abrasion and corrosion, the specimen in seawater experiences the least wear when the impact is minimal. However, under severe impact, no significant change in wear rate is observed for the three fluids, as the wear mechanism shifts from material removal to predominantly plastic deformation. Under impact, the effect of corrosion is reduced to a minimal factor [5].

Many of the metal phases in the experiment were found to have approximate compositions of Al4(La, Ce), Al8Mn4Ce, and reduced b-Mg17Al12 phases. All of the metal phases were noble when compared to the α-Mg matrix, according to SKPFM analysis. The a-Mg matrix surrounded by b-Mg17Al12 or Al8Mn4Ce was susceptible to spot corrosion, according to in situ EC-AFM

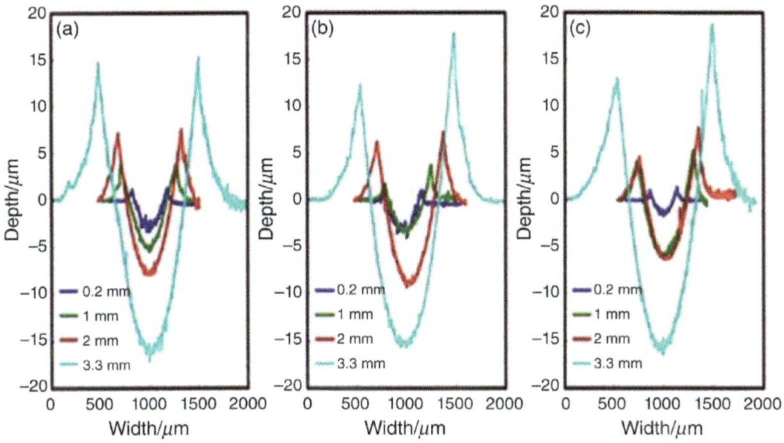

FIGURE 8.11 Comparison of wear scars in three different liquids (a) ethanol, (b) deionized water, and (c) seawater) at different impact heights. For each scenario, the average of three consecutive results was used. To facilitate comparison, the wear scar was shifted to the center [5] [open access].

observation of the early stages of corrosion. In particular, both AZ91 and AZ91 alloys containing (La, Ce) MM contained the b-Mg17Al12 phase. Figure 8.12 illustrates how both elongated and dispersed b-Mg17Al12 grains act as effective microcathodes [16].

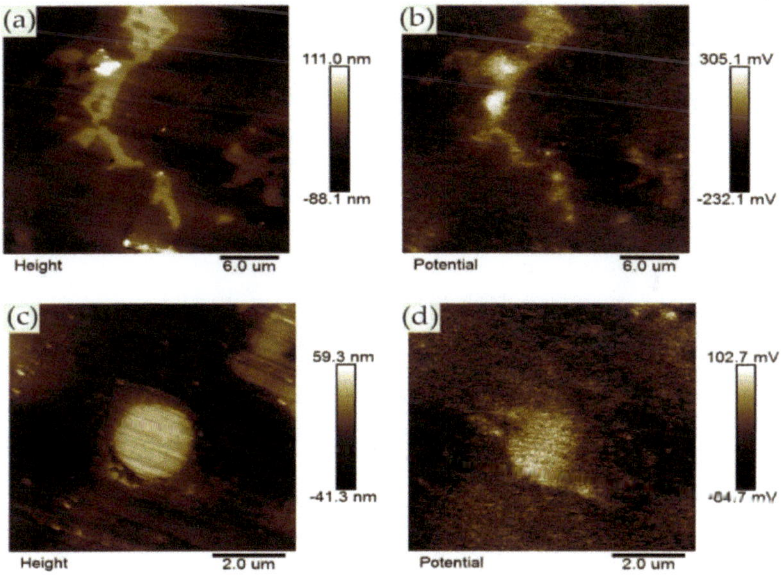

FIGURE 8.12 The polished AZ91 alloy (a) surface topography and (b) relative Volta potential map of elongated b-$Mg_{17}Al_{12}$; (c) surface topography and (d) relative Volta potential map of granular-like b-$Mg_{17}Al_{12}$ [1] [open access].

8.5 POWER INVESTED IN CREATION OF SURFACES

The ability of solid materials and objects to restore their previous shape after removing loads or external forces that cause deformation is called elasticity. When an object regains its original size and shape after the load is removed, it is called elastic. Different materials behave differently for different physical reasons, which are influenced by their microscopic structure. For example, the stretching of polymer chains under the influence of applied force provides the elasticity of rubbers and polymers. Conversely, under the influence of external pressure, the crystal lattice cells that make up the structure of metal material undergo changes in size and shape, which determine the elasticity of the metal.

The two factors that determine the elasticity of material are the elastic limit and elastic modulus. Materials that require large loads to produce significant deformation or those that are difficult to deform have a high modulus of elasticity. Steel strip is one example. Materials that easily deform under load, such as elastic bands, have a low modulus of elasticity. The material deforms permanently if the stress under load is too great, at which point it takes on a different shape and size rather than returning to its original state. The stress value above which a material ceases to behave elastically and instead permanently deforms is known as the elastic limit. The material with the lowest elastic modulus is the most elastic among materials with the same elastic limitations [2, 8–10, 16, 18, 20]. Surface engineering processes are constantly evolving, providing tremendous opportunities for a wide range of applications across multiple industries. From consumer products to electronics, medical devices, aerospace, automotive components, product performance and durability are greatly influenced by the adaptability of metals to their surface quality and process. Research into new materials, improved creative processes and integration of surface technology into other manufacturing processes are enhanced by ongoing research and development in this field. Three-dimensional representation of the boring bar used by PAFANA (Figure 8.13) becomes much more substantial because it raises with the material's strength. In any model of chip creation, then, labor necessary to generate a new surface must be taken into account in hard machining while balancing forces and/or energy.

8.5.1 Requirements for Rigid Processing

Models created using multibody tools in a real-time virtual environment require a trade-off between computational efficiency and accuracy. When complex multibody models come into contact with other parts of system or with external surfaces, they are often deformed to some degree. Different modeling assumptions may be applied, resulting in different types of models depending on the intended use of virtual environment. Although finite segments and methods offer computationally efficient ways to represent both large and small deformations, they are unable to convey the stresses and strains experienced by deformable components. Finite element analysis is a reliable method for describing the deformation of a system, including stress and strain fields, but requires more

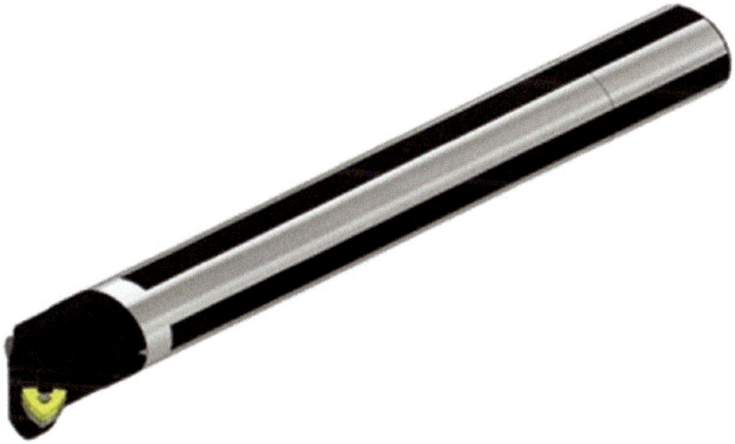

FIGURE 8.13 Three-dimensional representation of boring bar used by PAFANA [13] [open access].

Appropriate contact models must be included in systems where touch or impact scenarios occur [6, 7, 14, 15, 17]. However, the driving force (friction changes) in the majority of sliding systems is frequently unknown or needs to be discovered by modeling. Frictional vibrations can, at most, be minimized its frequency can be raised out of the undesirable range by altering the dynamics of the system.

8.6 OPERATIONS FOR HARD-MACHINING

The final manufacturing procedures of parts and components typically involve complex manufacturing such as boring, drilling, and the required dimensions and surface finishes typically require stringent manufacturing standards and conditions.

8.6.1 Hard Turning

Hard turning has the obvious benefit of eliminating grinding processes. The same machine, which "softly" spins the part before hardening it, enables this productivity and maximizes the equipment's effectiveness. However, some stores commit the error of initially selecting the incorrect tool insert for the job (one that is less expensive). Tool steel, bearing steel and case-hardened steel are examples of commonly used hard materials [11, 12], this is an untrue since these materials have hardness's that are substantially lower than 47 HRC, which means that the mechanism for how chips form during the operation, including the materials used for tools, differs greatly the best process predictability, from a metallurgical perspective. As mentioned previously, unsupported workpieces should generally have a length-to-diameter (L/D) ratio of no more than 4:1. This recommendation reduces the likelihood of failure or distortion during machining and helps maintain structural stability. In many manufacturing applications, including machining and fabrication, where unsupported components may be more prone to bending or instability, maintaining this ratio is essential. Workpiece integrity and performance can be improved by manufacturers by maintaining an acceptable L/D ratio [6, 7, 13–15].

The level of turning accuracy is determined by the rigidity of machine. Most modern machines have the rigidity needed to handle tasks that require heavy turning. In many cases, the overall condition of a car is more important. However, machine tool rigidity becomes an increasing concern as required part tolerances and surface finishes increase. The best option is to use specialized equipment and turning centers designed for serious machining. However, the secret to success is not the rigidity of machine, but system. Unfortunately, the industry does not fully understand this simple concept. To maximize system rigidity, all overhang, tool and part extensions must be minimized, such as by eliminating spacers and shims. As already stated, the use of coolant is one of major considerations when designing a turning process. Because dry cutting has obvious financial advantages, most turning operations are done dry. Parts can become hot, making it difficult to measure and handle components. Additionally, a surge in cherry red chips can lead to a number of other problems. Thus, high-pressure coolant through the tool is ideal for cooling the workpiece and hot chips while maintaining small chip size and shape control when coolant is required [8, 18].

In light of this, if coolant is used, for obvious reasons it should be low-concentration and water-based. When crushed by hand, the chilled chips practically fall apart, proving that the chips produce the right amount of heat. Depending on the required precision and surface polishing, the correct tool material, cermet, or ceramic) must be selected to ensure process efficiency. Different tools, cutting inserts and tool manufacturers offer different brands. Optimal machining conditions, tool holders and insert shapes only complicate the multi-parameter optimization task [15–18].

8.6.2 Reaming and Hard Boring

Reaming is a widely used method for making holes in manufacturing. When it comes to machining smaller holes, reaming is more practical and cost-effective than internal grinding and fine boring. Boring is a machining technique in which a pre-made hole is enlarged using a cutting tool. In situations where deformation and dimensional changes caused by heat treatment create undue constraints in "soft stage" of machining, hard boring will be a more economical alternative that reduces waste. A brief description of hole machining includes boring, reaming, drilling and honing. Adding Additional Cutting Tools More Cutting Tools Select "More Cutting Tools" and we will be here to support you every step of the way through the metal cutting process! In addition to being important surfaces on components such as discs, rings, couplings, brackets and bushings, holes are also common surfaces encountered in machining process. Cutting holes is a more complex task than cutting cylindrical surfaces while maintaining the same standards of accuracy and surface finish, resulting in lower productivity and higher costs.

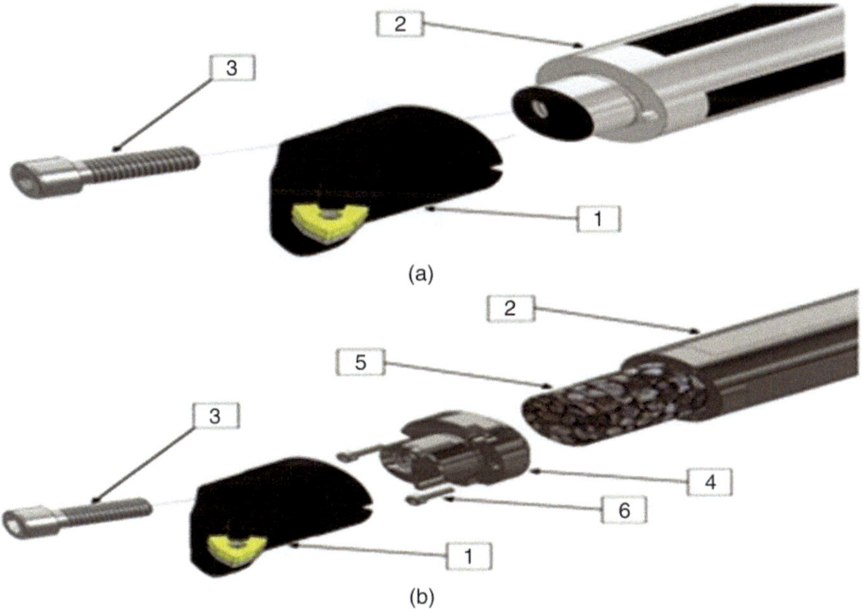

FIGURE 8.14 Comparison of modified design (b) and original design (a). 1 head, 2 rods, and 3 M10x30 mounting screws four fasteners, five filler cores, and six fastener mounting screws [13] [open access].

Figure 8.14 shows contrasting the updated design (b) with the original design (a). Head (1), shank (2), and fastening screw M10 × 30 (3). Fixing screws for fasteners are numbered four, five and six. Multi-bladed tools are used to achieve high performance [13]. However, single-edge tools with backing plates are used when the hole being machined must have a high degree of accuracy. Clamping tiny inserts onto tiny boring bars becomes more challenging for holes smaller than 12 mm in diameter. To solve this problem, a soldering tool is used [9, 10, 20].

8.6.3 Milling of Hard

Businesses frequently concentrate on creating dies and dies, which are used to cut materials like P20, H13, W5, and S7. Solid rods and cavities have traditionally been produced from these materials using electrical discharge devices. These materials can now be produced directly into hardened materials thanks to the development of new technologies over time. This is often accomplished by utilizing new toolpath techniques and hard milling processes. Prominent mold manu facturers have realized that, in an increasingly competitive global environment, adjusting to new technologies may be essential to their survival. The deployment of a system consists of the system's most important element is the machine. The machine needs to be made for hard milling and also has some high-speed machining center features [21].

FIGURE 8.15 Various types of CNC milling machines [13] [open access].

Solid milling requires a fundamental machine design as well as specific machine parts such as the motor, spindle and various types of CNC milling machines (Figure 8.15). It is a crucial to have a study base with adequate vibration-dampening properties. Because polymer concrete bases often have vibration dampening that is 6 to 10 times better than cast iron, they are a viable option for high-speed, heavy-duty milling. Digital drives' quick acceleration and deceleration capabilities and superior contouring accuracy aid in cutting down on cutting tool wear. Depending on the application, each of these tools has different strengths and limitations. End mills made of solid carbide are often highly expensive, precisely ground and coated. Cutter sets with interchangeable carbide inserts are a second category of carbide cutters. They are not ideal for harsh milling tasks. Ceramic replacement inserts are a third type of strengthened ceramic inserts with whiskers [13].

Faster cycle times and fewer operations per item are advantages of employing a milling system with changeable ceramic inserts. Using interchangeable ceramic inserts, ceramic cutters can safely and quickly grind everything from big face mills to small diameter end mills. Modern whisker-reinforced ceramics may run at speeds much above the point at which they melt since they have a melting temperature of over 2000°C that fails with carbide tools.

8.6.4 Broaching of Hard

The broaching operation is the most cost-effective way for mass-producing parts with complex features in the field of metal cutting where high production rates and great component stability are required. One motion can be used to draw the part's profile. While exterior broaching is intended to machine the surface profile, internal broaching begins from a hole that has already been pre-machined. Both horizontal and vertical directions are open to firmware. Both linear and helical cutting movements are possible. It is feasible to use one of two rigid broaching techniques with this tool configuration:

- The cutting speed during typical broaching procedures is known to let between 0.5 and 6 m/min. Since that time, researchers and professionals have focused on rigid broaching's effectiveness and dependability, as well as productivity and surface quality.
- When it comes to the quality and dependability of broaching operations, coated carbide broaches, intended as side-cutting tools, perform well. Obtain IT profile quality.
- When utilizing carbide broaches, an efficient cutting speed is identified as a crucial component in the broaching process. When compared to green (soft) feed, the cycle time at this speed is shorter since the feed stroke lasts between one and two seconds.

8.6.5 Operations for Manufacturing Hard-Gear

Cutting, forming, grinding and heat treating are the four main steps in gear manufacturing process required to produce gears that meet these requirements. Using techniques such as hobbing, forming, milling and broaching, gear cutting produces precise tooth shapes on gears. One of the most important technologies in production of gears is an abrasive processing. Over the past 20 years, among the methods of abrasive processing of hardened gears, operations with cutting speeds, 0.3–5 m/s have gained an importance. Further study has been done to make these low-speed grinding operations more productive. The development of new abrasive tools is an integral part of this study, requiring a large amount of empirical research and, therefore, a significant investment of time and money [11, 12]. The method allows you to change the geometric surface of side surfaces of teeth to improve the surface quality. In addition, the noise level of the gearbox is reduced due to characteristic surface roughness of final gears. The most common process and wear pattern of milling tools is development and propagation of cracks caused by thermal fatigue (high processing temperatures). The origin and wear patterns of coated tools are still the subject of recent research as they have a direct impact on tool life. Coating wear studies can help develop new coatings and improve processes.

Recent research has shown that using these methods can extend the life of coatings, and improve surface quality and overall process efficiency. However, machining with rigid gear machines places unique demands on the tool, machine and overall process. Understanding these prerequisites is necessary for the successful implementation of this process. High-speed steel is the material most commonly used in industry today to make cooktops. In addition, final machining cannot be done on a high-speed steel milling machine.

Even these relatively soft materials, but this can go so far before additional finishing with other dental instruments such as razors is required. The main reason is the incorrect angle of lateral surface along tooth profile. To improve the characteristics of finished surface, the cutting wedges of form-cutting tool are often created with an inclination angle of 3–4° [2, 19, 20, 22–24]. Since the tooth profile must be maintained within a limited tolerance when a cutter is sharpened repeatedly, substantially smaller flank angles are used in hob tools. Due to high yield strength and high elasticity of processed materials, when working with hard materials, excessive friction occurs on side surface. The workpiece and flange stick together due to high temperatures caused by this friction, reducing surface quality and shortening tool life. A carbide cutter with replaceable carbide inserts is shown in Figure 1.10 next to the cutter used for tiny modules.

8.6.6 Skiving

8.6.6.1 Turning of Hard

Turning is a very effective and simple machining method that produces high-quality surfaces for rotating body parts made from a variety of materials. Processing efficiency for each unique situation, the best combination of workpiece and tool materials and cutting parameters is determined. circumstances, geometric characteristics of cutting tool, condition of apparatus and technical data. Machine parts consisting of hardened steel annealed and special cast iron and difficult-to-cut materials are successfully machined. The use of cubic boron nitride (CBN) tools for fine turning of workpieces composed of these materials often avoids the need for grinding, the traditional method of producing surfaces with a roughness less than Ra = 1.25 μm. Currently known as "hard turning", this processing method involves machining materials with a hardness greater than 47 HRC [2, 19, 22–24].

Figure 8.16 demonstrates an overview of prototype boring bar [13]. Polishing hardened gears is known as gear hobbing. Using this method, errors in the gear

FIGURE 8.16 An image of the prototype boring bar [13] [open access].

ends and profiles associated with heat treatment are eliminated. The end result is increased transmission accuracy. Hardened gears with a hardness up to 63 HRC can be processed with carbide cutters [22–24]. The development of extremely sophisticated automated machine tools has made this possible, making the approach efficient and practical [13].

8.6.6.2 Skiving of Gear

Figure 8.17 shows an illustration of a comparison between the machined surface's look with the original PAFANA tool (a) and PC prototype tool (b). The body of worm gear tool is made of high-speed steel, with carbide cutting inserts soldered onto it. Each side surface of the gear is sharpened by removing a 0.5 mm thick allowance in one pass. Benefits of this technology include the ability to change gear flanks, such as cone and crown, over a range of 100 m or more in just one pass, as well as dramatic reductions in tooling costs and cycle times. The character has achieved DIN 8 quality, which is the most accurate indicator of process quality [13].

Many scientists believe that the main difference between sharpening teeth with worms and cutting teeth is the negative rake angle of tool in the hand, caused by displacement of front surface of tooth [3, 25]. The carbide cutter appears to be protected from impact and sudden overload by a negative rake angle. However, this is not entirely true as process modeling is sometimes too confusing, and important variables are ignored because the geometry of tool teeth in machine tool that is not taken into account [21, 26, 27].

By altering the surface characteristics of solid materials, surface engineering is an important field that makes significant contributions to the manufacturing and development of consumer products. It is a critical to prevent the deterioration of industrial tools and components caused by harsh operating conditions. Surface degradation has a significant financial impact, but advanced surface engineering

(a) (b)

FIGURE 8.17 An illustration of how the machined surface looks while using the original PAFANA tool (a) and PC prototype tool (b) [13] [open access].

techniques provide a way to reduce these costs, and improving the performance and durability of materials. Advances in this field will further enhance the efficiency, sustainability and ingenuity of industrial processes.

8.7 PROBLEMS

1. Why is the industry so concerned about the failure of DLC coating degradation investigation using triboelectric imaging and metal charge measurements?
2. How can the phase and properties of high-entropy FeCoCrNiAlTix alloy coatings on Q253 be developed using laser cladding?
3. How can impact corrosion affect structural steel with a yield strength of 235 MPa in different fluids?
4. Why operations for manufacturing hard gear?
5. If two bodies with different mechanical forces come into contact, what causes the transfer of atomic deposits?

REFERENCES

1. Chen, H., Qin, Z., He, M., Liu, Y., & Wu, Z. (2020). Application of electrochemical atomic force microscopy (EC-AFM) in the corrosion study of metallic materials. *Materials*, *13*(3), 668.
2. Krupanek, K., Staszczyk, A., Sawicki, J., & Byczkowska, P. (2018). The impact of nozzle configuration on the heat transfer coefficient. *Arch. Mater. Sci. Eng.*, *3*, 16–24.
3. Altintas, Y. (2012). Manufacturing Automation: Metal Cutting Mechanics, Machine Tool Vibrations, and CNC Design, Cambridge University Press, Cambridge, UK.
4. Fotovvati, B., Namdari, N., & Dehghanghadikolaei, A. (2019). On coating techniques for surface protection: A review. *J. Manuf. Mater. Process*, *3*, 28.
5. Fuentes, G.G. (2015). Surface engineering and micro-manufacturing. *Micromanuf. Eng. Technol.*, *2*, 459–486.
6. Xiao, W., Zi, Y., Chen, B., Li, B., & He, Z. (2014). A novel approach to machining condition monitoring of deep hole boring. *Int. J. Mach. Tools Manuf*, *77*, 27–33.
7. Khoroshailo, V., Kovalov, V., & Predrag, D. (2016). Improving of vibration resistance of boring tools by big diameter holes tooling on lathe. *Procedia Technol*, *22*, 153–160.
8. Ferdous, W., Bai, Y., Almutairi, A., Satasivam, S., & Jeske, J. (2018). Modular assembly of water-retaining walls using GFRP hollow profiles: Components and connection performance. *Compos. Struct.*, *194*, 1–11.
9. Żółtowski, B. *Badania Dynamiki Maszyn*; Wydawnictwo MAKAR: Bydgoszcz, Poland, 2002.
10. Pawłowski, W. *Wibracyjne Szlifowanie Wgłębne Wałków*; Zeszyty Naukowe Politechniki Łódzkiej, Nr 654; Rozprawy Naukowe; Wydawnictwo Politechniki Łódzkiej: Łódź, Polska, 2010
11. Ciniero, A., Le Rouzic, J., & Reddyhoff, T. (2017). The use of triboemission imaging and charge measurements to study DLC coating failure. *Coatings*, *7*(8), 129.
12. He, B., Zhang, N., Lin, D., Zhang, Y., Dong, F., & Li, D. (2017). The phase evolution and property of $FeCoCrNiAlTi_x$ high-entropy alloying coatings on Q253 via laser cladding. *Coatings*, *7*(10), 157.

13. Kępczak, N., Bechciński, G., & Rosik, R. (2024). Operating properties of deep hole boring tools with modified design. *Materials, 17*(7), 1551.
14. Orgiyan, A., Oborskyi, G., Ivanov, V., Tonkonogyi, V., & Balaniuk, A. Features of flexural-torsional oscillations of cantilever boring bars for fine boring of deep holes with small diameters. In Grabchenko's International Conference on Advanced Manufacturing Processes; Lecture Notes in Mechanical Engineering; Springer: Cham, Switzerland, 2022; pp. 98–108.
15. Product Catalogue PAFANA. Available online: https://pafana.pl/wp-content/uploads/2021/04/11-133_noze_tokarskie_skladane_96dpi_07_2020.pdf (accessed on 13 May 2022).
16. Agavriloaie, L., Oprea, S., Barbuta, M., & Luca, F. (2012). Characterisation of polymer concrete with epoxy polyurethane acryl matrix. *Constr. Build. Mater, 37*, 190–196.
17. Broschüre Mineralguss EPUMENT. Available online: https://www.rampf-group.com/fileadmin/rampf-gruppe.de/media/machine_systems/downloads/Mineral-casting-EPUMENT-EN.pdf (accessed on 18 May 2022).
18. Ferdous, W., Manalo, A., & Aravinthan, T. (2017). Bond behaviour of composite sandwich panel and epoxy polymer matrix: Taguchi design of experiments and theoretical predictions. *Constr. Build. Mater, 145*, 76–87.
19. Kilikevičius, A., Rimša, V., & Rucki, M. (2020). Investigation of influence of aircraft propeller modal parameters on small airplane performance. *Eksploat. Niezawodn. Maint. Reliab, 22*, 1–5.
20. Pawłowski, W., Kaczmarek, & Louda, P. (2016). Theoretical and experimental modal analysis of the cylinder unit filled with pur foam. *Eksploat. Niezawodn. Maint. Reliab, 18*, 428–435.
21. Alammari, Y., Sanati, M., Freiheit, T., & Park, S.S. (2015). Investigation of boring bar dynamics for chatter suppression. *Procedia Manuf., 1*, 768–778.
22. Ondra, V., & Titurus, B. (2019). Theoretical and experimental modal analysis of a beam-tendon system. *Mech. Syst. Signal Process, 132*, 55–71.
23. Product Data. (2006). DeltaTron® Accelerometers—Types 4514, 4514-001, 4514-002, 4514-004, 4514-B, 4514-B-001, 4514-B-002 and 4514-B-004, Brüel&Kjær. Available online: https://www.bksv.com/media/doc/bp2066.pdf (accessed on 7 June 2022).
24. Product Data. (2005). Impact Hammers—Types 8206, 8206-001, 8206-002 and 8206-003, Brüel&Kjær. Available online: https://www.bksv.com/media/doc/bp2078.pdf (accessed on 7 June 2022).
25. Ma, J., Xu, J., Li, L., Liu, X., & Gao, M. (2022). Analysis of cutting stability of a composite variable-section boring bar with a large length-to-diameter ratio considering internal damping. *Materials, 15*, 5465.
26. Chen, F., Lu, X., & Altintas, Y. (2014). A novel magnetic actuator design for active damping of machining tools. *Int. J. Mach. Tools Manuf., 85*, 58–69.
27. Li, W., Zheng, H., & Feng, Y. (2023). Optimization of cutting parameters for deep hole boring of Ti-6Al-4V deep bottle hole. *Materials, 16*, 5286.

9 Health and Safety Issues in Emerging Surface Engineering Techniques

9.1 INTRODUCTION

In certain situations, engineering materials need to have improved surface properties to increase their reliability and service life for various industrial applications. Surface engineering is the process of improving the surface and subsurface properties of a material while maintaining the bulk properties of processed material unchanged, either by applying a surface coating or by surface treatment [1–6]. Using these procedures, it is possible to reduce maintenance costs that arise due to frequent failure of engineered components. Surface engineering techniques have been used for a very long time in a wide range of applications to create a specialized and customized surface that responds quickly and effectively, ensuring an acceptable design life of engineered artifacts. Anodizing, salt bath nitriding and traditional surface engineering techniques, like any manufacturing method (coating), pose risks to public health, safety and environment. These processes use high levels of energy (e.g., heat and electricity) as well as highly toxic chemicals and hazardous solid, liquid or gaseous by-products that may pose a risk to environmental health, safety and health of workers. Hazardous elements are classified based on the type and level of risk they pose. First, high levels of exposure over a short period may have acute health effects due to hazardous components. Second, there may be serious long-term effects from repeated or prolonged exposure to hazardous materials as a result of surface engineering procedures. Due to increasing stringent laws, international and national health and safety requirements, several new surface engineering processes have been developed (e.g., chemical vapor deposition, CVD; plasma nitriding and physical vapor deposition, PVD). Reducing environmental, health and safety risks through using of toxic chemicals and high voltage, automated computer-controlled systems and enhanced engineering controls are some of major factors driving the development of new surface engineering technologies [7]. Although new hazards have been added during this development, the health and safety issues associated with traditional procedures have been reduced via new processes. High vacuum, new by-products (e.g., gas phase nanoparticles), energy sources (e.g., laser and plasma) and new feedstocks (e.g., hazardous gases and nanoparticles) are among the risk areas. Since the type and severity of hazards determine the level of health and safety hazards, it is essential to evaluate the new types of hazardous elements

DOI: 10.1201/9781003584735-9

included in the evolving process and take appropriate steps to control their impact on the environment and public health. This chapter provides a summary of various health and safety issues associated with new procedures for surface engineering. Actions to be taken to address health and safety issues have been discussed.

9.2 SURFACE ENGINEERING PROCESSES

As shown in Figure 9.1, the various surface modification methods can be divided into main categories; mechanical, chemical, electrochemical and thermal methods. Moreover, mixed treatments are mixtures of two or more treatments from each of main groups. The main advantages of each treatment are integrated into these combined treatments to provide a finished material with excellent mechanical and microstructural surface qualities, thereby significantly extending the service life of tool and/or workpiece under service-like operating conditions. Surface engineering can be broadly classified as microstructural modification, which is the process of changing the microstructure by applying an appropriate heat treatment. In contrast, compositional modification involves changing the surface composition. Of all composite surface engineering techniques discussed so far, the metal coating is one of the most practical and accessible methods [8]. Surface engineering, known as advanced surface engineering, is an important

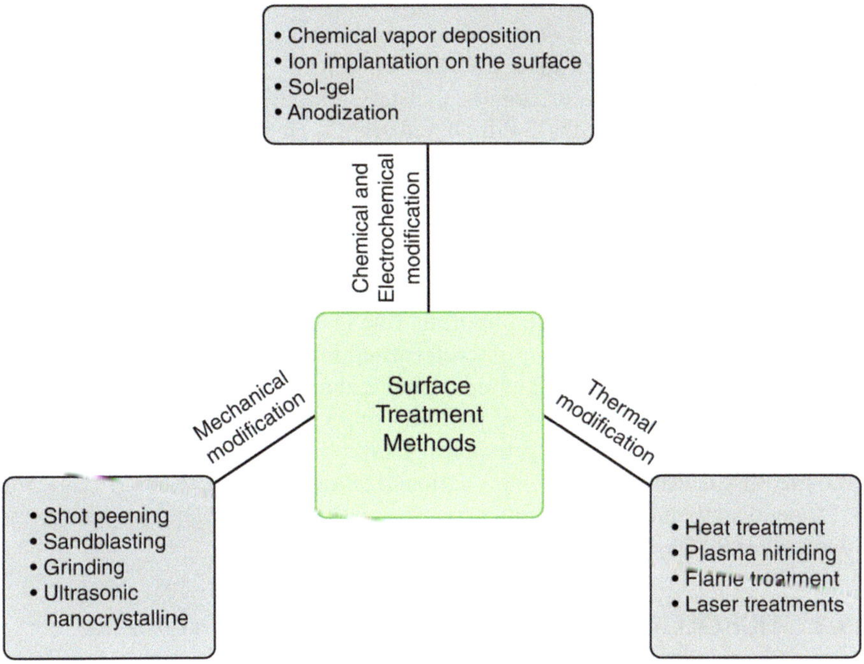

FIGURE 9.1 Diagram illustrating most popular methods of surface modification [1] [open access].

knowledge-based industry that is critical to the production of consumer goods and many manufacturing processes. Surface engineering includes all technologies that can be used to modify the surface of a solid to make it perform better or have new characteristics. Protecting industrial components and manufacturing tool surfaces from severe wear, oxidation, corrosion and friction have been the focus of surface engineering in recent decades. These phenomena, which are commonly seen as catalysts for surface degradation, result in huge manufacturing costs each year, mainly due to component scrapping and tool repurposing or replacement. It is estimated that surface degradation can result in losses up to 4% of gross domestic product (GDP) in wealthy economies [1].

The protection of industrial components and manufacturing tool surfaces under severe wear, oxidation, corrosion and friction conditions has focused on surface engineering in recent decades. These phenomena, which are commonly regarded as catalysts for surface degradation, result in enormous manufacturing costs each year, mainly due to component scrap and tool rework or replacement. In rich economies, surface deterioration is expected to cause losses up to 4% of GDP. This major impact is due to several variables, such as higher maintenance costs, shorter equipment life, and the need to replace tools and components more frequently. The above losses underscore the need to implement effective surface engineering and material protection tactics to increase the strength and functionality of industrial assets, thereby enhancing economic efficiency and sustainability. In addition to reducing financial losses, addressing surface deterioration promotes the development of manufacturing technologies and techniques. These qualities amount to more than USD280 billion per year in the United States. In addition, some estimates suggest that in Germany alone, the manufacturing industry spends up to USD1–2 billion per year [1] on petroleum-based lubricants to avoid wear. In the United States alone, annual wear and friction costs are projected to be over USD100 billion [2]. The formation of residues from galvanic processes, such as hexavalent chromium, results in additional costs for surface protection. The development of micro- and nanofabrication technologies has presented new and exciting challenges for surface engineering. Shortly, surface modification processes will play an important role in enabling the industrialization of several technologies, including microforming, microfabrication and micro-nanotexturing. In addition, new and emerging technologies require new functional surfaces that could provide new products that can outperform those currently available based on classical principles. Examples include (1) biomaterials that require sophisticated surface biofunctionalization techniques; or (2) fabrication of functional membranes and other functional coatings for renewable energy production for H_2-based fuel cells [8].

9.3 TRIBOLOGY AND SURFACE CONTACT PHENOMENA

Tribology, or study of wear, lubrication and friction are an important and ongoing goal in all manufacturing processes. It is widely recognized in mechanical engineering that these surface degradation mechanisms are responsible for many manufacturing failures. Tribology is the study of physical and chemical interactions of two surfaces in relative motion, as well as their contact interactions and resistance to environmental degradation [3–5]. Aluminum disks exhibited significant wear when lubricated. Aluminum discs experienced less wear than other materials, even though the wear process is essentially the same for all specimens. This variability can be explained by the unique properties of aluminum, which can reduce damage under some conditions, such as its low hardness and tendency to produce more gentle wear debris. Understanding these differences in wear behavior is essential to maximizing lubrication tactics and extending component life in a variety of applications. Further research into the subtle interactions between aluminum discs and lubricating oil may yield important information for enhancing wear resistance. Abrasive wear is observed in each situation, especially after lubrication with BOA and BO + PIL combinations. All wear marks have exhibited adhesive wear in addition to plowed grooves within, which indicate abrasive wear. Plastically deformed materials are folded at the boundaries of marks. Three specimens lubricated with 4-dodecyl benzenesulfonate-based PIL, longer alkyl chain family of PILs has exhibited less plastically deformed material in narrower wear tracks compared to specimens lubricated with an equivalent p-toluenesulfonate-based PIL [5]. Figure 9.2 shows optical micrographs of the wear traces on the surface of an aluminum disc after it has been combined with mixtures of oils and chemiluminescent oils (BO and BOA). It shows an independent pattern of exploitation of the interaction between the aluminum surface and the oils used under a specific operation and differences in mixing and amplitude of the fingerprints that have appeared over a long period can be observed, reflecting the influence of the physical and chemical oils used on the corrosion resistance and environmental performance of materials used in industrial applications.

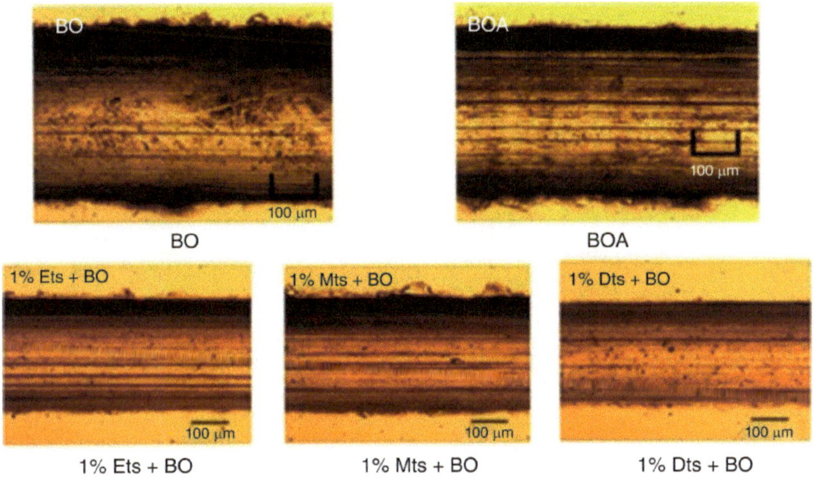

FIGURE 9.2 Optical micrographs of wear marks on aluminum disk after experiments with blends and biodegradable oils (BO and BOA) [5] [open access].

Understanding the benefits of surface engineering and how it can improve the surface performance of tools and components requires a comprehensive understanding of fundamentals of friction, wear and lubrication. Tribology in general [3, 4] and micromanufacturing tribology in particular [5, 6] are the subject of a large body of literature. A comprehensive literature update will allow the reader to understand better surface-related manufacturing failure processes and preventive measures from this study and references therein. Understanding the fundamentals of wear, lubrication and friction is necessary to appreciate the benefits of surface engineering and its role in improving the surface performance of tools and components (Chapter 10). Tribology in general [2, 3] and micromanufacturing tribology in particular [4, 5] are the subject of a large body of literature. The reader will gain a deeper understanding of surface-related failure processes in manufacturing and preventive methods by carefully studying these articles and references included therein [8]. Figure 9.3 shows the behavior of the kinetic friction coefficient for a 100Cr6/8620 metal system lubricated with mineral oil and CO/SO. The graph in Figure 9.3 shows how the two oils behave similarly under typical load conditions. In addition, the CO/SO binary mixture was found to provide the highest protection against friction at 2.5 N of normal load, with a friction coefficient that was 19% lower than that of mineral oil at the same load. However, at higher loads, the friction coefficient of the CO/SO mixture - which was 3% at 5 N and 6% at 7.5 N - was slightly higher than that of mineral oil [5].

It is clear from the figure that adding 1% Eds to the base oil (BO) results in a wear depth of 10 μm. According to this result, the addition of Eds improves the lubrication properties and successfully reduces the wear depth compared to using the base oil alone. The reduced wear depth means improved lubricity and protection of lubricated surfaces, indicating the potential advantages of using Eds as an additive in lubrication applications. Further investigation may elucidate how

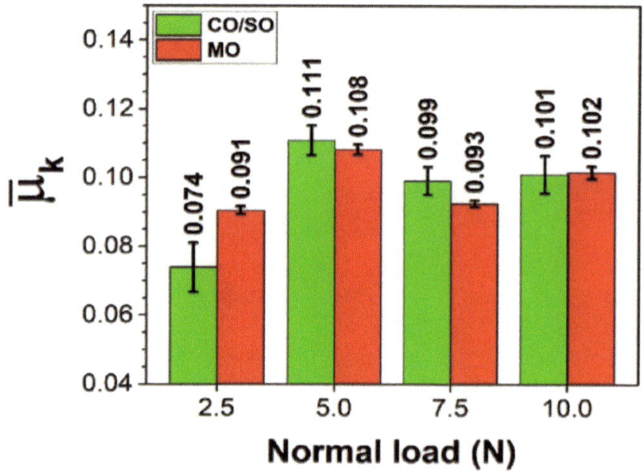

FIGURE 9.3 Comparison of the mean kinetic friction coefficient on 100Cr6/8620 metallic system under lubrication with CO/SO and mineral oil, at 70°C [9] [open access].

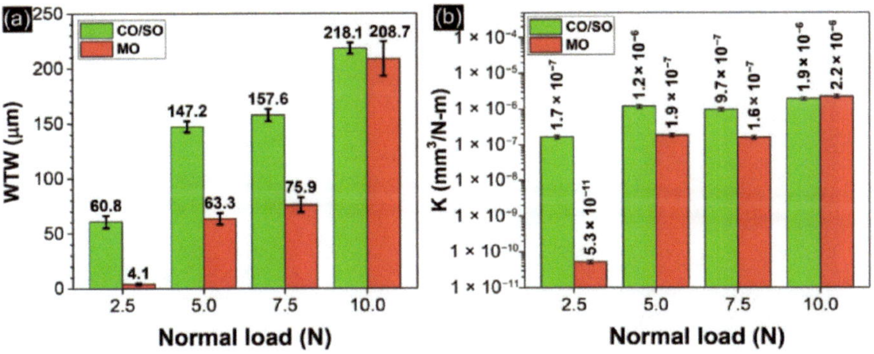

FIGURE 9.4 Comparison of the (a) wear track width (WTW) and (b) wear rate (*K*) value of the AISI 8620 steel disks lubricated with CO/SO mixture and MO, at 70 °C [9] [open access].

Eds enhance the overall lubrication effectiveness and reduce wear. This reduction indicates a noticeable reduction of 60% compared to commercially available BOA and about 67% compared to pure BO. Scanning electron microscope (SEM) images and EDS analysis are performed on worn aluminum surfaces to learn more about wear mechanisms and chemical interactions between lubricants and aluminum surfaces. The wear parameters used to evaluate the wear response of 100Cr6 steel discs lubricated with mineral oil and CO/SO are shown in Figures 9.4, 9.5 and 9.6. It is clear from the results of Figure 9.4 that both lubricants

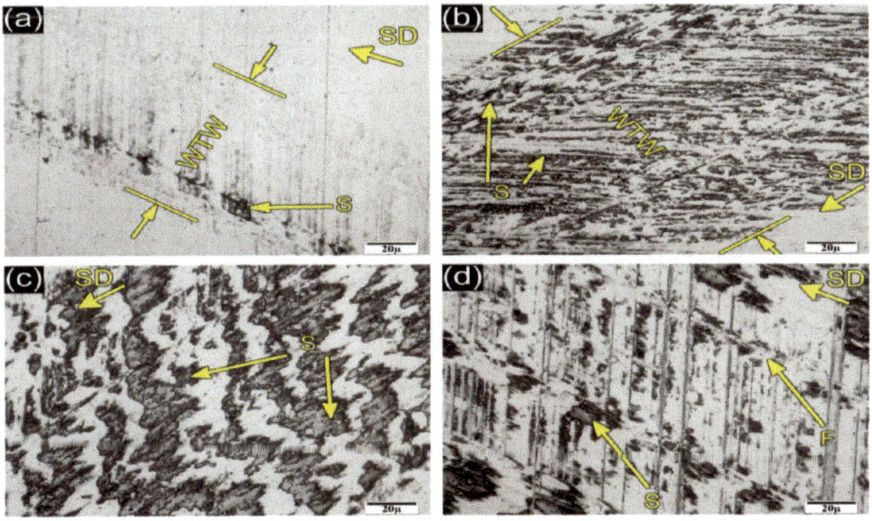

FIGURE 9.5 Optical micrographs of worn surfaces of AISI 8620 steel lubricated with CO/SO under (a) 2.5 N, (b) 5 N, (c) 7.5 N, and (d) 10 N, at 70 °C. Note: SD means sliding direction; S means spalling; and F means furrow [9] [open access].

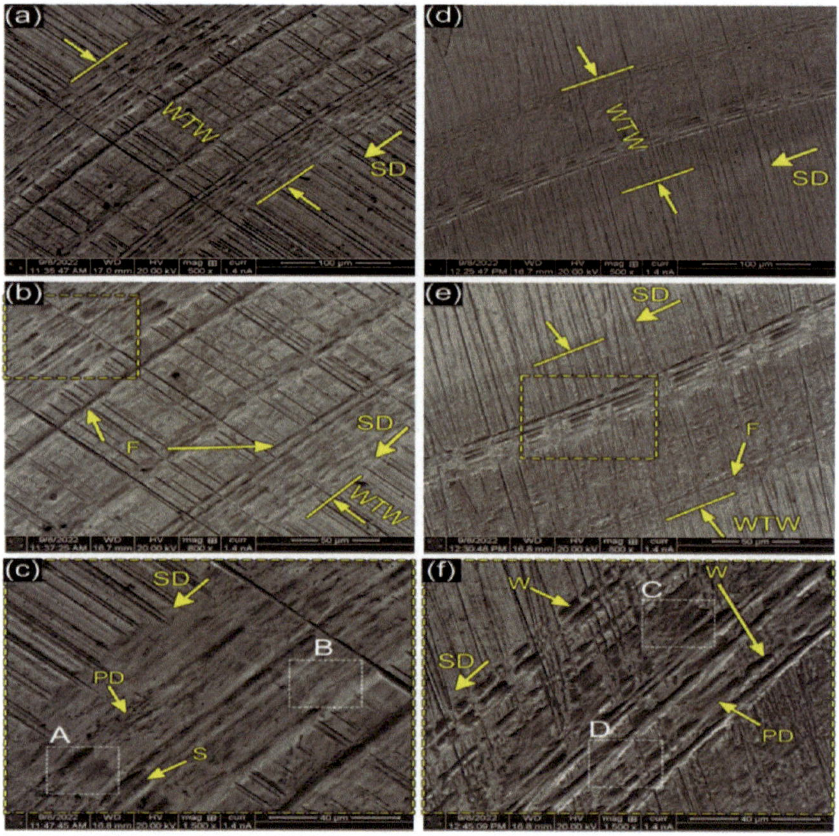

	Spectrum A		Spectrum B		Spectrum C		Spectrum D	
	Wt%	σ	Wt%	σ	Wt%	σ	Wt%	σ
Fe	91.7	0.4	91.3	0.4	91.3	0.4	90.3	0.4
C	7.1	0.3	6.9	0.3	7.6	0.3	6.7	0.3
O	0.3	0.1	1.1	0.2	0.2	0.1	2.1	0.2
Cr	0.7	0.1	0.5	0.1	0.7	0.1	0.7	0.1
Si	0.2	0.1	0.2	0.1	0.2	0.1	0.2	0.1

FIGURE 9.6 SEM micrographs of worn surfaces of AISI 8620 steel lubricated with (a–c) CO/SO mixture, and (d–f) mineral oil at 10 N and 70°C. Notes: (c, f) are magnifications of the regions in yellow dashed boxes in (b, e), respectively; A–D dashed boxes are the regions of the EDS analyses; SD means sliding direction; S means spalling; F means furrow; PD means plastic deformation; and W means wedge wear [37] [open access].

showed an increasing wear trend with increasing load. At 2.5 N, it was found that the mineral oil and CO/SO mixture lubricated systems showed a significant wear variation. At this load, the mineral oil produced a wear track width (WTW) that was 90% smaller than the CO/SO mixture, and the wear rate was four orders of magnitude smaller. However, as the typical load increased, the variation steadily decreased and the readings became closer, such that the wear track width under mineral oil lubrication was only 5% smaller at 10 N, and the wear rate was on par with the CO/SO mixture. As can be seen, the wear response of the CO/SO lubricants was similar to that of the MO lubricants, except for the 2.5 N load. The corrosion rate values, which indicate systems with good corrosion behavior, were between 10−6 and 10−7 [5].

9.4 CHARACTERIZATION TECHNIQUES

It is necessary to use specially developed sophisticated characterization approaches to understand the principles of surface functionality and strategies for its modification. Mechanical engineers can more effectively solve surface-related problems during prototype design, simulation or process testing if they have a solid foundation in surface characterization [3–5]. Related characterization methods focus on the chemical composition, mechanical properties (e.g., hardness, fracture toughness, friction coefficient, wear rate), thermochemical stability (oxidation, corrosion) and surface topography (roughness, texturing) in the specific case of manufacturing tools and component surface protection. However, many pores are created during the production process and polishing sandpaper left scratches on the surface after grinding and polishing [6].

The weight percentage of each element fell within the corresponding ranges, and there are no defects in the elemental composition of alloy or material composition. It is important to emphasize that many characterization methods currently in use have been adopted as validation standards by national and international standards organizations such as International Organization for Standardization (ISO) and American Society for Testing and Materials (ASTM) (www.iso.org and www.astm.com, in that order). The purpose of this chapter is to provide the reader with a brief yet comprehensive introduction to the field of surface engineering and its relevance to microfabrication. The basics of most sophisticated surface modification techniques will be covered in the first section, with an emphasis on technologies that may be more useful in microfabrication due to their unique features. The final segment covers several case studies where surface engineering is important [6]. The surface roughness is measured at 9.8 nm Ra and 13.2 nm Rq. The coating thickness was about 600 nm.

9.5 ADVANCED MECHANICAL PROPERTIES OF COATINGS

The bonding property of h-BN coating on the X750 alloy is measured by performing a micron scratch test on the coatings at a constant speed. SEM images and acoustic emission signal of the scratches are shown in Figure 9.7. It is shown that the critical load is 2.03 N, at which the emission signals began to fluctuate. The criterion chosen to determine the bond strength between the coatings and the substrate is this critical load. Microscopic observation and change in the acoustic emission signal are used to calculate the critical load of the coatings [6].

The coefficient of lubricant efficiency (CLE) concerning wear and friction is shown in Figure 9.8. Overall, it is clear that the CO/SO mixture reduced wear and

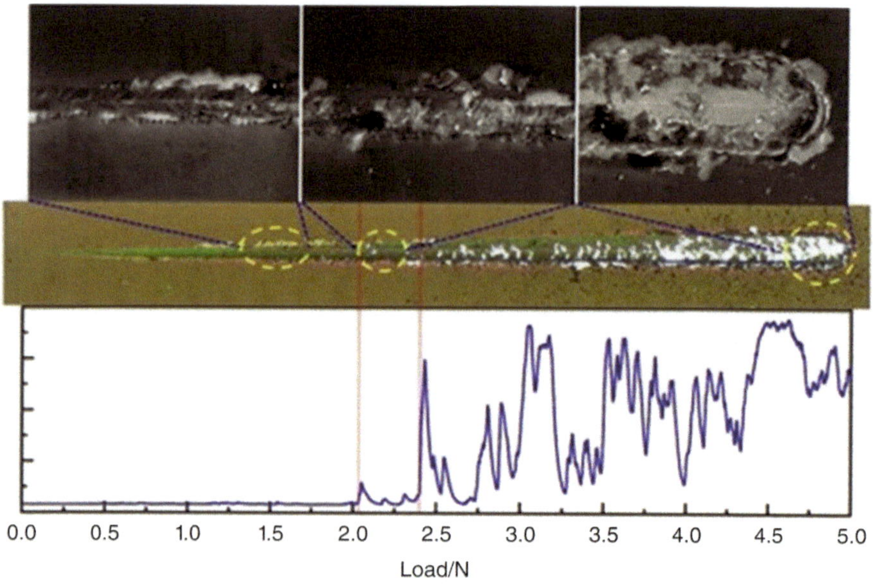

FIGURE 9.7 Critical load, corresponding AE signal diagram and scratch morphology of h-BN coatings [6] [open access].

friction to a level similar to that of mineral oil. At 5 N (81.9%), the CO/SO friction efficiency was the lowest, while at 2.5 N (92.2%) it was the highest. The wear reduction efficiency of both oils is very close to 100%. The mineral lubricant has an average wear and friction reduction efficiency of 85.8% and 99.6%, while the biodegradable lubricant has an average of 85.9% and 99.4% [6].

Figure 9.9 shows the microstructure of AZ91 surface magnesium alloy with and without Gd. The microstructure of AZ91 surface magnesium alloy consists of α-Mg

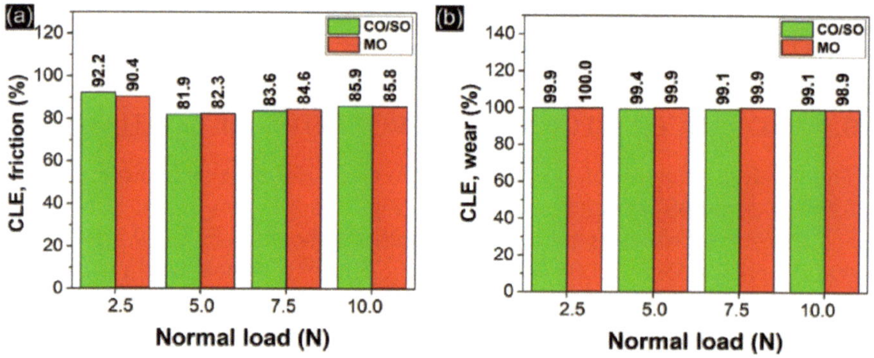

FIGURE 9.8 Coefficient of lubricant efficiency (CLE) in terms of friction (a) and wear (b) for the CO/SO mixture and mineral oil at 70°C [9] [open access].

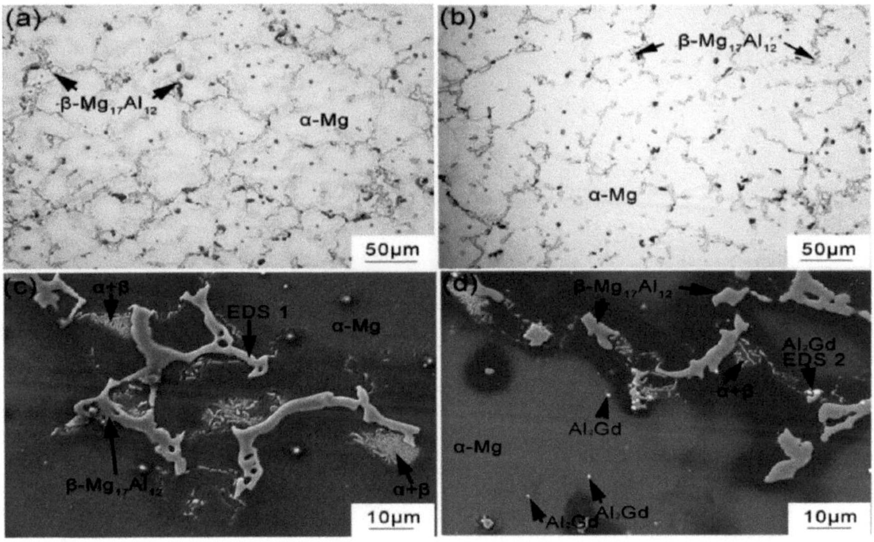

FIGURE 9.9 (a): Microstructure of AZ91 alloy. (b): Microstructure of AZ91 + 0.5 Gd alloy. (c): Another view of AZ91 alloy. (d): Another view of AZ91 + 0.5 Gd alloy [10] [open access].

matrix and coarse-mesh β-Mg17Al12 phase, which are deposited as separate eutectic structures at grain boundaries, as shown in Figure 9.9a,c. In the pores of β-phase, a small amount of α-Mg appears. The margins of the β-phase are covered with a large number of lamellar eutectic structures with alternating α-Mg and β-Mg17Al12. Figure 9.9b shows the optical microstructure of 0.5%Gd alloy, respectively. Upon the addition of Gd, the amount of β-phase is greatly reduced and it breaks off from the lattice to form asymmetric blocks and strips [10].

The worn topography is shown in Figure 9.10 from room temperature to 800°C. There is no thermal expansion effect at room temperature. The coating surfaces are dense and smooth and appeared to accumulate and break down when stress is applied. The metal has a higher coefficient of thermal expansion than coatings. The coating partially cracks as a result of thermal stress [6].

The surface morphology of as-deposited and laser-scanned Au films is investigated using AFM measurements. The topography images of 18 nm (5a–5c) and 39 nm (5d–5f) thick Au films on the quartz substrate are shown in Figure 9.5 for before and after green and UV laser scanning. AFM images and electrical measurements are in good agreement. The scratches and grooves on the worn surface disappeared and were replaced by a large smooth area due to the plastic deformation mechanism. At the same time, a layered structure developed at the edge of the specimen with the surface metal slowly coming out from the metal edge (Figure 9.9a). There was also a lot of chip-like worn debris (Figure 9.9c). These morphological changes demonstrated that severe plastic deformation and flaky wear prevailed in the wear process under the load of 100 N [10].

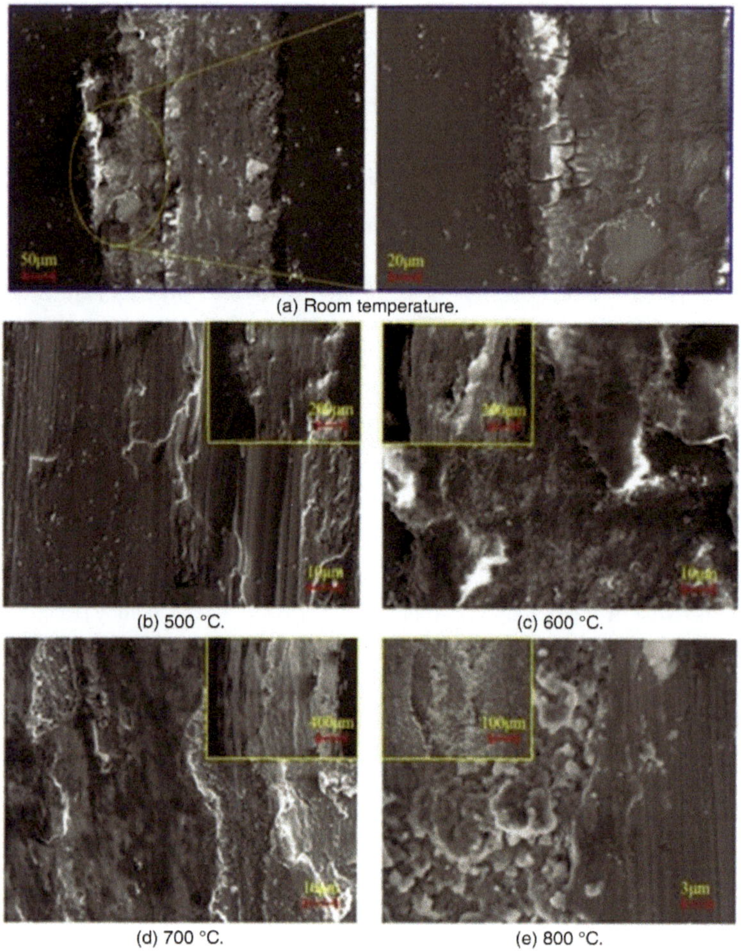

FIGURE 9.10 SEM Images of Worn Coating Surface on X750 Alloy: (a): SEM image showing the worn surface of X750 alloy at a lower experimental load, revealing initial wear characteristics. (b): SEM image of the worn coating surface of X750 alloy at a slightly higher load, showing progressive wear patterns. (c): SEM image taken at a higher load, illustrating further wear and surface degradation. Part (d): SEM image displaying the worn coating surface at an elevated load, highlighting more significant damage and wear features. (e): SEM image at the highest tested load, showing the most severe wear on the X750 alloy surface, with noticeable wear increase [6] [open access].

The experimental apparatus was subjected to vibrations due to the relatively high experimental load of 120 N, which affected the accuracy of the results. As a result, as shown in Figure 9.11, only the data for loads between 15 and 100 N were kept. The surface AZ91 alloy without Gd showed a gradual linear increase in wear rate with increasing load for loads ranging from 15 to 60 N. The wear rate of AZ91 alloy was significantly accelerated to about 1.5 times 60 N when the load reached 100 N, indicating a change from a stable and slightly corroded state

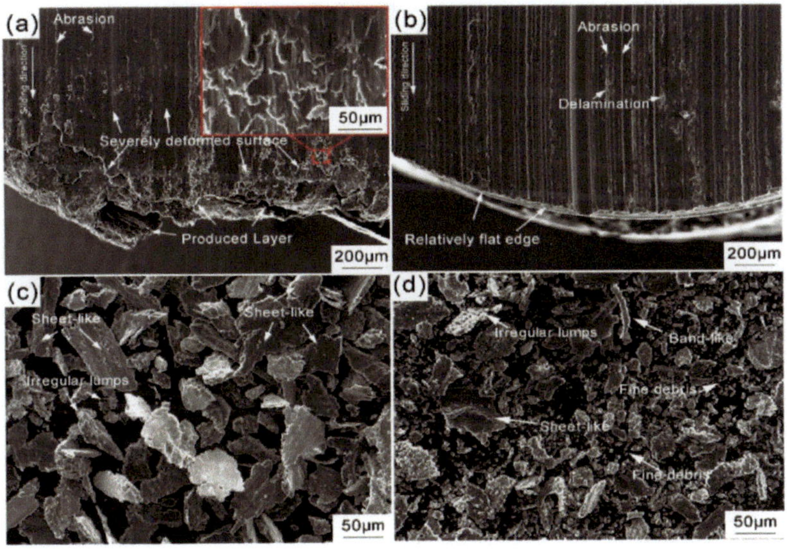

FIGURE 9.11 Wear Edge Morphology and Debris Morphology of Surfacing Magnesium Alloys Under a 100 N Load: (a): Wear edge morphology of AZ91 alloy under a 100 N load, showing the surface wear characteristics.Part (b): Wear edge morphology of AZ91 + 0.5 Gd alloy under a 100 N load, illustrating differences in wear patterns compared to the AZ91 alloy. (c): Debris morphology of AZ91 alloy under a 100 N load, highlighting the debris generated during the wear process. (d): Debris morphology of AZ91 + 0.5 Gd alloy under a 100 N load, showcasing the difference in debris formation between the two alloys [10] [open access].

to an unstable and severely corroded state. The wear rate of the alloy under 15 N stress increased somewhat with 0.5% Gd. However, the wear rate of AZ91 + Gd alloy was significantly higher between 30 and 100 N than that of AZ91 alloy. Furthermore, during the experiment, the wear rate of the AZ91+Gd surface alloy showed a steady linear progression pattern with no significant inflection point as the load increased. Figure 9.12 shows the results of friction and wear tests on magnesium alloys under various loads. In particular, Figure 9.12a shows the wear rate, which shows how the material loss varies with the applied load. This helps to evaluate the wear resistance of the alloy under various conditions. [11].

The temperature and hydrogen production are monitored throughout the reaction. The Figure 9.13 illustrates the theoretical maximums. There is not a single source of reference. They are computed using the reaction equation and weight quantities of 0.81 g and 0.81 g (Venpure TM SF powder, Rohm and Haas) [11].

The morphology of worn coating surface on text surface is shown in Figure 9.14. The overall morphology of texture is visible, indicating that it is not polished. The presence of concave texture retained the excess debris, which reduced the CoF and prevented sharp fluctuation of friction curve. The width of the wear scars is still small at room temperature, reaching its maximum at 700°C, and its depth is small at 800°C. In addition, no cracking of the coating is found at any temperature. When compared with the expansion crack of h-BN coating without texture at

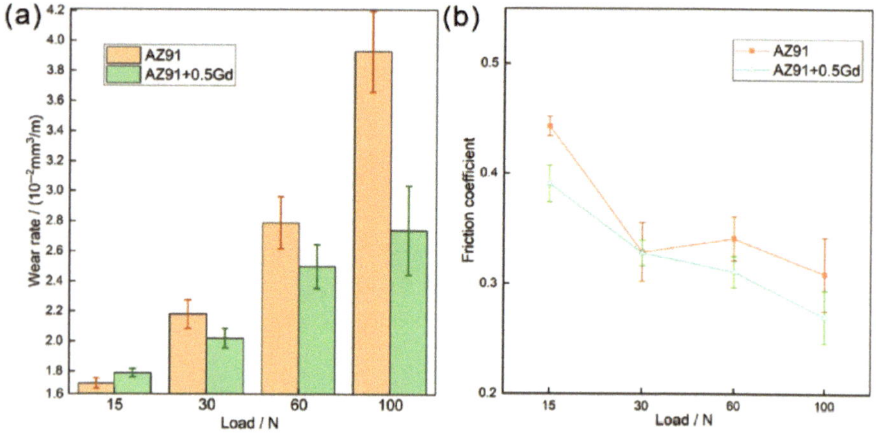

FIGURE 9.12 Results of friction and wear tests of surfacing magnesium alloys under different loads: (a) wear rate, (b) friction coefficient [10] [open access].

500°C, it is found that the texture could stop the cracking through thermal mechanisms, which is certainly beneficial for high-temperature friction of coating [6].

Figure 9.14 of platinum-coated microspheres show accelerated hydrogen evolution and a significant increasing in temperature, although this is still within theoretical limits. In the next stages, alternative catalysts are being investigated to facilitate the reuse of high-gradient catalytic (HGM) microspheres. Furthermore, several layer designs are eventually put into practice to maximize sustainability and performance. By increasing the lifetime and efficiency of catalytic processes, this strategy hopes to open the door to more productive uses in hydrogen production and related technologies [6].

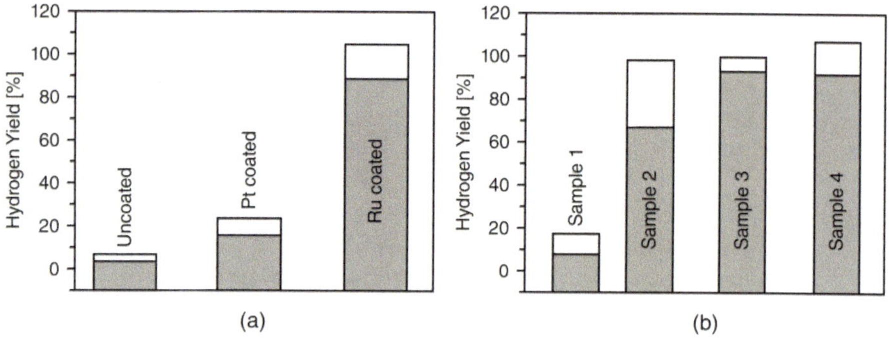

FIGURE 9.13 Hydrogen production equation in reaction (1) for differently coated HGMs: Microspheres coated (a) and layers of Ro on different substrates (b). Light gray areas in the bar chart represent measurement errors [11] [open access].

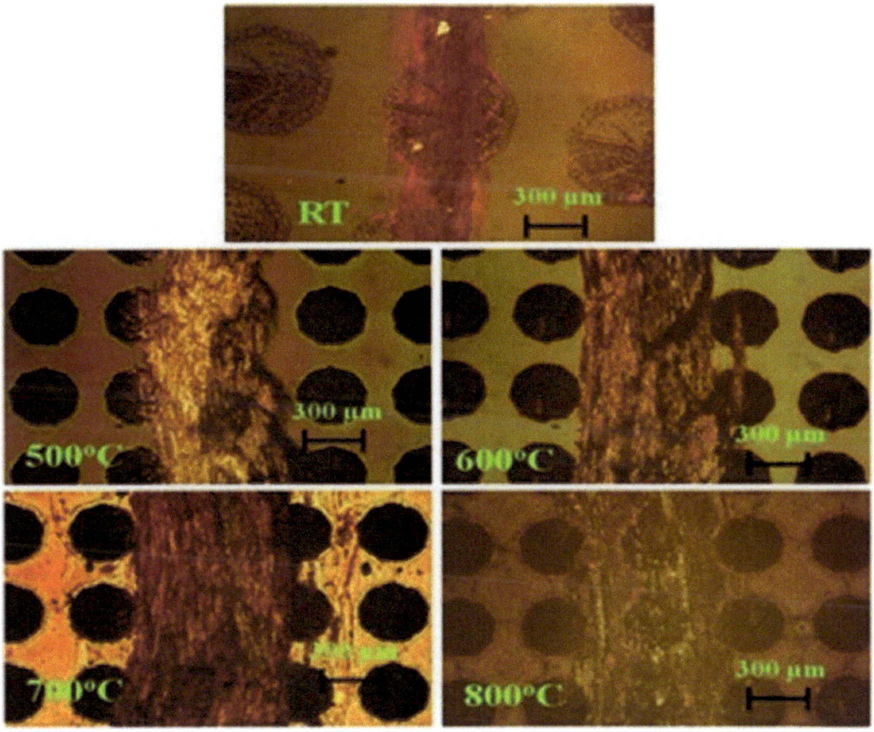

FIGURE 9.14 Surface relief of worn textured surface [6] [open access].

The mechanism of the antifriction effect of coating on textured X750 alloy is shown schematically in Figure 9.15. The friction coefficient of X750 alloy is relatively high despite its good antifriction properties. As a result, the alloy is prepared for h-BN coatings with a lamellar structure. The tribotest results have showed that high-temperature superlubricity is achieved and coating is particularly useful in improving friction properties. On another hand, the friction coefficient changes during sliding. The purpose of texture design is to reduce stress and collect wear particles, which improved wear resistance and high-temperature friction stability [6].

9.6 STRATEGIES FOR MINIMIZING HEALTH AND SAFETY IMPACTS

The processing sector is known for its complex structure and potential hazards, which require the implementation of effective control measures to ensure the health and safety of people and protect the working environment. The objective of this study is to evaluate the impact of interventions on health, safety and environment (HSE) outcomes associated with the processing industry. Several intervention

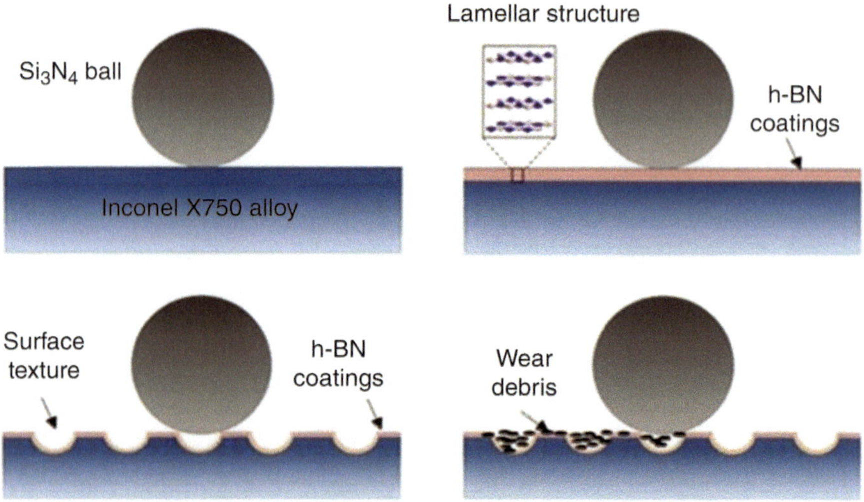

FIGURE 9.15 Friction mechanism's schematic diagram [6] [open access].

tactics have been implemented to address issues related to fires, explosions, product spills, occupational health hazards and mechanical failures. These strategies include engineering controls, administrative controls, personal protective equipment (PPE) and behavioral treatment. These measures aim to reduce the likelihood of illness, injury, environmental accidents and property damage. Research shows that interventions have beneficial short-term effects on HSE outcomes such as increased worker safety, reduced hazards and improved safety behavior [6]. In addition, extensive evaluations have shown consistent improvement and risk reduction. The most successful strategy for improving HSE outcomes is found to combine multiple types of interventions including behavioral, administrative, PPE and engineering controls. The results highlight the importance of carefully considering and implementing process industry interventions to ensure the safety of people, maintain the working environment, protect facilities and improve overall safety performance [12].

Significant developments and advances have occurred in industrialization since its inception in the 18th century. The next significant change came with the invention of electricity in the 19th century, which allowed for the widespread distribution of energy from centralized sources. This is followed by the introduction of equipment powered by regional steam generation, which freed manufacturing from the constraints of manual labor (Industrial Revolution 1.0). Machines could operate faster with less weight thanks to electricity (Industry 2.0). The 20th century has presented the introduction of motorized assembly lines, and as electronics developed, manufacturing became increasingly automated and productivity-oriented (Industry 3.0). By creating more flexible, safer and more ergonomic machines, automation has created the opportunity to improve industrial process optimization and productivity [1].

Understanding the impact of interventions on health, safety and environment (HSE) performance in the manufacturing sector is facilitated by using systems theory [13–17]. Businesses can manage HSE interventions with a holistic strategy and improve overall performance by implementing ideas from systems theory [18]. This analysis explores how systems theory can be applied in practice to assess how interventions impact HSE performance in the manufacturing industry. The components of systems theory are:

- *Component interdependence*: The interdependence of parts that make up a system is emphasized by systems theory. These relate to different components of system rather than individual actions in context of HSE interventions in manufacturing industry [19–23]. Examples of such interventions include process improvements, equipment adjustments, training programs and communication system upgrades. It is vital to recognize the interdependencies between these elements when assessing the potential impact of changes on HSE performance [7, 24].
- *Feedback loops*: These are important in systems theory and provide a useful tool for assessing the effects of actions. Interventions in the process industries can provide both favorable and unfavorable feedback on HSE performance. When an intervention improves a specific area of HSE performance, positive feedback occurs, which leads to a range of positive outcomes [8]. For example, the implementation of new safety training program may increase employee awareness and reduce accidents. On another hand, unfavorable feedback occurs when an intervention unintentionally produces negative outcomes. For example, while the introduction of new equipment may improve operational efficiency, it may unintentionally increase the risk of exposure to hazardous materials. It is possible to detect unintended consequences and improve actions to achieve the intended HSE outcomes by understanding and monitoring these feedback loops [25].
- *Emergent features*: Emergent features that result from the interaction of system components are recognized by systems theory. When considering specific initiatives in isolation, it is not possible to predict the emergent impacts on HSE that may arise in a process industry context [8, 26]. For example, incorporating interventions such as process changes, training programs and strengthening the organization's safety culture can significantly reduce the number of incidents and create a safety culture that permeates the entire organization. Organizations can develop comprehensive HSE plans and identify synergistic impacts by understanding the emergent characteristics of these activities.
- *Adherence to systems theory*: Examining system constraints and hazardous materials unintended consequences can be detected and actions can be improved to achieve intended HSE outcomes by understanding and monitoring these feedback loops [25].

One of the major obstacles to band transfer in semiconductor nanocrystals (NCs) is structural disorder. However, there is a lack of systematic studies on how structural disorders affect electrical performance. Here, we measure the effect of disorder on electron transfer in semiconductor NC assemblies using a tightly coupled model that explicitly includes active and localized disorder. If mobility edges exist, they are determined by wavefunction delocalization and density of states analysis. Phase diagrams of transition from localization to delocalization for several superlattice configurations provide an overview of obtained results. It should be noted that the critical disorder at the turning point is mainly determined by short-range arrangement of the superlattice, not by its long-range arrangement. In addition, we have calculated the binding energy between adjacent crystals at the nanoscale level. Calculations show that there is no direct correlation between the magnitude of electronic coupling and red shift of exciton peak in optical absorption spectrum. We find that in the specific case of PbSe nanocrystals bound by ethanedithiol bonds, the disorder is too large to allow nonlocal states. Finally, we use a phenomenological model to study the effect of valley degradation and separation connectivity within NCs as well as between neighboring NCs.

The exact nature of the coupling of individual valleys is unclear and will determine the effect of multiple valleys on charge transfer. We have calculated the coupling strength and found that the delocalization of wave function is slightly affected by the presence of multiple troughs. In general, we expect belt transmission to be possible with an appropriate surface passivation and bonding. Understanding the process of energy loss and defect creation in implanted materials has been the subject of much research and development. However, ion implantation leads to the appearance of some defects in materials, most of which are electrically active. It is also known that during the Si-ion implantation of SiO_2. Due to their role in the formation of NC states and sub-oxide boundaries, these defects are essential to the process [26, 27].

It has been suggested that doping with phosphorescent light can reduce the density of interfacial defects [28]. It is expected that new physical and chemical properties, different from those of comparable bulk materials, would be discovered by nanoparticles smaller than 100 nm. In addition, nanoparticles will become a component of several applications across a wide range of industries. These applications use electronic, magnetic, optoelectronic and electrical materials. In this study, we summarize the research conducted on expected and current advances in nanoparticle electrical devices. Although the device's potential uses have not been proposed, we briefly mention research efforts to control the shape and array of nanoparticles as a new class of materials [29]. Because they act as charge transfer traps and significantly reduce the material's ability to carry a charge, these defects limit using of NCs in complex electronic devices.

Currently, several electrical characterization methods are used to study the charge transfer properties of NCs. The most commonly used method is current-voltage (IV) measurement. Characteristics of charge transfer features [7], it has been shown that the size and spatial position of NC affects its transport qualities. The effect of Si-NC diameter on charge transfer [24], which attributed to the effect of filtration-conduction mechanism. Memory devices are another application of CNCs embedded in an oxide matrix. Charge trapping in metal oxide semiconductor (MOS) devices [25]. Any electronic device requires the passivation of these defects associated with the interaction of substrate and oxide matrix. The vacuum annealing technique, which is often used to crystallize NCs (embedded in an oxide matrix), is another factor that contributes to increase the trap density at interface.

However, most interfacial defects that arise during vacuum annealing can be eliminated in the resulting gas or hydrogen environment by low-temperature annealing (450°C) [7]. Although there may be additional challenges associated with the new processes and/or chemicals required to create such nanoscale structures, the using of nanotechnology in electronics to miniaturize electronic components does not in itself pose any threat to human health. However, it is important to note that there are applications of nanotechnology in electronics and electrical products that directly raise environmental and human health concerns. It is the process of creating electronic components or surface coatings of electrical products using artificially created nanoparticles in "nanomaterials". Materials designed and produced with structural characteristics having at least one dimension of 100 nanometers or less are called nanomaterials. Many different nanomaterials are currently used in electronics on a commercial basis or for development and research purposes. Silver nanoparticles for surface coatings, as well as carbon nanotubes and quantum dots for electronic and electrical equipment, are among the most widely used nanomaterials.

Ruthenium (Ru) has showed an excellent performance when tested with other catalysts as illustrated in Figure 9.16. Reaction analysis has revealed values within the error bars of theoretical limit for both hydrogen production (Figure 9.16a) and mercury gain. However, when reactive oxygen (RO) is administered directly, the catalyst is found to be completely removed from the microspheres after one application. As a result, downstream research has focused on creating different layered structures to suppress adhesion failure caused by corrosion. The goal is to improve the robustness and efficiency of catalytic systems, which will ultimately increase their suitability for using in the synthesis of hydrogen and other chemicals [30].

The 100× objective of Raman microscope focuses on 488 nm laser beam onto material, creating laser-annealed regions of less than 1 μm in size. These regions are easily visible in the camera photographs (Figure 9.16a), indicating changes in the absorption and/or reflectance of film. Figure 9.17b shows the transmission, PL and Raman patterns across a typical laser-annealed spot [30]. Temperature-induced Si–SiO_2 phase separation and ordering, as well as the development of Si–nc, result in a significant increase in Raman signal and a decreasing in transmittance. Through careful manipulation of experimental parameters, the thermal electron release of DLTS from NCs can be determined [26]. An Arrhenius plot is

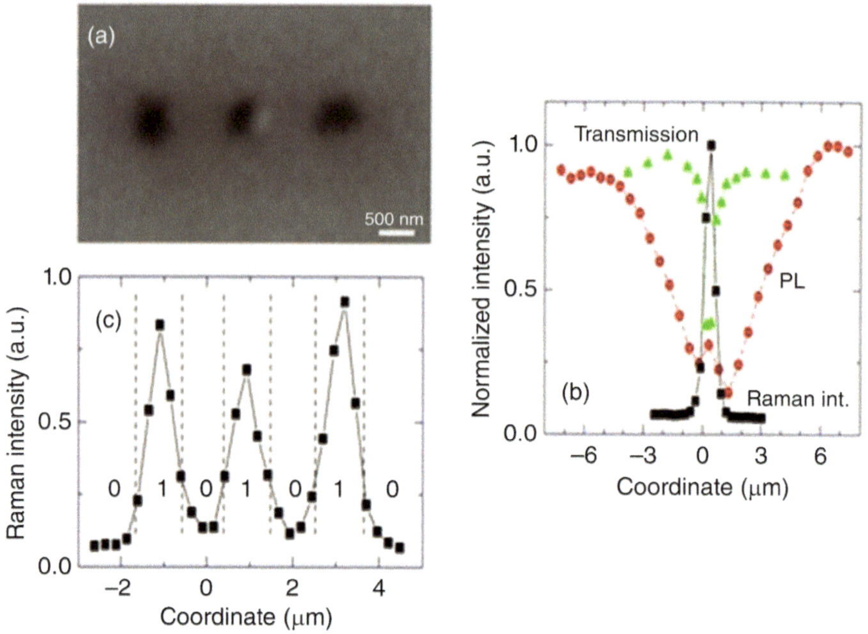

FIGURE 9.16 Digital data storage on free-standing 2 μm thick Si/SiO_2 SL. (a) Camera image of three spots created by annealing the 488 nm laser for about 10 milliseconds. (b) Cross-sections of the Raman, transmission and 1.5-eV PL signals through a standard laser spot. (c) Cross-section of Raman intensity through locations shown in panel (a), where digital data points are shown. 500–540 cm^{-1} range is integrated for the Raman intensity [30] [open access].

used to estimate the electron emission activation energy of the NC trap based on DLTS analysis. The difference between the two lowest energies of confinement states [25, 26].

9.7 SEPARATION OF BIOLOGICAL MATERIALS

In the late 1980s, biodegradable polymers have begun to be used in modern biomedical applications. Currently, one of the most interesting areas of biomaterials research is the study of biodegradable biomaterials. Medline has recorded a significant growth in publications in this area of research over the past 10 years. Biodegradable materials are expected to promote the healing process of damaged tissue and eventually disappear. These new biomaterials have three promising applications: cardiovascular, orthopedic and pediatric. Coronary artery stent is one of the interesting biodegradable implants for cardiovascular applications. This small implant opens the narrowed artery until the vessel can be repaired and serves as a temporary scaffold. When biodegradable stents are first introduced in 2001, researchers are intrigued. Several models are currently undergoing preclinical testing in humans, but none are yet ready for clinical use [7, 8, 18, 24].

Developing ideal biodegradable mineral biomaterials requires attention. This chapter provides an overview of research on biodegradable mineral biomaterials. It examines in detail the properties of ideal biodegradable mineral biomaterials and proposes methods for their development and difficulties in creating and characterizing biomedical interfaces. Thanks to new developments in other fields of science and technology, biomedical research is constantly growing. One area of great interest in this research is the control of biomolecular and cellular interactions, which can be achieved by modifying the surfaces of materials of various types and sizes. At the same time, advances in biomedicine are driven by understanding the complex properties of biomedical interfaces. It is important to note that advances in vacuum science and technology offer many new approaches to the modification and characterization of matter, specifically designed for biological systems [28]. The following paragraphs present four separate cases of bio-interface preparation where vacuum procedures are necessary for modification or additional analysis. These include bulk (mineral) biomaterials, supported (porous) nanomaterials, primary calcium phosphate (CaP) nucleation phases, and an additional example highlighting recent advances in our knowledge of nanoparticle conjugates. Activation of nanoporous structures on their surface. A variety of biomedical processes applied in both therapeutic and diagnostic approaches find strong support in nanoporous materials. By comparing materials to their size-matched counterparts, the aspect ratio of these surfaces has important implications for controlling biomolecular interactions. By investigating how surface features influence the behavior of biomolecules, scientists can optimize material properties for use in biotechnology, medicine and materials science. This comparison helps designers create materials that support specific biomolecular interactions, as well as highlight differences in reactivity and adhesion at the nanoscale. These advances could lead to improved tissue engineering, drug delivery systems and biosensor performance, ultimately improving outcomes in a range of biomedical applications.

For example, modification of surface chemistry has already been used to impart specific biomolecular activities to silicates [24], titanium oxides [25] and porous alumina [25]. In this regard, there is still controversy regarding the extent of early reactions of biomolecules/cells to nanoscale rough surfaces; Some scientists report greater affinity, while others highlight their antifouling properties [26]. In summary, Sr has been described as a species distributed on YSZ and, on another hand, in La-Sr perovskites including transition metals. The perovskite lattice is represented by the first state, which is identified by XPS spectrum with peaks at BEs between 131.4 and 132.5 eV. The latter state indicates the separation of secondary oxide phase and peaks at BEs between 132.8 and 133.8 eV (Figure 9.17). These two archetypal chemical states are

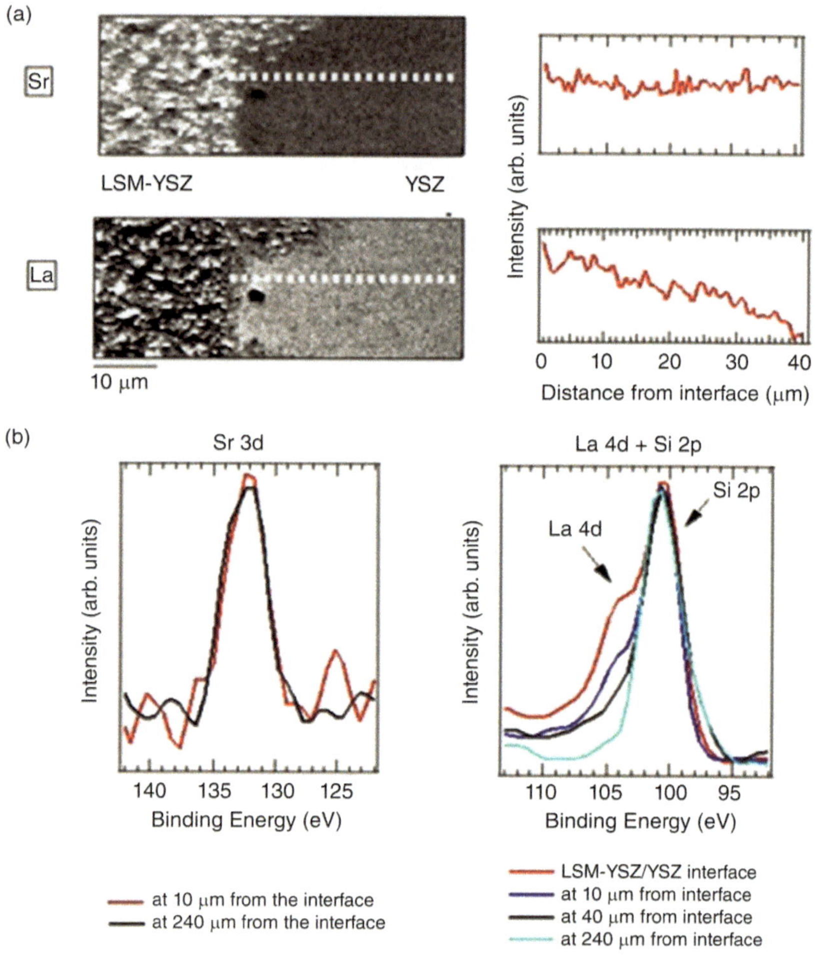

FIGURE 9.17 (a) Elemental concentration profiles and SPEM maps at energies of Sr 3d and La 4d obtained using a clean cell under 0.05 mbar O2 at 924 K. The interface between the YSZ electrolyte (right) and LSM-YSZ cathode (left) is region that is photographed. (b) μXPS spectra of Sr 3d and La 4d obtained along the dashed line in panel (a) at indicated distances from the cathode/electrolyte contact [11] [open access].

usually found together in the studied volume in situations documented in the literature [11].

10.1% of this organic support layer has disappeared. Figure 9.18 illustrates the elemental composition and morphology of corrosion products under different corrosion times in 110 g/L sodium aluminate solution containing 3 g/L S2O32− and 5 g/L S2−. To give a more comprehensive observation of different morphologies of corrosion products, Figures 9.18a, d, g, j, m are annotated to indicate the surface areas with squares, which are enlarged in

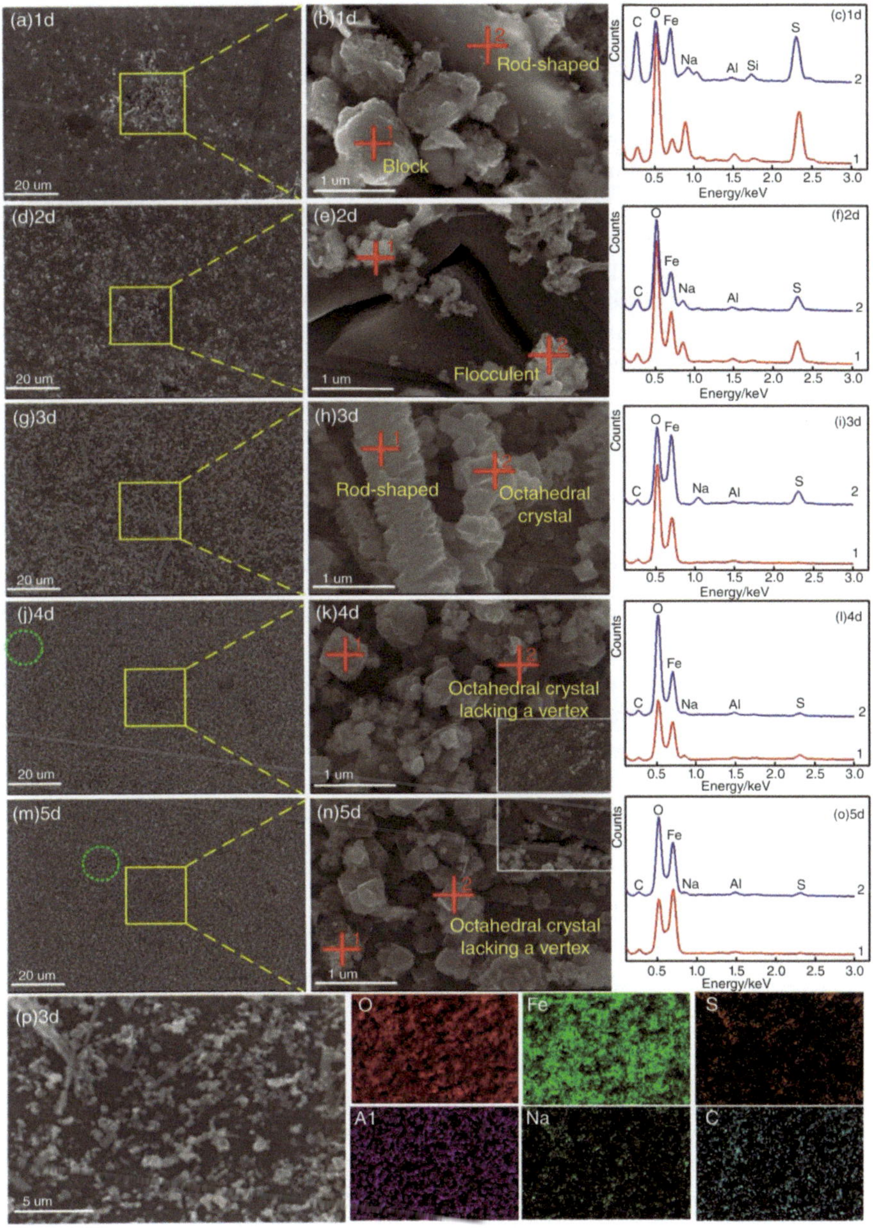

FIGURE 9.18 SEM morphologies (a, b, d, e, g, h, j, k, m, n), element distribution (p) and EDS analysis (c, f, i, l, o) of corrosion products on surface of Q235 steel with different corrosion times: (a–c) 1 day; (d–f) 2 days; (g, h, i, p) 3 days; (j–l) 4 days; and (m–o) 5 days [31] [open access].

Figures 9.18b,e, h, k, n. O, Fe, Na, Al and S accounted for the majority of corrosion products (Figure 9.18c, f, i, l, o, p) [31].

However, the elemental contents changed significantly with increasing of corrosion time. The surface coverage is increased with increasing of corrosion time, and shape of corrosion products changes dramatically. After one day of corrosion, a considerable amount of corrosion products is localized on the steel surface (Figure 9.18a), which made the rusting phenomenon obvious. Large blocks or flakes of corrosion products are locally formed on surface after 2 days when the amorphous corrosion products locally accumulated and crystallized slowly [31] (Figure 9.18d). With the growth of coating after two days, it is obvious that the steel surface has significant cracks and delamination (Figure 9.18b, d). The surface coverage increases significantly when the corrosion is maintained for three days (Figure 9.18g). The growth rate of crystalline particles is higher than nucleation rate due to long corrosion time [8]. Most, but not all, metallic biomaterials are alloys used in prosthetics and abutments, including total joint replacements and orthodontic brackets. All these metals have high corrosion resistance, which is critical, good biocompatibility (immune response), suitable mechanical properties, good osseointegration (or vascular integrity) and reasonable price [25]. One way to improve this property is to create alloys containing components that increase corrosion resistance. However, the addition of chromium (≥11% in stainless steel production) can stop the formation of rust or decomposition of metal. Likewise, adding nickel to steel promotes the development of austenitic phases, which effectively prevent corrosion. Mo reduces pitting corrosion as an alloying component [26]. Recently, "third-generation biomaterials", a new class of biomaterials that take advantage of device biodegradability, have been developed. When it comes to mineral materials, biodegradability means that mineral implant dissolves in biological environment. middle. Corrosion thus has beneficial uses. At a pH level of around 7, very few minerals are present.

However, depending on the cells, zinc can be cytotoxic, this can be taken into account when performing a specific transplant. The two biodegradable metals currently in use are iron and magnesium, but magnesium has a relatively high hemolysis rate and dissolution rate, both have been improved in vitro by doping [25, 26]. Also, localized corrosion (pits, cracks, galvanic connections) can result from melting of the metal. However, the selection of magnesium as a soft metal that biodegrades and corrodes the body's electrolyte environment has potential applications in two medical fields: bone fusion [26] and coronary stents [29]. The creation of metallic implants from biodegradable materials that will eventually degrade and regenerate new tissue is a huge technological challenge. Therefore, current biomaterial research is focused on the rediscovery of highly corrosion-resistant Mg-Mg alloys, which offer several advantages such as avoiding long-term problems and the need for implant removal, especially in orthopedic applications. In this study, the surface of Inconel X750 alloy is first given a circular texture of 400 μm in diameter and then coated. The texture is established by the resulting matrix structure, where the circular center is located at 600 μm and surface texture density is 35%. SEM examination of coating reveals a smooth surface with small protrusions surrounding the annular roughness. To minimize variations in the coefficient of friction (CoF) caused by the accumulation of large chips and to prevent disruption of coating lubrication, this texture is designed to retain abrasive chips in a lubricating manner.

Therefore, research is underway to create third-generation biomaterials, such as biodegradable stents that take advantage of metal corrosion rather than traditional formulas. The early stages of CaP nucleation are studied using HAXPES. Another popular topic is layer-by-layer deposition techniques, which will facilitate the production of biomaterials for regenerative medicine and tissue engineering in the coming years. The periodic addition of elemental and molecular components to the sample allows these methods to more precisely control the final composition and characteristics [28].

This example shows how a calcium phosphate (CaP) core can be formed using cyclic electrochemical deposition (CED), a process specifically designed to form a hydroxyapatite (HAP) core, and also shows silicon (PS/CaP) pores [28]. The spectrum of PSi–CaP interface, which highlights how complex the interactions are at this intersection. Critical information about phosphorus binding can be found in P 1s spectra of CaP contact (Figure 10.21), but a reference source for these data is currently unavailable. The peak found at 2149.0 eV is caused by P–O/P=O bonds that are associated with CaP and located at 2148.5 eV. In addition, the P0 peak is located at 2145.5 eV, suggesting unique binding properties at interface. This analysis highlights the importance of understanding the chemical interactions at PSi–CaP interface for biomaterials and tissue engineering applications [25].

Advances in vacuum and pressure science are often necessary for further discoveries. There has been a noticeable uptick in the study of Hoover science, technology and applications where the International Union of Vacuum Sciences is underrepresented. The achievements of Hoover's science and technology, as well as some modern developments in several scientific fields, are the main topics of this study. Interdisciplinary teams from fields such as electrical engineering, biology, medicine, physics and chemistry will use creative thinking to solve new problems.

This chapter shows that interventions have a significant impact on environmental, safety and well-being factors in the process industry. To highlight the importance of interventions, it is important to recognize that they are ongoing processes that require ongoing monitoring and evaluation, rather than one-off fix. Businesses need to put the health of their employees and environment first. Standard operating procedures, training initiatives and using of technology have showed to reduce workplace accidents, illnesses and injuries. This strategy supports the process industry's efforts to create a sustainable ecosystem and safer working conditions for employees. To ensure the long-term sustainability of initiative, it is a vital to consider its economic feasibility. Businesses must carefully consider the costs of starting and maintaining initiatives, as well as the financial benefits and payback periods. This assessment ensures that projects are both financially viable and capable of maintaining long-term viability. Regulatory compliance is a critical component that cannot be ignored. Businesses must comply with environmental laws and guidelines to avoid fines, legal problems, reputational damage and business interruptions. Businesses need to plan for future legal requirements and create sustainable solutions that both meet and exceed these requirements [29, 32–36].

In addition, integrating advanced technologies is essential to improve the efficiency and longevity of treatments. Manufacturing industries are constantly changing, so keeping a close eye on technological advances gives businesses a special chance to reduce waste and improve environmental performance by using fewer resources. The path to long-term sustainability is to continually integrate green measures into daily operations while maintaining the optimal reliability of these factors. This ensures the continued profitability of the industry in the face of changing and new challenges and has a positive impact on public health, safety and environment [33–38].

The fact that this study offers valuable information should not be overlooked. This theoretical study also has some significant drawbacks, such as a dearth of empirical data, difficulties in making broad generalizations, a lack of context knowledge, and difficulties with quantitative measurement. The limitations are discussed as follows:

- *Insufficient empirical data*: Theoretical studies may limit direct access to empirical data from the real world, as they rely heavily on existing theories and models. This limitation may make it difficult to use real industrial data to test theoretical propositions.
- *Difficulty in generalization*: Theoretical studies' focus on general principles and abstract concepts makes difficult to immediately relate the results to specific process industry sectors or initiatives.
- *Limited contextual understanding*: Complex contextual elements specific to the process industry, such as its specific challenges, regulations and operational nuances, may not be adequately captured by theoretical research.
- *Difficulty in measurement*: It may be difficult to measure theoretical entities statistically, making difficult to operationalize these ideas for empirical testing.

To ensure practical relevance and applicability in the future, the authors consider adding a case study to this study and engaging industry stakeholders to address these limitations. This strategy will focus on specific interventions within a given processing industry to gain a deeper understanding of their impact on the environment, health and safety. Using a mixed methods research approach to gain deeper insights is another possible avenue. This may involve collecting qualitative information through interviews and quantitative information through surveys. Using multiple research methodologies promises to enhance the sustainability of results and increase their impact on commercial procedures [31, 37, 38].

Surface engineering is a proven technology that is widely used in various sectors to improve component performance. However, it comes with inherent health, safety and environmental hazards, just like any other industrial operation. Stricter health and safety laws have led to the development of new surface engineering methods that have reduced the risks associated with traditional approaches but have introduced new hazards. To cope with these ever-changing hazards, careful risk analysis, strict safety protocols and continuous innovation that strikes a balance between safety and innovation are necessary. Industry and regulators must work together to ensure the safe and sustainable development of surface engineering technologies.

9.8 PROBLEMS

1. Explain the importance of surface engineering processes.
2. List several strategies for minimizing health and safety impacts.
3. How would relative wear rates change if the surface changed along a circular path?
4. What are advanced tribology and surface contact phenomena?
5. What are the main causes of problems in characterization techniques?

REFERENCES

1. Rezayat, M., Karamimoghadam, M., Moradi, M., Casalino, G., Roa Rovira, J.J., & Mateo, A. (2023). Overview of surface modification strategies for improving the properties of metastable austenitic stainless steels. *Metals*, *13*(7), 1268.
2. GVRI, The Global Process Industry Market Size, G. V. R. I., 2020.
3. Rahman, M., Haider, J., Hashmi, M.S.J., Hashmi, S., Batalha, G.F., Van Tyne, C.J., & Yilbas, B. (2014). 8.03-health and safety issues in emerging surface engineering techniques. *Comprehens. Materials Process.*, 35–47.
4. Khan, F., Rathnayaka, S., & Ahmed, S. (2015). Methods and models in process safety and risk management: Past, present and future. *Process Saf. Environ. Protect*, *98*, 116–147.
5. Guo, H., Stoyanovich, B., Pang, J., & Iglesias, P. (2023). Lubricating ability of protic ionic liquids as additives to a biodegradable oil for aluminum-steel contact: Effect of alkyl chain length and propensity to hydrogen bonding. *Lubricants*, *11*(8), 329.
6. Zeng, Q. (2023). High-temperature superlubricity performance of h-BN coating on the textured inconel X750 alloy. *Lubricants*, *11*(6), 258.
7. Hopkins, A. (2009). Safety, Risk and Culture: An Integrated Approach, Routledge.
8. Reason, J. (2008). The Human Contribution: Unsafe Acts, Accidents and Heroic Recoveries, Ashgate Publishing.
9. García-Miranda, J. S., Aguilera-Camacho, L. D., Hernández-Sierra, M. T., & Moreno, K. J. (2023). A Comparative Analysis of the Lubricating Performance of an Eco-Friendly Lubricant vs Mineral Oil in a Metallic System. *Coatings*, *13*(8), 1314.
10. Chen, Q., Yu, Y., Sun, J., Jing, C., Zhao, Y., & Wang, J. (2021). Investigation of the wear behavior of surface welding AZ91 and AZ91+ Gd alloys under variable loading conditions. *Crystals*, *11*(5), 554.

11. Sharif, A., Farid, N., Vijayaraghavan, R.K., McNally, P.J., & O'Connor, G.M. (2021). Femtosecond laser assisted crystallization of gold thin films. *Nanomaterials*, *11*(5), 1186.
12. Broughton, E. (2005). The Bhopal disaster and its aftermath: A review. *Environ, Health: Global Access Sci. Source*, *4*, 1–6.
13. Dorfman, P., Fucic, A., & Thomas, S., Late lessons from Chernobyl, early warnings from Fukushima, in: Late Lessons from Early Warnings: Science, Precaution, Innovation, 2013, pp. 464–489. http://www.eea.europa.eu/publications/late-lessons-2/late-lessons-2-full-report/late-lessons-from-early-warnings.
14. International Nuclear Safety Advisory Group. IAEA Report INSAG-7 Chernobyl Accident: Updating of INSAG-1, International Atomic Energy Agency, Vienna., 1992.
15. Singh, B., Jukes, P., Poblete, B., & Wittkower, B. (2010). 20 years on lessons learned from Piper alpha. The evolution of concurrent and inherently safe design. *J. Loss Prev. Process. Ind*, *23*(6), 936–953.
16. NOSC, Deep water – The Gulf Oil Disaster and the future of offshore drilling, National Commission on the BP Deepwater Horizon Oil Spill and Offshore Drilling, N. O. S. C., Washington, DC, 2011.
17. Paltrinieri, N., Øien, K., & Cozzani, V. (2012). Assessment and comparison of two early warning indicator methods in the perspective of prevention of atypical accident scenarios. *Reliab. Eng. Syst. Saf*, *108*, 21–31.
18. Hale, A.R., & Borys, D. (2013). The Handbook of Safety Management: A Practical Guide to Developing Safety Programs that Work, CRC Press.
19. HSE (2015). Process Safety Management: Leading Framework, HSE Books, London.
20. OSHA, Safety and health programs, Retrieved from, O. S. a. H. A., 2018, https://www.osha.gov/shpguidelines/.
21. Hallowell, M.R., Gambatese, J.A., & Behm, M. (2008). Applying the PDCA cycle to accident/incident investigation and analysis. *J. Construct. Eng. Manag*, *134*(11), 866–875.
22. Arshad, U., Abbas, M., Waheed, A., Zafar, S., & Siddique, R. (2020). Investigating the impact of safety audit on industrial safety performance in chemical industry. *SN Appl. Sci*, *2*(6), 1–10.
23. Checkland, P. (1999). Systems Thinking, Systems Practice, Wiley.
24. Hale, A.R. (2010). Safety, management and safety science., *Saf. Sci.* 48 (6), 623–627.
25. Leveson, N. (2011). Engineering a Safer World: Systems Thinking Applied to Safety, MIT Press.
26. Hse, H.S.E., Managing for Health and Safety, HSG65), 2015.
27. Johnson, C.D. (2009). Managing Safety: A Guide for Executives, CRC Press.
28. Checkland, P., & Poulter, J. (2006). Learning for Action: A Short Definitive Account of Soft Systems Methodology and Its Use for Practitioners, Teachers, and Students, Wiley.
29. Wilson, J.R., & Corlett, E.N. (2005). Evaluation of Human Work: A Practical Ergonomics Methodology, CRC Press.
30. Nikitin, T., & Khriachtchev, L. (2015). Optical and structural properties of Si nanocrystals in SiO2 films. *Nanomaterials*, *5*(2), 614–655.
31. Quan, B., Li, J., & Chen, C. (2021). Effect of corrosion time on the synergistic corrosion of Q235 steel in sodium aluminate solutions. *Metals*, *11*(5), 753.
32. Goh, Y. M. (2024). Introduction to workplace safety and health management: A systems thinking approach (Vol. 8). World Scientific, Singapore.

33. Cox, T., Jones, D., & Innes, R. (2019). Developing and implementing a human factors toolkit to improve health and safety in the UK offshore oil and gas industry. *Saf. Sci*, *118*, 19–28.
34. Azadeh, A., Farokhnia, M., & Maghsoudloo, A. (2021). Risk-based health, safety, and environment management (HSE) framework for petrochemical industries, *Journal of Process Safety Progress*, 8(3).
35. Reason, J. (1997). Managing the Risks of Organizational Accidents, Ashgate Publishing.
36. Amalberti, R., & Deblon, F. (2011). Human factors engineering and patient safety. *Qual. Saf. Health Care*, *20*(6), 486–490.
37. Hollnagel, E. (2009). The ETTO Principle: Efficiency-Thoroughness Trade-Off, Ashgate Publishing.
38. Guldenmund, F.W. (2000). The nature of safety culture: A review of theory and research. *Saf. Sci*, *34*(1–3), 215–257.

10 Recent Developments in Surface Engineering

10.1 INTRODUCTION

Surface and subsurface processes play a decisive role in the functional properties of solids. Due to the complexity of this phenomenon, an interdisciplinary approach is required in various fields, including thin films, nanostructures, electronic materials, surface engineering, pure and applied surface sciences, and bio-interfaces. Because of nonequilibrium gas states typically form surfaces and subsurfaces, researchers with expertise in plasma science and engineering are often involved in studying these phenomena. In addition, under vacuum or low pressure, the functional properties of solids or liquids often change. In fact, high or very high vacuum conditions are used in the most advanced surface and subsurface characterization techniques, so vacuum science and technology play a critical role [1–4]. It also acknowledges ongoing progress in the scientific fields mentioned above. This chapter presents the achievements of recent years, selected by the heads of scientific departments of IUWTA. The previous audit report is issued in reference [5] and highlighting from the above areas as well as other areas of science and technology.

10.2 NANOSTRUCTURED SMART MATERIALS

In nanotechnology, self-assembly and self-organized (or self-ordered) manufacturing methods are recognized as the most elegant and cost-effective route to the production of complex and functional nanostructured materials [6, 7]. Several top-down and bottom-up methods based on chemical, electrochemical, sol–gel and hydrothermal processes involving self-organization have been explored for their synthesis [2, 8]. For example, highly porous metal oxide nanostructures such as anodic aluminum oxide (AAO), nanotubular titanium oxide (TNT) and porous silicon (pSi) have been successfully fabricated by taking advantage of self-ordering phenomenon that occurs during electrochemical anodization metals [2–5]. Comparing this fabrication method with traditional lithography-based fabrication methods, several advantages can be highlighted, it is less expensive, widely available, allows downstream manufacturing with nanoscale resolution and provides access to high aspect ratio structures [9]. The possibility of using electron beam surface engineering to modify the composition and properties of Co-Cr-Mo alloys is investigated by Valkov et al. [10]. Adjusting the engineering variables achieved the highest cooling rate without surface burnout. Three-dimensional (3D) atomic force microscopy (AFM) images of sample surfaces are

 DOI: 10.1201/9781003584735-10

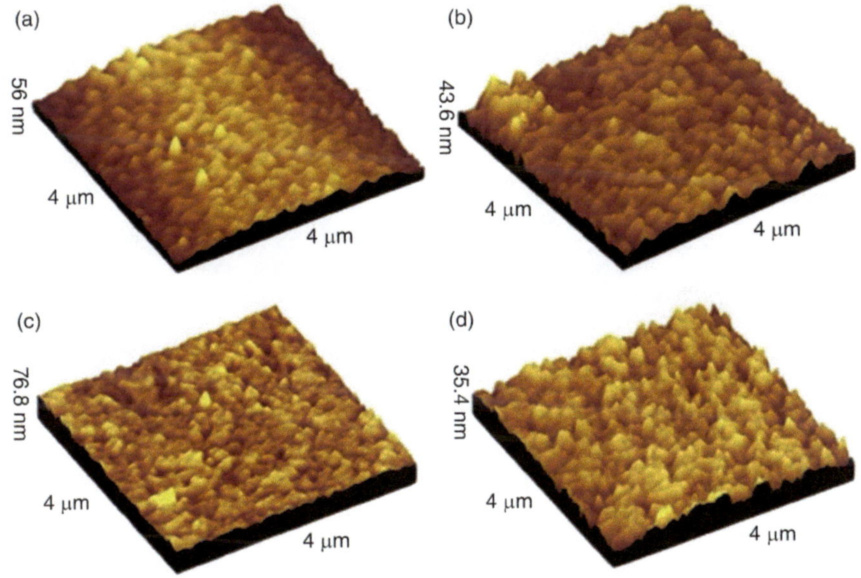

FIGURE 10.1 AFM images showing (a) pure ITO, (b) HfCx-treated ITO, (c) HfOx-treated ITO and (d) HfO_2-treated ITO anode. The samples are measured using scan size, 4×4 m^2 in non-contact mode [4] [open access].

shown in Figure 10.1. The substrate morphology is strongly depending on the electron beam machining (EBM) process [4].

AFM images of variously treated indium tin oxide (ITO) surfaces are shown in Figure 10.1: (a) pure ITO, (b) treated with HfCx, (c) treated with HfOx and (d) treated with HfO2. When 4×4 μm^2 area is used as the scanning size, corresponding root mean square (RMS) values of samples are as follows: 1.38 nm for HfCx-treated ITO, 0.8 nm for HfOx-treated ITO and 1.0 nm for HfO_2-treated ITO. The results demonstrate that HfOx treatment improves the interface properties between the organic hole transport layer (HTL) and the ITO anode as shown by lower RMS value of HTL. Figure 10.1 depicts AFM images furthermore, the chemical reaction between HfCx and HfOx on the ITO anode is revealed by near-edge processing, confirming its importance in improving the interface quality for OLED applications. For example, the initial procedures are as follows: (1) placing the pretreated workpiece into an electrochemical mold with hard oxidation parameters of 1.34-9.38 A/dm2, 75 V, −4 °C, and 40–100 min; (2) transferring the oxidized workpiece to a deacidification bath (H2O, 60 °C, 5 min); (3) transferring the deacidified workpiece to a sealing bath (H_2O, 100 °C, 40 min); and (4) polishing the surface of the workpiece with 0.3 μm Al_2O_3 polishing powder. The actual images of the surface color transition from silver to black on AISI 1045 steel and thermal spraying of aluminum during hard oxidation at (a) 1 min, (b) 10 min, (c) 20 min, (d) 0.5 h, and (c) 1 h are shown in Figure 10.2c 1 h, and (f) 1.5 h at −4 °C with a final voltage of 75 V [5].

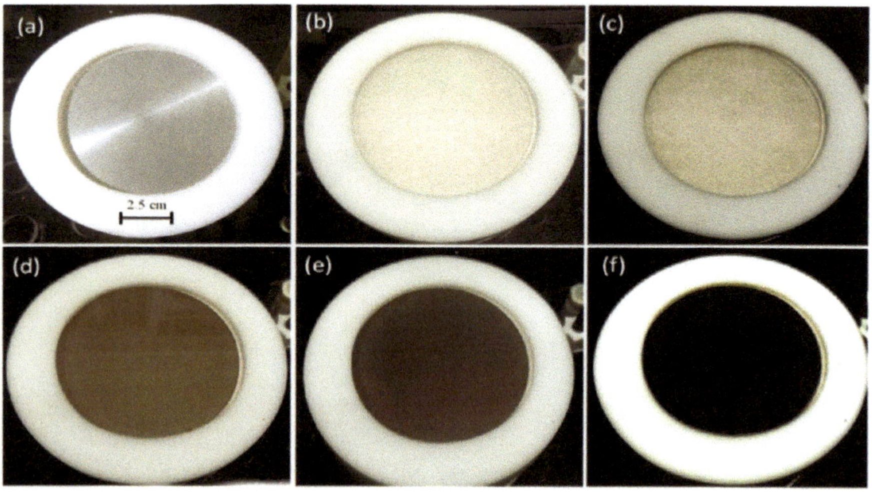

FIGURE 10.2 Actual photos of surface color change on AISI 1045 steel/thermal sprayed Al during hard anodization in (a) 1 min, (b) 10 min, (c) 20 min, (d) 0.5 h, (e) 1 h, and (f) 1.5 h [11] [open access].

10.3 ADVANCED METHODS FOR DETERMINING SURFACE PROPERTIES

When two or more immiscible materials with different mechanical and physical properties are combined, a composite is created and the final material has better mechanical properties. A wide range of substances of biological or synthetic origin fall under these criteria. Teeth, bones and plants are some examples of composite materials. Synthetic composites include reinforced concrete, glass fiber reinforced plastics and rubber fillers such as silica or carbon black. Currently, the term "composite" is generally used to refer to materials consisting of a continuous phase known as a matrix and one or more fibrous components. Thus, the term "carbon composites" refers to composites made from carbon fiber in general. In 1960s, the commercial production of carbon fibers began, leading to creation of carbon-based composites. Depending on the type of matrix, four main classes of carbon composites can be distinguished: carbon-ceramic composites, metallic carbon composites, carbon-polymer/plastic composites and carbon-carbon composites. We provide a brief overview of important advances in spectroscopic techniques made possible by development of new light sources such as synchrotron radiation facilities, which are tunable and modulated X-ray sources, and their use in chemical imaging.

Even if it is a possible to create intelligent nano organisms with amazing capabilities, their potential would be severely limited by the lack of sophisticated tools to characterize them at various length scales, from few micrometers to tens of nanometers. As a result, the electrical and structural descriptions of complex materials

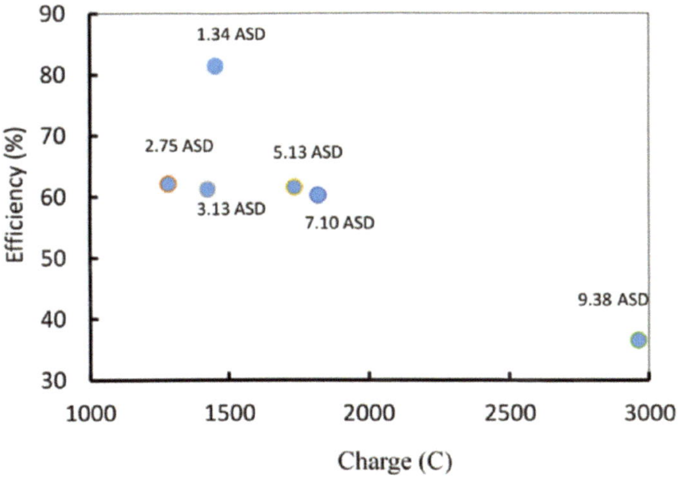

FIGURE 10.3 Plot of anodization efficiency related to charge with various current density [11] [open access].

become precise and refined. More precisely, their surfaces and subsurfaces have proven to be very complex, especially when phase separation involves new materials. These include but are not limited to, Mott-doped insulators, magnetic fields, composite materials containing nanocrystals or microcrystals grown on flexible or insulating precursors, and nano- or micro-patterned objects. Therefore, there has been a remarkable and rapid expansion of modern microscopic techniques. Even with angstrom spatial resolution achieved, coupling energy spectroscopic techniques with nanometric lateral resolution remains challenging [3–5]. Figure 10.3 shows a graph of the efficiency related to the charge and experimental current density (ASD). In general, the oxidation efficiency increases as the oxidation current density decreases. Due to the different sample sizes, the current applied and current density were calculated in practical oxidation. Time (which indicates cost) is another critical element in practical oxidation.

Plasma therapy has improved the surface shape at nanoscale, but the benefits are noticeably smaller. Most researchers have found that the presence of plasma increased surface defects and hardness values of some carbon materials, including fibers and nanotubes.

Furthermore, they suggest that increased duration of therapy may have a counter-effect, such as texture smoothing [5]. An incident photon beam focused on a sample allows the beam to be distributed throughout the material in XRD. A chemically rich characterization of material can be created by measuring photoelectron emission from each illuminated spot. In addition, virtually comprehensive records of core levels (CL) and their chemical composition can be obtained using scanning electron microscopy (SEM). However, the analysis of electronic states in the valence band using SEM becomes more challenging at photon energies, 250 eV where the valence state cross-section decreases significantly. This limitation highlights the need for advanced methodologies to effectively study electrical properties and improve our understanding of material behavior at the nanoscale.

The influence of inhomogeneity and confinement on local or partially nonlocal electronic states of valence band of surfaces and near-surface regions of modern materials is a major area of interest. Understanding electronic structures are important due to their electronic states (with binding energies below 20 eV) directly affect solid thermal and mechanical properties, chemical bonding and electrical transport [12–14]. Although the heat-treated composite film did not exhibit distinctly different nanoparticle sizes at elevated temperatures, > 300°C, this newly developed composite material may find application in various surface coatings. Figure 10.4 shows the results of morphological analysis performed on bare Ag and nanocomposite (Ag+ SiC) sheets. The bare Ag sheet is transformed into nanoparticles that are a quite large in size and structure, as shown in Figure 10.4a,b. On another hand, the composite coating demonstrated resistance to transformation into noticeably different-sized nanoparticles (Figure 10.4c, d). These results indicate that this composite material can withstand temperatures that are somewhat higher than those required to produce a more typical optical coating [5].

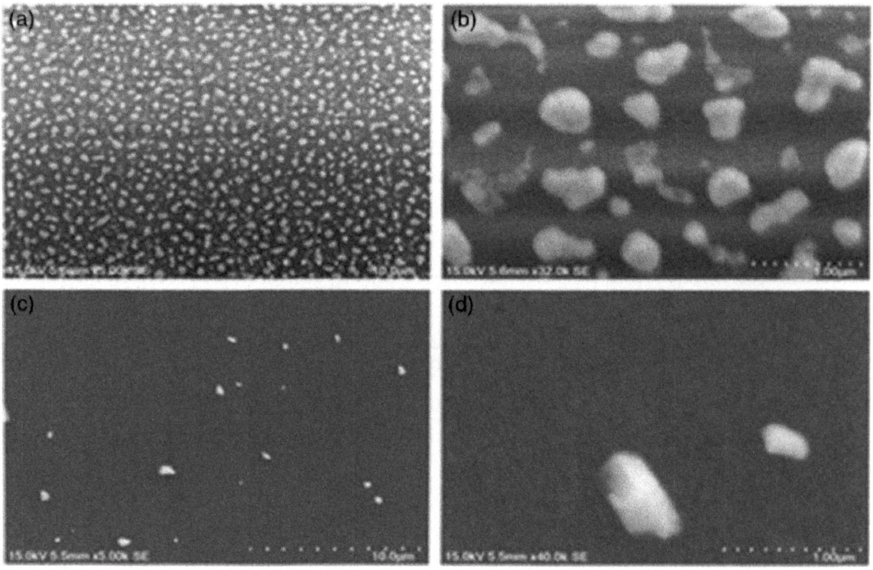

FIGURE 10.4 Morphological study of pure silver and pure silver mixed with 5% SiC. SEM images of thin Ag coating are obtained after annealing at 350°C for one hour (see a, b), and images of thin Ag + SiC composite coating are obtained after exposure to same temperature (c, d). Light deposits or contaminants on surface could cause surface defects in photographs (c, d). Reprinted from [110] is licensed under a Creative Commons Attribution Model (CCBY 4.0) license [5] [open access].

Infrared microscopy and AFM have been effectively combined to create a new type of image that contains chemical information at nanoscale. Biomaterials research has benefited greatly from using this innovative tool [15–17]. The polymer matrix restricts the temperature at which carbon fiber-reinforced polymers can be used. Alloys and light metals like aluminum can be utilized as a matrix. Carbon–carbon composites, in which the matrix and fiber are both composed of carbon, are the result of substituting carbon for organic resins. In an inert atmosphere, these composites exhibit remarkable refractory characteristics. However, in order to be used in hot air, they must be shielded from oxidation. In addition, ceramics and glasses are utilized as carbon composite matrices. In particular, the development of standard synchrotron beam (BL) is explored, in which the high energy and spatial resolution of angle-resolved photoemission spectroscopy (ARPES) could be combined with the nanoscale lateral resolution that many had long considered completely irrational. These reserves are due to very stringent angular and spatial DOF requirements associated with detecting typical ARPES geometries, as well as the difficulty of using small, thin Fresnel zone plates to focus XRD. Figure 10.5 illustrates how the geometric shapes of interface surfaces

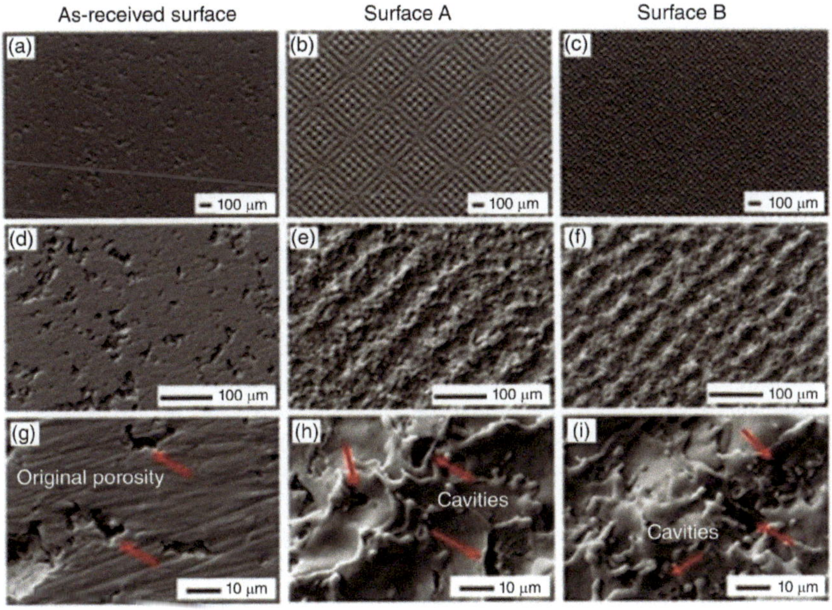

FIGURE 10.5 SEM images of processed substrates. The following surfaces are analyzed: (a) as-processed surface with 50 resolution; (b) surface A with 0.5 mJ ET and 50 resolution; (c) outer surface B with 5 mJ ET and 50 resolution; (d) surface before laser processing with 250 resolution; (e) outer surface A with 0.5 mJ ET and 250 resolution; (f) outer surface B with 5 mJ ET; (g) SEM scan showing the original porosity; However, (h) shows more cavities due to penetration of corrosion ions and (i) shows less cavities, which prevents the penetration of corrosion ions to the metal surface [5] [open access].

differ between the laser-cut and as-received interfaces. Significantly broken, wavy shapes of recycled and hardened material are visible on both treated substrates [5].

Since surface B is treated with higher energy applied to material (ET = 5 mJ), its roughness appears slightly finer than surface A. The square pattern created by laser beam mobility cross-hatching method is visible with increasing incident light intensity. Upon closer inspection, both treated substrates showed voids ranging from 1 to 5 nm in size at a higher resolution of 1500. Numerous metals including Ti, Zr, Co, Cr, Ni and their alloys serve as the basic building blocks for orthopedic and dental permanent implants. Unfortunately, infections caused by corrosion, excessive stress, or lack of osteogenesis are often the cause of failure of these biomaterial implants. To meet the extremely stringent standards of biomaterial surfaces for specific biomedical purposes, several surface modification techniques are widely used including ion implantation, thermal spraying, ultrasonic nanocrystal surface modification, glow discharge plasma, electrophoretic deposition, physical vapor deposition and coatings. These techniques help prevent corrosion of biomedical implants by enhancing interfacial adhesion and encouraging proactive behavior.

10.4 THE LATEST DEVELOPMENTS IN THE FIELD OF PACKET SYSTEM DEVICES

Large sites use the most powerful spectroscopic techniques available. Use of synchrotron radiation sources. Conventional beams effectively combine high transmission beam optics with one or more input devices. The effectiveness of the sprayed aluminum in preventing electrolyte penetration into the steel surface at varying current densities can be evaluated using micrographs of the interface between the carbon steel substrate and the aluminum spray, which are shown in Figure 10.6. It should be noted that the hot-rolled and thermally sprayed aluminum film may contain defects. An excessively high current density may result in a poor anodizing film with low hardness. (a) 1.34 ASD (A/dm2), (b) 2.75 ASD, (c) 3.13 ASD, (d) 5.13 ASD, (e) 7.10 ASD, and (f) 9.38 ASD were the current density values applied during oxidation. The hardness values were 524 compared to those of uncoated AISI 1045 steel (170 ± 8 HV0.1). Under current densities of 1.34, 2.75, 3.13, 5.13, 7.10, and 9.38 ASD, respectively, and ±23, 443±17, 421±20, 408±20, 353±14, and 302±16 HV1.0 [6].

These two properties are important for the efficient development of surface spectroscopy techniques and, more broadly, for characterization of new materials. In addition, with this installation, electronic analyzers with high power and angular resolution can be used without problems. Currently, these efficient photoelectron analyzers on the market can accurately detect chemical transformations with high resolution, detect resonant photoemission, detect fundamental levels with high resolution, and perform photoelectron diffraction studies. The surface roughness values of two-dimensional (2D) topography of PEDOT

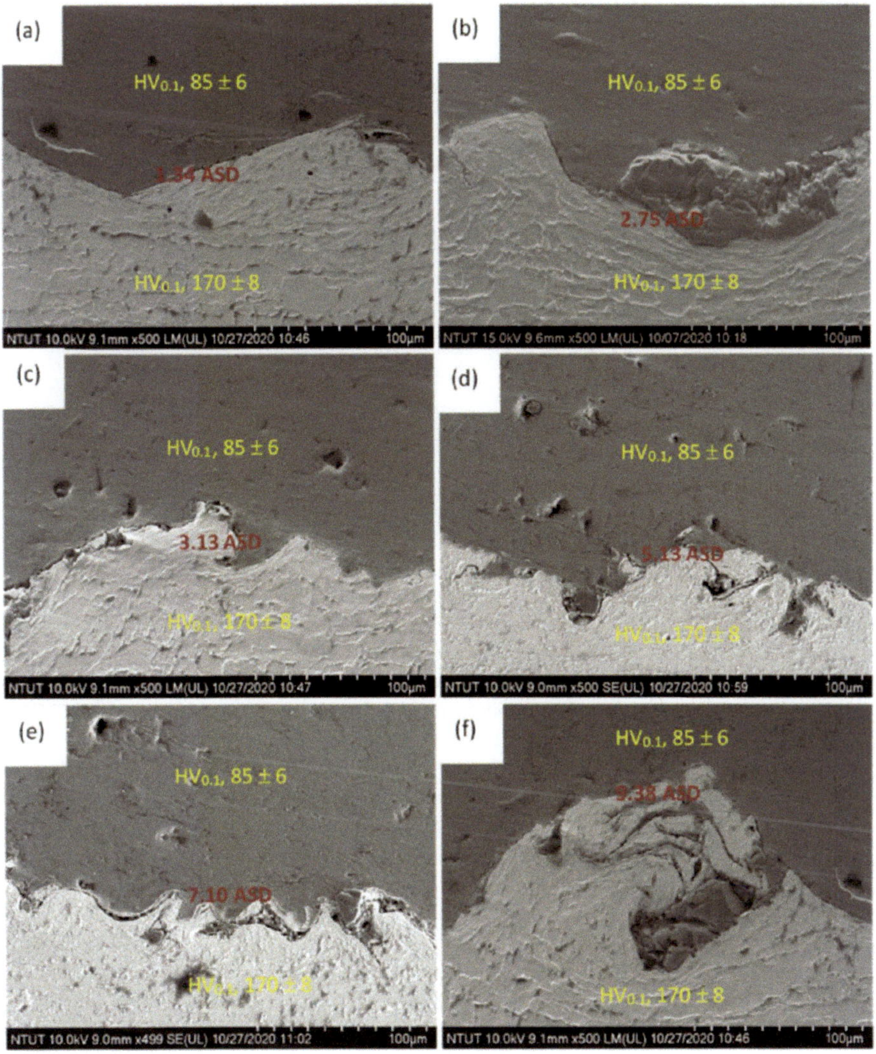

FIGURE 10.6 SEM images of AISI 1045/thermally sprayed Al interface after application of current densities: (a) 1.34 ASD, (b) 2.75 ASD, (c) 3.13 ASD, (d) 5.13 ASD, (e) 7.10 ASD, and (f) 9.38 ASD. Microhardness values ($HV_{0.1}$ for AISI 1045 and $HV_{1.0}$ for hardened surface) were given for comparison [11] [open access].

(poly(3,4-ethylenedioxythiophene)) film and PEDOT + CsI are measured to be 1.41 nm and 1.23 nm, respectively. These results indicate that the PEDOT:PSS film is successfully aligned and smoothed by the addition of CsI. The 2D and 3D topography results show consistency (Figure 10.6b, e). Moreover, the height profiles of each grain are obtained by AFM, as shown in Figure 10.6c, f. All these results indicate that the surface of PEDOT + CsI film is smoother and

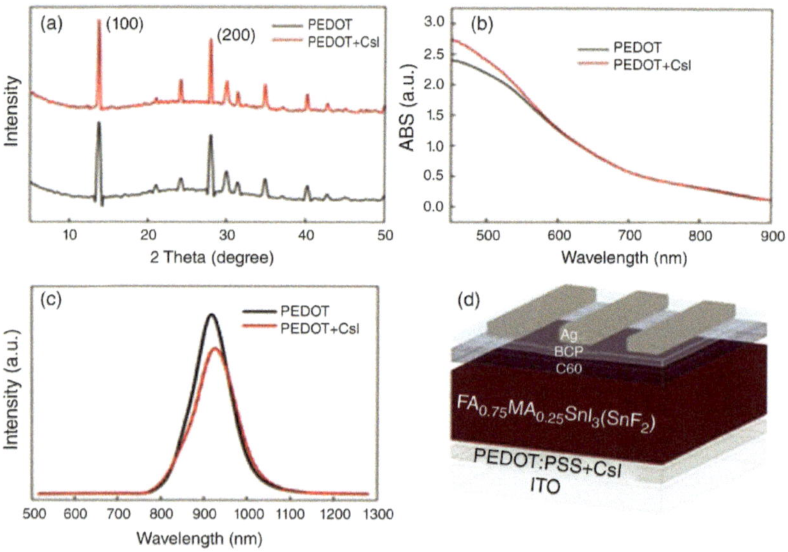

FIGURE 10.7 XRD and Spectral Analysis of Perovskite Films Deposited on PEDOT and PEDOT + CsI Films: (a): X-ray diffraction (XRD) pattern of perovskite films deposited on PEDOT and PEDOT + CsI films, showing that the addition of CsI does not significantly affect the bulk crystallization or composition of the perovskite film. (b): UV-visible absorption spectra of the perovskite films, comparing the absorption characteristics of films deposited on PEDOT and PEDOT + CsI films. (c): Photoluminescence spectra of the perovskite films, illustrating the light emission properties of films with and without CsI doping. (d): Sn-CsI-based PSC (perovskite solar cell) device configuration, showing the structural setup for testing the performance of perovskite films in solar cells [6] [open access].

more uniform, which should improve the interface contact and help with charge selection while suppressing recombination. How different concentrations of CsI affect the top layers of PEDOT:PSS perovskite is also investigated. X-ray diffraction (XRD) of perovskite films deposited on different PEDOT:PSS films (Figure 10.7a) shows that the almost identical locations and intensities of diffraction peaks for both films indicate that the addition of doped CsI to the PEDOT:PSS layer does not significantly affect the bulk crystallization or composition of the perovskite film. On another hand, it is documented that CsI dopant could create new perovskite phases by reacting with metal halide at PEDOT: PSS/perovskite interface. The passage of charge carriers could be facilitated and interface defects could be passivated by current interface phase [6].

The SEM images of Sn-based perovskite films prepared with different SnF_2 concentrations are shown in Figure 10.8. The addition of SnF_2 can improve the shape and crystallization of Sn-based perovskite films. Severe recombination occurs due to the pure perovskite film without SnF_2, which has a rough surface shape and inconspicuous grain boundaries (Figure 10.8a). The grain boundaries are visible when 0.02 M SnF_2 is introduced into the perovskite precursor solution, although the surface remained rough (Figure 10.8b) [6].

FIGURE 10.8 Top-view SEM images of Sn-based perovskite films prepared at four different SnF_2 concentrations: 0 M, 0.02 M, 0.03 M, and 0.04 M [6] [open access].

The addition of CsI can improve the dispersion of PEDOT: PSS in an aqueous solution to provide a more consistent film. Spectroscopy can be used to study metals, semiconductors, and even insulators with excellent lateral resolution and energy. BL units used for micro- and nanoimaging must operate with incredibly stable parameters. Improving functional performance and quality of surface characterization depends largely on sophisticated and accurate nanoscale measurement techniques. The two basic methods are hard measurement methods and soft measurement methods. Coordinate measuring devices, roundness tests, surface roughness measurements, interferometric techniques, confocal optical microscopy, scanning microscopy and nano-computed tomography are examples of advanced soft measurement techniques. However, to address characterization problems caused by surface texture and measurement methods, a new technical committee is being formed within ISO. This study reviews the main advanced soft measurement techniques obtained through optical, tactile and other instrumentation methods, as well as their classification schemes and application to engineering surfaces. In addition, upcoming developments are presented and discussed in this section to show proposed solutions to major problems that require scientific research. For nanodot research. Analytical rooms are equipped in the experimental stations. These defects are often depicted in POM plots as a sudden offset or blurry line that intersects the bands; in contrast, AFM image in Figure 10.9 offers a much more detailed perspective on the specifics of these

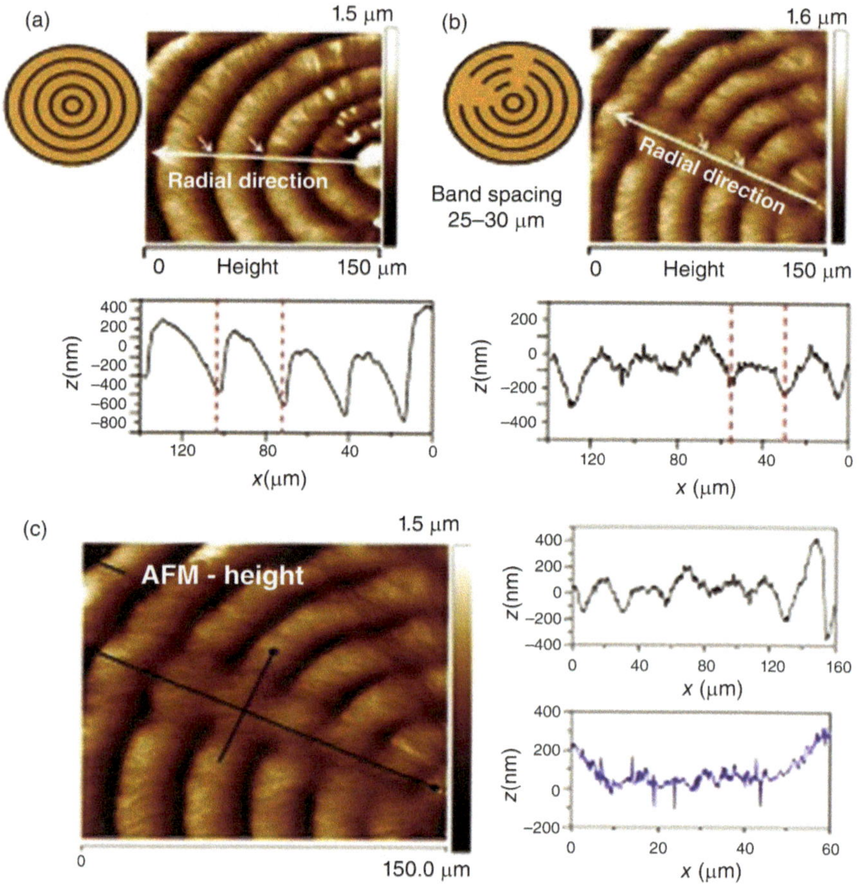

FIGURE 10.9 Comparison of AFM images and height profiles for P12T spherulites crystallized at 110°C, 2 wt% (film thickness: 1.5–2 μm), showing (a) dislocation-free bands versus (b) dislocation segments and (c) height profiles in the defect region of extinction ring bands [7] [open access].

defects than POM images. In addition to being a slightly offset intersection of two circular/lobed band sections that meet at an offset distance, the defect appears as a sharp convex arc within the band [3–5]. The offset distance is 10 μm for the inner bands and 1 μm for outer bands [the indicated sections in the AFM plots]. The offset is more significant for inner bands but becomes progressively smaller for outer ones. The area between the bands is seen as a "defect" in the ring bands. Since there is a periodic depletion or shortage of molten P12T chains to pack into the crystals, the defective sections may be the result of abrupt directional changes in the crystals. According to the results of height profiles on normal strips free of defects, a larger vertical difference of 750–1000 nm and a more orderly change in height up and down are observed [7].

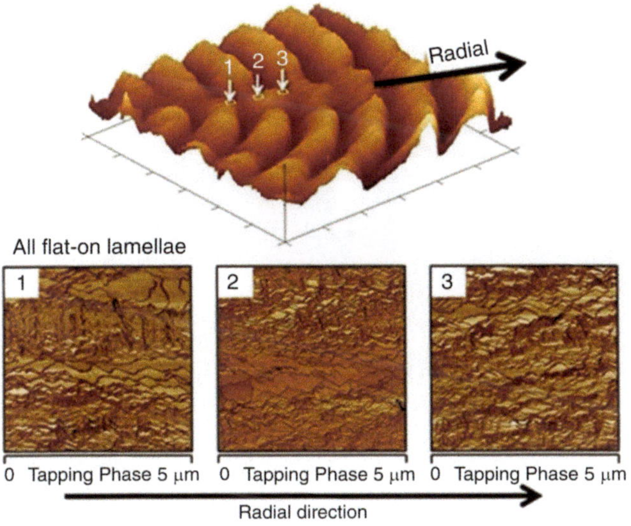

FIGURE 10.10 P12T crystallized at Tc = 110°C with dislocated segments in petal-shaped annular bands having AFM 3D profiles and phase images (film thickness ~1.5–2 μm). For enlarged examination, points 1-2, 3 represent three different locations of dislocation defects [7] [open access].

AFM analysis is performed on the dislocation band offset regions. Three locations (spots 1, 2, 3) of dislocation segments in petaloid ring bands of P12T crystallized at Tc = 110°C are depicted in Figure 10.10 (top) with 3D AFM profiles and (bottom) with corresponding enlarged phase images. The corresponding phase images of these three locations are displayed as enlarged micrographs, where the numbers denote the inner band, intermediate band and outer band along the band defect line. Region 1 represents the inner band, which has a dislocation offset of 8–10 μm. The regular band in the adjacent region-1 consists of a systematic stacking of single crystal plate assembly into a terrace-like structure according to AFM image; however, the offset in dislocation spot displays highly irregular crystals, which are completely smeared. The medium displacement dislocation (4-5 μm) is found in region 2. Except for displacement, which is smaller in region 2, both regions show similar characteristics. In comparison, region 3 consists of smallest displacement dislocation (1 μm or less). In region 3, the displacement region and the surrounding outer band regions show a much more ordered assembly of single crystal. Although not as much as in regions 1 or 2, the crystal assembly nevertheless shows some slight discontinuity due to concave protrusion at dislocation site [7].

Thus, the upper left corner of spherulites is the focus of defect analysis. The height profiles of dislocated regions are compared with the profiles of regular bands. The height profiles are compared with the nearby regular bands along the dislocated bands, either circumferentially or radially across the dislocated regions.

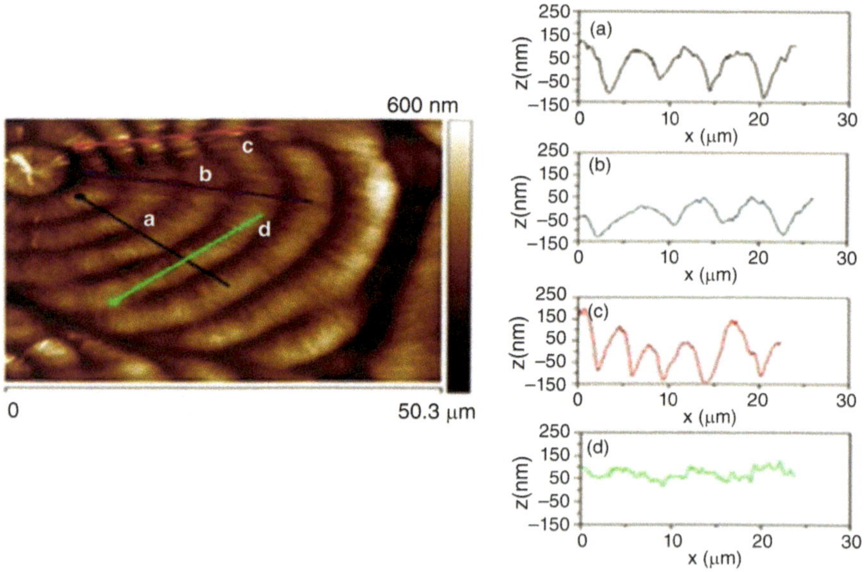

FIGURE 10.11 (a) Line profile showing 200 nm height variation. (b) Line profile with 100 nm height variation. (c) AFM profile showing bands with dislocations. (d) Inferred additional line profile. (e) Not mentioned in the provided context. [7] [open access].

Figure 10.11 shows the height pattern and height profile analysis of regular bands with localized dislocations in four cross-sections of lines a, b, c and d. This is illustrated by AFM height profile diagrams. The height profile radially across the regular bands is depicted by line a (black line), which shows a periodic height fluctuation, 200 nm. The blue line, or line b, represents a smoothed fluctuation in height profile, 100 nm [7].

In contrast, Figure 10.12a illustrates that the interband region, which is considered a "defect" between the annular bands, is where highly deformed and trapped lamellae abruptly drop into a crack with a vertical height difference of 260 nm. Periodic dehydration or lack of available molten P12T chains to pack into crystals may be the source of defect zones where the crystals drop down and abruptly change direction. Thus, the lamellae in spherulites grow in a discontinuous manner rather than continuously, with the length of each interruption varying depending on Tc, film thickness, top cover glass, until the spherulites collide, resulting in spherulites with extinction bands. On another hand, screw dislocation growth, where the crystals change their orientation, may be the source of intraband defect [7].

The SEM image of the surface of the AISI 1045 carbon steel substrate after hard oxidation and thermal spraying of aluminum is shown in Figure 10.13. The stack structure of AISI 1045 carbon steel, thermal sprayed aluminum, and the hard oxidized film is shown in Figure 10.13a. The SEM plot of AISI 1045 steel is shown in Figure 10.13b. Fe, Al and O are the representative elements of

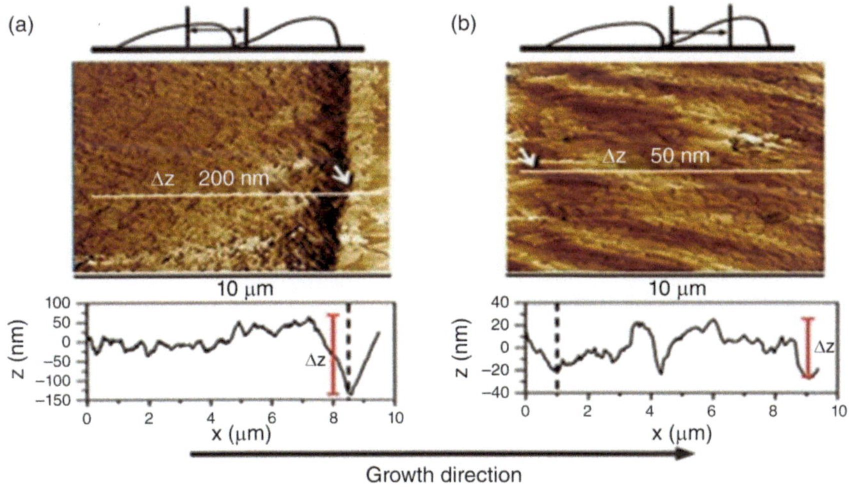

FIGURE 10.12 AFM analysis performed on interband regions. (a) Gradual increase at beginning of subsequent band development cycle; (b) deposition into crack at end of growth cycle (260 nm vertical) (film thickness = approximately 1 μm) [7] [open access].

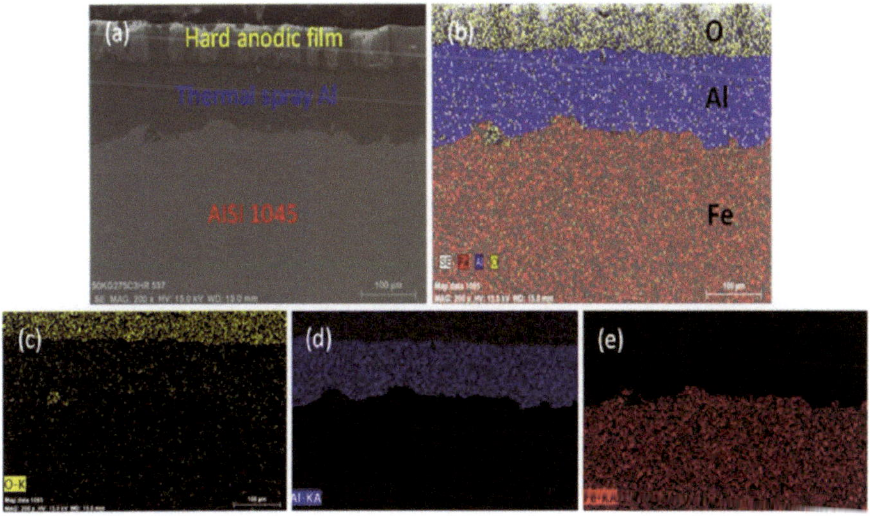

FIGURE 10.13 SEM Images of AISI 1045/thermally sprayed Al/hard anodic film interfaces; (a) hard anodic film on thermally sprayed Al/AISI 1045, (b) SEM mapping of AISI 1045/thermally sprayed Al/hard anodic film with (c) O mapping, (d) Al mapping, and (e) Fe mapping, respectively [11] [open access].

steel, thermal spray and hard oxidized film, respectively. The small amount of O present at the Al-Fe interface in the image can be attributed to the local oxidation of the surface of the carbon steel substrate during the arc spraying procedure. The SEM plots of individual oxygen (O), aluminum (Al), and iron (Fe) in Figure 10.13b are shown in Figure 10.13c–e [7].

The trend of platelet morphology change across the stripe is generally similar for P12T films of different thicknesses (1–2 μm vs. 3.5 μm). Larger and more widely spaced single crystal platelets, whose packing becomes more strained near the peak maximum as the stripe packs are characteristic of initial expansion. However, the lamella shape changes for thicker P12T films (~3.5 μm) in this study are more noticeable compared to the thinner films. Each lamella in region 1 is at least 3 μm wide, and lamella thickness is several tens of nanometers. The radial direction is indicated by white arrow in these phase images.

The angle-resolved photoemission spectroscopy microscope has a special preparation chamber with gas lines and a fast entry chamber that makes it easy to quickly transfer samples from room temperature to ultra-high vacuum (UHV) environments. The system consists of two chambers; one for camera and another for analysis, which contains the optics for focusing on nano-spots and the electron analyzer. To provide accurate scanning of samples at nanoscale, interferometric control is used to reduce mechanical vibrations and thermal drift. The interferometric system compensates for parasitic displacement and stabilizes temperature fluctuations to less than 0.1°C (Figure 10.13).

The working principles and applications of nanoARPES microscope and conventional spectroscopic instruments, including X-ray diffraction (XRD), are important points of distinction. While the major focus is on studying the crystal structure by analyzing the photoelectron emission, nanoARPES measures XRD and provides precise insights into the electronic structure and surface states of materials. Because of this feature, nanoARPES is particularly useful for studying complex materials and interfaces at nanoscale, providing new perspectives that may be missed by conventional techniques. The advantage of k-microscopy is that samples are positioned at nanoscale in the x, y and z axes due to they are mounted on high-resolution dish. It is possible to scan both the azimuthal angle and polar angle automatically within a range, 90 degrees. Spectroscopic mapping and nanodot spectroscopy are two main operating modes of k microscopes currently on the market. Therefore, the precise capture of XRD spot location on the sample in real-time using interferometry is automatically converted into a digital signal that adjusts the scanning position of sample through a feedback system [13, 14, 18].

Surface metrology has gained great importance in many scientific and industrial fields. The study of surface measurement and dimensions is becoming increasingly common in a variety of settings, including research and data processing according to standardized rules. Therefore, surface characterization using sophisticated and precise nanometric techniques is critical, especially in the fields of industrial engineering, criminology, tribology and biotechnology. This diversity has led to development of increasingly complex measurement techniques, each with its specific application, which is used to analyze surface characteristics using pen, optical and non-optical methods [19]. In the area of dimensional and geometric specifications and product inspections, ISO Technical Committee TC-213 has established Working Group WG-16 to address standardization of surface texture (3D) and measurement methods, as well as review existing standards for traditional profiling methods (2D), including Here are the properties of the tool.

The task of this working group is to develop recommendations for three main methods for measuring surface roughness [20]. The line profiling method has used a high-resolution sensor to detect peaks and valleys in a terrain surface for obtaining a quantitative Z(X) profile of surface. The method of defining linear features in three dimensions is extended by cadastral topography techniques, which often involve quantitative topographic photography or reconstruction of a series of parallel projections of patterns. It is very important to remember that some spatial projection methods can determine the Z height based on the X and Y coordinates.

Visible topographic images, on another hand, contain a series of parallel Z(X) profiles with relatively random heights in direction Y. It allows movements of more than few millimeters to ensure precise sample placement. In addition to the lateral movements, this rough top block also has polar and azimuthal angular drivers mounted to adjust the angle at which the excited photoelectrons exit and angle at which the samples are azimuthally.

10.5 NEW DIRECTIONS IN THE STUDY OF SURFACE SCIENCES

There are difficulties in surface analysis caused by the emergence of new types of nanostructures, which are essential for the development of future semiconductor chip manufacturing and offer opportunities for solving biological problems, especially those related to the structure and function of protein molecules. Research into "nanostructured surfaces" is expected to require the development of new surface analysis methodologies, as well as the introduction of new techniques, especially when considering molecular systems sensitive to radiation damage or biological systems in aqueous environments. Specific proposals are made for potential developments and energy loss spectroscopy scanning is one example of a new approach to surface research.

The ideal 2D system is graphene, an isolated monolayer of graphite. The new electrical system demonstrates several distinctive transportation features. This substance and its derivatives have been the subject of a large amount of research. A variety of techniques have been used to study single, including Raman scattering, infrared diffraction, transmission electron microscopy and others. The relationship between CL and reactivity has been demonstrated by excellent ARPES and nano-ARPES studies.

AFM probe scans the sample topography in the first round of electrophoretic force microscopy (EFM) phase. In the second round, DC bias is given to the probe or graphene sample along the same scanning line. The probe can re-trace

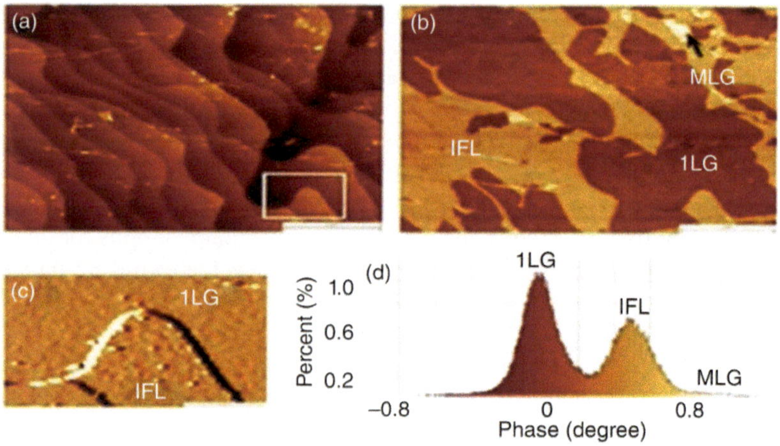

FIGURE 10.14 Typical area of epitaxial graphene on 4H-SiC(0001): The following images are shown: (a) Topography (z-range: 13 nm); (b) corresponding EFM phase image (z-range: 2°); (c) topography deviation error signal indicating adsorbates on the interface layer (IFL) compared to pure 1LG; (d) histogram of EFM phase image, which displays the 1LG (which accounts for 40% of the total area) and IFL regions. MLG accounts for less than 2% of the total area [21] [open access].

the topography using this method at a constant lift height above the surface. We perform EFM at room temperature and at increasing temperatures in our analysis (Figure 10.14) to produce potential difference maps of epitaxial graphene on 4H-SiC (0001). It is possible to clearly distinguish between graphene layers of different thicknesses thanks to the quantification of these potential difference maps [21]. The graphene layers of our samples on carbon-terminated single-crystal silicon and silicon carbide are typically generated using one of following techniques; single-layer (1) multilayer graphene grown on both overlapping and non-overlapping layers, (2) overlapping graphene produced by chemical vapor deposition (CVD) on copper foils, and (3) exfoliated graphene incorporated into artificial heterostructures [22]. This comprehensive approach highlights the adaptability of graphene fabrication methods and advances our understanding of its electrical properties on different substrates.

Figures 10.15 and 10.16, respectively illustrate the torsional resonance conductivity atomic force microscopy (TRCAFM) current map and torsional morphology of graphene grown at 2000°C in argon environment at on-axis (6H-SiC(0001)) and 8° off-axis (4H-SiC(0001)). A comparison of shape maps between the two images reveals that smoother surface (RMS = 2.4 nm) is obtained in on-axis sample, but significant gradual aggregation is found in off-axis sample. Moreover, the current maps are significantly different. The graphene layer on-axis substrate shows a uniform current over the entire scanned area, indicating homogeneity at nanoscale, as can be shown by comparing Figures 10.15b and 10.16b [21]. Wrinkles in the graphene coating appear as bright lines in the shape map [9] They

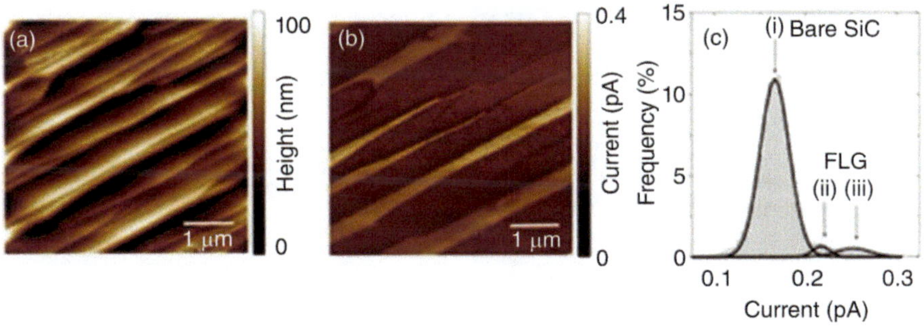

FIGURE 10.15 (a) AFM image, (b) AFM conductivity map, (c) Histograms, showing three investigated regions: A region devoid of graphene (curve (i)). A region covered by graphene (curves (ii) and (iii)). [21] [open access].

arise as a result of mismatch in thermal expansion of SiC and graphene during sample cooling and high current density [21].

The Moon and Mercury are the most suitable and well-studied examples of airless rocky bodies in the Solar System; However, asteroids and moons of Mars are subjected to similar interactions; The main difference is the effect of gravity. Although the absence of dense atmosphere on Moon is known before the 1970s, the Apollo missions provided the first evidence of the presence of Ar and He in lunar exosphere. Sodium and potassium are discovered by land-based methods over the next ten years. A potential material for the next generation of quantum Hall resistance standards is epitaxial graphene on silicon carbide. Due to their excellent breakdown current density and high accuracy, quantum Hall resistance standards based on epitaxial graphene on SiC have already outperformed the most advanced GaA standards currently in use. To evaluate the quality of available material, 100 arrays of Hall rods connected in parallel on top of epitaxial graphene are fabricated (Figure 10.17). At magnetic fields from 7 to 9 T, one out of four devices have showed a quantized resistance that corresponded to an

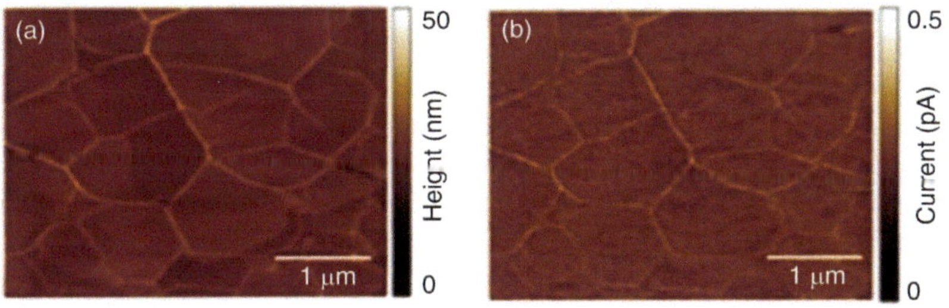

FIGURE 10.16 TR morphology (a) and current map (b) were obtained on the axial sample [28] [open access].

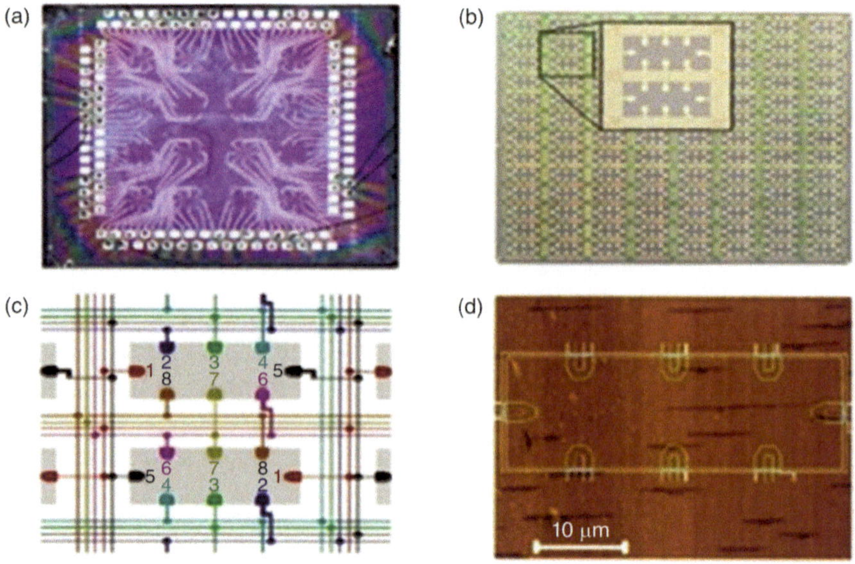

FIGURE 10.17 (a) Photomicrograph of entire chip. (b) Photomicrograph of 100-Hall bar array. (c) Part of schematic of the array. There is a lot of redundancy in arrangement of inter-correction wires. (d) Graphene substrate in standard AFM phase image with Hall bar drawn on top. The bilayer regions are the dark areas [21] [open access].

accurate R-K/200 value within a measurement accuracy of 10(-4). The main reason for the poor performance of other arrays is uneven doping [21].

Lunar sensors have been detected near the Moon by WIND spacecraft's Ultra thermal Ion Spectrometer (STICS) and more recently by Kaguya observations. The successful ground-based observation of Na during the arrival of an interplanetary coronal mass ejection (CME) at Mercury, recorded by MESSENGER on September 20, 2012, confirms a strong connection between exospheric Na variability and surface plasma forcing. Indeed, the distribution of Na in the exosphere shifted during this observation, possibly as a result of the compression of daytime magnetosphere by thick iCME plasma [23]. When stacked, the carbon atoms of two adjacent layers are ideally aligned in the same lateral directions. Each layer in ABC layout is offset horizontally from the layer below it by an amount. Three-quarters of the diagonals of lattice unit cell are stacked on top of each other, and the rotated graphene layers are stacked so that the rotation angles between them differ from AB stacking. There are differences in the structural and electrical properties of graphene obtained on Si and C faces of SiC. Continuous carbon accumulation layers AB (Bernal) are shown. They are smeared with continuous diffraction rings on face C, but with preferential rotation angles, indicating the presence of islands with different azimuthal orientations or adjacent layers rotated. The source link is not found. The surfaces and orientation of the islands vary greatly. For each islet type, size, distribution and substrate assessment are determined.

The most interesting result is that the speed of Bernal islands in multilayer graphene grown on faceted SiC crystals is equal to the speed of graphene grown on faceted SiC crystals [24, 25]. On another hand, the films in first scenario consist of islands of different thicknesses and orientations, while second, one direction is detected.

The new state of matter is called a topological insulator (TI). These materials have gapless surface states, where strong spin-orbit coupling locks the electron spin perpendicular to its velocity, as well as a full dielectric gap in the bulk. The interesting areas of physics, complex spintronics devices and quantum computing [22] the main methods used to study 3D TI and topological surface states. Topological nanobodies are shown in Figure 10.9. There is no reference source. NanoARPES promises to be a new tool for studying electrical structures [26].

Latest developments in ultra-thin layer production of innovative low-dimensional physics have studied to date include anomalous quantum Hall effects in single-layer graphene and massless Dirac fermions in materials down to the unit cell thickness. Another class of materials that exhibit new physical phenomena due to d-electron interactions is layered transition metal dichalcogenides. Typical transition metal chalcogenides are MoS_2 and its derivatives. In addition, individual atomic levels can also be combined in a specific order to create engineered heterostructures. The first, highly complex examples of these heterostructures (often called "van der Waals") are created and studied, revealing specific properties and new phenomena.

The unique features of van der Waals family materials, resulting from the ability to thin a 3D crystal into 2D layers, have attracted particular research interest in recent years. After the "appearance of graphene" [15], this family is replenished with semiconductors (MoS_2 [23], $MoSe_2$), metals ($TaSe_2$ [27]) and superconductors ($NbSe_2$ [28])), insulators (e.g. hexagonal boron nitride [17]), semiconductors (e.g., MoS_2 [22]). In addition, new interfacial phenomena such as moiré physics [15] or band structure alignment [21] or proximity effects [16] are created when different van der Waals materials are combined by stacking individual layers vertically to create so-called heterostructures. It is desirable to tune and exploit their properties to develop new tools and new applications. Interestingly, graphene on the graphene template is a flexible material with a high density of surface and edge nucleation centers of the class of transition metal dichalcogenides (TMDs) is obtained [8].

MoS_2 is perhaps the most researched TMD. However, in recent years, $MoSe_2$ has emerged as a better choice in a number of applications. For example, $MoSe_2$ has been shown to exhibit greater activity in the hydrogen evolution reaction [29], charge carrier mobility [14] and internal quantum efficiency [16] compared to MoS_2. Moreover, $MoSe_2$/graphene heterostructures have an almost ideal bandgap for solar cells [18], making them a promising option for use in green technologies, including photovoltaics, low-power electronics and the hydrogen economy [19]. This is due to the exceptional properties of graphene and the success of its production: from small flakes when mechanically laminated to large (centimeter) sheets when applied to polycrystalline copper foil. According to theory, individual strips and sheets of silicene, it will remain stable in both small folded and flat configurations and exhibit essentially the same electrical properties as graphene in both scenarios.

The expected electrical configuration of silicene shows bands at the corners of Brillouin zone diagram that create Dirac cones. Silicon electronics. Using ARPES, the electrical properties of silicene, a promising material are tested and theoretically predicted. However, there is still much debate about how to fully understand the experimental data for these types of 2D materials using different theoretical frameworks [21, 23]. Microelectronic systems using nanomechanical systems (MEMS) or microelectromechanical systems, and due to their great adaptability, nanoelectromechanical systems (NEMS) have attracted the attention of scientific and technical communities.

Many other important factors affect the stability and efficiency of modern technologies. Exotic matter features arise from correlations induced by electronic interactions, which are crucial to many modern condensed matter problems and become even more important at lower dimensions. Classic examples of binding effects that appear under certain conditions on the surface of large crystals are the 2D electron gases found on the surfaces of oxides such as $SrTiO_3$. Due to the inherent inertness of graphene and the high vapor pressure of corresponding dichalcogenides, most heterostructures have been generated by CVD [20] at pressures that often exceed UHV or even higher bottom conditions. Preparation using peel and stack [27], limits the ability to characterize growth in-situ. However, as we have showed previously for MoS_2/Au(111), understanding the nucleation and growth dynamics is crucial to understanding their effect on the structure and electrical properties [28]. However, a comprehensive description of in-situ growth processes of TMD/graphene heterostructures is rare in the literature and limited to a small number of real-time scanning tunneling microscopy (STM) nucleation process descriptions [22], and XPS studies of MoS_2 [22] is an effective method to monitor the growth process in-situ [16]. It can acquire images across space as well as full field-of-view images at a rate of one to several images per second with a lateral resolution of only a few nanometers. For imaging, low-energy electron diffraction (LEED) uses elastically scattered electrons with input kinetic energies ranging from 0 to several tens of electrons. This ensures maximum surface sensitivity [17]. In addition, thanks to the flexibility of the optoelectronic mount, LEED patterns can be measured by imaging the back focal plane of the lens. This provides structural information about the periodic arrangement of atoms and surface symmetry, among other things, at the atomic level due to the diffractive nature of backscattering. By limiting the illuminated area using microscopic diffraction (μLEED), this information can be obtained locally using an electron beam.

MEMS and NEMS are attracting interest from the science and technology sector. Their small size allows them to have much higher resonant frequencies at high-quality factors, much lower mass and much higher force sensitivity than any equivalent electrical circuit. MEMS is a small device that combine mechanical and electrical components, often less than 1 mm in length but greater than 100 nanometers in length. While, NEMS is a nanoscale device that integrate mechanical and electrical components and has a typical length less than 100 nm. These are integrated electrical and mechanical components called miniaturized intelligent systems that include sensing, processing, and/or actuation functions. The acronym MEMS appeared for the first time in the United States. "Micromachining" is the name used in Japan, and "microsystem technology" (MST) is the term commonly used in Europe.

The phrase "micro/nanodevices" is often used. The terminology used in MEMS/NEMS is now widely used and includes mechanical, optical, fluidic, electrical and/or biological activity. Micro/nanoelectromechanical systems (MOEMS/NOEMS) are MEMS/NEMS used in optical applications. RF MEMS/NEMS or RF-MEMS/RF-NEMS is the term used to describe MEMS/NEMS for electronic applications. BioMEMS/BioNEMS is a term used to describe MEMS/NEMS in biological applications. MEMS and emerging nuclear power circuits, similar to the influence of semiconductor technology, information technology, or cellular and molecular biology are expected to have a significant impact on our lives [15, 22, 23, 26]. System applications include electromechanical, electrical, chemical, biological and information/communications. Moreover, understanding the related problems entails the interrelationship between electronic and magnetic instability, the influence of constraints and chaos in electronic structure, etc. The main strength of many university surface science laboratories is fundamental research into principles of superconductivity of greatest concern are unconventional superconductivity. Electricity eliminates carbon emissions from industrial and household appliances while providing flexibility, efficiency and ease of use. What to improve government efficiency and effectiveness, oversight and management must be improved.

The expected progress is driven by advances in materials production, transportation transformation and energy storage, with the latter being the weakest element. The surface science community has developed several ways to overcome new technical obstacles, including developing systems, devices and components that take advantage of innovative materials and physical principles. With the support of national and international funding agencies, scientists and industry partners are developing technological advances that are part of this innovation synergy. Nanotechnology is fundamental to improving the efficiency of light conversion in the field of nanophotonics for energy production as both nanoparticles and nanostructures improve light harvesting and conversion [16].

In many laboratories, the three main components of biophysical activity are measurement, modification and manipulation of cells. Many projects are being developed that require collaboration in chemical, biological and physical research. This interdisciplinary research is beginning to take shape and is being conducted in several laboratories. For example, ambient pressure photoemission (APPES) can reduce artifacts and provide information about the surface chemistry of biomaterials under near-real-life conditions, solving the problem associated with nanoparticle solution preparation [22]. Surface science research now extends beyond the laboratory, increasing its impact on industry and delivering stronger, more reliable materials. However, the future Surface science has many challenges. The most plausible theory is that he will leave.

Almost all elements of life will be unaffected and widespread adoption is expected by 2025. Future 2D materials such as graphene, silicene and others will enable structures that are much lighter but stronger than is possible with carbon fibers. These materials will be frequently used in many new products requiring precise surface properties. The main challenge is how to create complex materials at low cost while maintaining their stability and ease of production control. Knowledge of atomic structure is the way to achieve this goal, the structure of surface and processes occurring, and how they differ from the corresponding features of bulk matter. It is therefore essential that surface science activities contribute to these new research topics [20]. Surface Physics marks the end of the past 25 years and beginning of next millennium in surface physics. In the context of condensed matter physics, this study aims to identify guiding directions for surface physics for the future. The bottom line is that surface sciences and surface physics in particular, must undergo a cultural revolution to become a science-oriented field rather than a technology-oriented field. Surface physics has many exciting prospects, but to take advantage of them, our culture and more importantly, the way we train the next generation of surface scientists must change. We highlight some of the opportunities that arise in complex, highly interconnected, low-dimensional systems to highlight the critical role played by surface physics that can be exploited in modern condensed matter physics.

10.6 USE OF SCANNING MICROSCOPES TO STUDY NANOSTRUCTURED MATERIALS

Here are some of the key innovations that contributed to the dominance of the field; in 1980s, the STM, AFM and scanning probe microscopes (SPMs) are developed to allow targeted manipulation objects. Single atoms and visualization at the atomic level, as well as exploration of a range of physical properties at the nanoscale. The discovery and control of production of three small and low-dimensional materials. We summarize some recent advances in SPM in the areas of nano photovoltaics characterization, the study of graphene-based nanostructures and atomic-scale imaging and manipulation. Finally, we show how nanoscience is penetrating even traditional, "legacy" businesses, such as papermaking and petroleum engineering, that are researching and using nanotechnology [21].

For example, a modified H-shaped STM tip can be used to separate or separate helical molecules. Since surface molecules are not strongly attracted to this tip, repulsive forces can be applied to perform manipulations. Modified systems are less accurate due to chiral effects, but high-resolution images can be obtained by removing hydrogen atoms, making chiral structures easier to see.

Studies reveal that while molecules with opposite chirality interact, molecules with the same chirality do not bind to each other [21]. Single atoms have been successfully modified using STM [22], suggesting that these limitations can be partially addressed using an automated STM-based vacancy modification method on a low-chirality Cl (2 × 2) superstructure on a Cu(001) surface [28]. Density functional theory (DFT) presents that for Cl atom to reach a vacancy in the superstructure, it must cross an energy barrier, 0.3 eV. Precise control over the passage of these energy barriers through data storage sites is made possible by automated STM manipulation. In addition to their sensing capabilities, smart materials can also interact with stimuli by activating them or taking corrective actions; Their self-healing properties make them very useful in biological applications. Smart materials such as chromogenic, ferroelectric, magnetic, piezoelectric, thermoelectric and shape memory alloys are prominent examples of nanotechnology developments that demonstrate the enormous potential of these technologies in a variety of sectors [4].

Technology and science rely heavily on advanced interdisciplinary research into nanoscale surfaces, structures and subsurface phenomena. The importance of nanoscale research is underscored by recent advances in atomic characterization, electrical structure analysis and creative applications. These developments not only expand our understanding of science, but also open the door to new technical advances, particularly in the fields of materials science, energy and medicine. Interdisciplinary collaboration will be critical to overcoming barriers and achieving progress in surface engineering and nanoscale phenomena as research advances.

10.7 PROBLEMS

1. Describe the rationale for recent developments in surface engineering and nanotechnology of flexible materials.
2. Provide several sophisticated methods for understanding the tribological properties of surfaces. Describe the possible consequences if cutting failure occurs in these areas, such as wear damage or lubrication failure.
3. Describe recent advances in packet system hardware.
4. Why is there growing interest in recent developments in nanomaterials research?
5. What motivates people to create smart materials using nanostructures?

REFERENCES

1. Cordeiro, J.M., Nagay, B.E., Dini, C., Souza, J.G., Rangel, E.C., da Cruz, N.C., Yang, F., van den Beucken, J.J., & Barão, V.A. (2021). Copper source determines chemistry and topography of implant coatings to optimally couple cellular responses and antibacterial activity. *Biomater. Adv*, *134*, 112550.
2. Wu, H., Xie, L., He, M., Zhang, R., Tian, Y., Liu, S., Gong, T., Huo, F., Yang, T., & Zhang, Q.; et al. (2019). A wear-resistant TiO_2 nanoceramic coating on titanium implants for visible-light photocatalytic removal of organic residues. *Acta Biomater.*, *97*, 597–607.

3. Ali, A., Chowdhury, S., Janorkar, A.V., Marquart, M., Griggs, J.A., Bumgardner, J.D., & Roach, M.D. (2023). A novel single-step anodization approach for pani-doping oxide surfaces to improve the photocatalytic activity of titanium implants. *Biomed. Mater.*, *18*, 015010.
4. Sohn, S., Lee, J., Park, K., Jung, D., Kim, H., & Yi, J. (2008). Enhanced performance of an organic light-emitting device by using an oxidant component during a surface reaction process by using atomic layer chemical vapor deposition. *J. Korean Physical Soc.*, *53*(6), 3416–3421.
5. Thakur, A., Kumar, A., Kaya, S., Marzouki, R., Zhang, F., & Guo, L. (2022). Recent advancements in surface modification, characterization and functionalization for enhancing the biocompatibility and corrosion resistance of biomedical implants. *Coatings*, *12*(10), 1459.
6. Sun, Y., Song, Y., Liu, M., & Zhang, H. (2024). Synergistic modulation of Sn-based perovskite solar cells with crystallization and interface engineering. *Molecules*, *29*(11), 2557.
7. Woo, E.M., Lugito, G., Yang, C.E., & Chang, S.M. (2017). Atomic-force microscopy analyses on dislocation in extinction bands of poly (dodecamethylene terephthalate) spherulites solely packed of single-crystal-like lamellae. *Crystals*, *7*(9), 274.
8. Visser, N.L., Verschoor, J.C., Smulders, L.C., Mattarozzi, F., Morgan, D.J., Meeldijk, J.D., van der Hoeven, J.E., Stewart, J.A., Vandegehuchte, B.D., & de Jongh, P.E. (2023). Influence of carbon support surface modification on the performance of nickel catalysts in carbon dioxide hydrogenation. *Catal. Today*, *418*, 114071.
9. Sun, X., Shao, X., Yi, J., Zhang, J., & Liu, Y. (2022). High-efficient carbon dioxide-to-formic acid conversion on bimetallic PbIn alloy catalysts with tuned composition and morphology. *Chemosphere*, *293*, 133595.
10. Kübler, M., Wagner, S., Jurzinsky, T., Paul, S., Weidler, N., Villa, E.D.G., Cremers, C., & Kramm, U.I. (2020). Impact of surface functionalization on the intrinsic properties of the resulting Fe–N–C catalysts for fuel cell applications. *Energy Technol*, *8*, 2000433.
11. Chiang, P. C., Chen, C. W., Tsai, F. T., Lin, C. K., & Chen, C. C. (2021). Hard anodization film on carbon steel surface by thermal spray and anodization methods. *Materials*, *14*(13), 3580.
12. Mncwabe, Z., Farahani, M.D., & Friedrich, H.B. (2020). Switching between oxidation types using molybdenum phosphate catalysts for paraffin activation using doped fe as surface acidity modifier and MoO_x as an oxygen insertion tool. *Catal. Lett*, *150*, 728–737.
13. Gao, X., Wen, Y., Tan, R., Huang, H., & Kawi, S. (2023). A review of catalyst modifications for a highly active and stable hydrogen production from methane. *Int. J. Hydrogen Energy*, *48*, 6204–6232.
14. Zhang, Y., Zhao, Y., Hu, W., Wang, X., & Yang, R. (2023). Microstructure and surface engineering through indium modification on Ni-rich layered cathode materials for enhanced electrochemical performance of lithium-ion batteries. *J. Alloys Compd*, *934*, 167862.
15. Mutharani, B., Ranganathan, P., Chen, T.-W., Chen, S.-M., & Vinoth Kumar, J. (2020). Thermoreversible switchlike electrocatalytic reduction of tizanidine based on a graphene oxide tethered stimuli-responsive smart surface supported pd catalyst. *Anal. Chem.*, *92*, 8965–8973.
16. Li, H., Ai, Z., Yang, L., Zhang, W., Yang, Z., Peng, H., & Leng, L. (2023). Machine learning assisted predicting and engineering specific surface area and total pore volume of biochar. *Bioresour. Techno.l*, *369*, 128417.

17. Aghazadeh, M., Ganjali, M.R., Morad, M.M., & Gharailou, D. (2020). Saccharide-capped superparamagnetic copper cations-doped magnetite nanoparticles for biomedical applications: A novel and simple synthesis procedure, in-situ surface engineering and characterization. *Curr. Nanosci.*, *16*, 770–778.
18. Hamidian, K., Rahimi, R., Hosseini-Kharat, M., & Alavi, M. (2021). Development of the molecular engineering of disazo dye sensitizers and TiO_2 semiconductor surface to improve the power conversion efficiency of dye-sensitized solar cells. *J. Photochem. Photobiol. A Chem*, *418*, 113408.
19. Sun, X., Jing, M., Dong, H., Xie, W., & Luo, F. (2023). CuO-ZnO submicroflakes with nanolayered Al_2O_3 coatings as high performance anode materials in lithium-ion batteries. *J. Alloys Compd*, *953*, 170137.
20. Wang, C., Gann, E., Chesman, A.S.R., & McNeill, C.R. (2019). Residual solvent additive enables the nanostructuring of PTB7-Th:PC71BM solar cells via soft lithography. *AIP Adv*, *9*, 065024.
21. Yazdi, G.R., Iakimov, T., & Yakimova, R. (2016). Epitaxial graphene on SiC: A review of growth and characterization. *Crystals*, *6*(5), 53.
22. Singh, N., Malik, A., Nohwar, S., Jana, R., & Mondal, P.C. (2023). Covalent surface modification of nickel ferrite nanoparticles for electrochemical supercapacitor performance. *N. J. Chem*, *47*, 5308–5315.
23. Zhang, Y., Gao, S., Zhao, T., Chen, D., Wang, R., & Yu, Y. (2023). Ge modified by metal–organic segments as anodes for Ge-air batteries with an alkaline gel polymer electrolyte. *Arab. J. Chem*, *16*, 105021.
24. Binding, S.C., Pernik, I., Gonçales, V., Wong, C.M., Webster, R., Cheong, S., Tilley, R.D., Garcia-Bennett, A., Gooding, J.J., & Messerle, B.A. (2019). Simultaneous functionalization of carbon surfaces with rhodium and iridium organometallic complexes: Hybrid bimetallic catalysts for hydroamination. *Organometallics*, *38*, 780–787.
25. Bai, J., Zhang, J., Eiler, K., Yang, Z., Fan, L., Yang, D., Zhang, M., Hou, Y., Guan, R., & Sort, J.; et al. (2022). Electrochemically fabricated surface-mesostructured cuni bimetallic catalysts for hydrogen production in alkaline media. *Nanomaterials*, *12*, 118.
26. Ramezani, M., & Ripin, Z.M. (2023). 4D printing in biomedical engineering: Advancements, challenges, and future directions. *J. Funct. Biomater*, *14*, 347.
27. Huang, J.Y., Fei, G.T., Xu, S.H., & Wang, B. (2023). ZnO–SiO_2 composite coating with anti-reflection and photoluminescence properties for improving the solar cell efficiency. *Compos. Part B Eng*, *251*, 110486.
28. Guo, C., Jia, R., Li, X., Tian, X., Chen, J., Ge, H., Huang, H., Yuan, L., & Xu, C. (2023). Influence of backside surface morphology on passivation and contact characteristics of TOPCON solar cells. *Sol. Energy*, *258*, 278–288.
29. Kottarathil, S., Narayanaru, S., Unni, S.M., Tamaki, T., Kuroki, H., Hareesh, U.N.S., Anilkumar, G.M., & Yamaguchi, T. (2022). Selective borohydride oxidation reactions of zeolitic imidazolate framework-derived bimetallic carbon alloy electrocatalysts for alkaline fuel cell applications. *ACS Appl. Energy Mater.*, *5*, 12571–12582.

Index

E

F

G

H

I

L

M

N

O